Simplicity and Complexity

Studies in Literature and Science

published in association with the
Society for Literature and Science

Titles in the series

Transgressive Readings: The Texts of Franz Kafka and Max Planck
by Valerie D. Greenberg
A Blessed Rage for Order: Deconstruction, Evolution, and Chaos
by Alexander J. Argyros
Of Two Minds: Hypertext Pedagogy and Poetics
by Michael Joyce
The Artificial Paradise: Science Fiction and American Reality
by Sharona Ben-Tov
Conversations on Science, Culture, and Time
by Michel Serres with Bruno Latour
Genesis
by Michel Serres
The Natural Contract
by Michel Serres
Dora Marsden and Early Modernism: Gender, Individualism, Science
by Bruce Clarke
The Meaning of Consciousness
by Andrew Lohrey
The Troubadour of Knowledge
by Michel Serres
Simplicity and Complexity: Pondering Literature, Science, and Painting
by Floyd Merrell

Simplicity and Complexity

Pondering Literature, Science, and Painting

Floyd Merrell

Ann Arbor

The University of Michigan Press

Published in the United States of America by
The University of Michigan Press
Manufactured in the United States of America
♾ Printed on acid-free paper

2001 2000 1999 1998 4 3 2 1

A CIP catalog record for this book is available from the British Library.

Library of Congress Cataloging-in-Publication Data

Merrell, Floyd, 1937–
Simplicity and complexity : pondering literature, science, and painting / Floyd Merrell.
p. cm. — (Studies in literature and science)
Includes bibliographical references.
ISBN 0-472-10860-3 (alk. paper)
1. Simplicity (Philosophy) 2. Complexity (Philosophy) 3. Simplicity in literature. 4. Complexity (Philosophy) in literature. 5. Science—Philosophy. 6. Science—Methodology. 7. Simplicity in art. 8. Complexity (Philosophy) in art. I. Title. II. Series.
B105.S55M47 1997
117—dc21 97-40774
CIP

Preface

Modernity is the transient, the fleeting, the contingent; it is the one half of art, the other being the eternal and the immutable.

—Charles Baudelaire

Prefaces are by and large of little use. So much has become quite obvious in recent years, and I will try to resist indulging myself inordinately with a lavish display of preliminary maneuvers.

Nevertheless, I should at least prime the pump that is expected to draw out notions of *simplicity* and *complexity* and *order* and *chaos* from literature, science, and art before proceeding. Let me begin this way. Look out the window, if you will, and you might see grass, flowers, shrubs, trees, birds, people, cars, buildings, smoke, clouds, sky, or whatever. If there is an inquisitive child nearby, a barrage of questions might be forthcoming, and you will be hard put to classify these items further, making what you think are necessary distinctions as you go along. This question-and-answer session takes place with a basic vocabulary of a few hundred words. Simple enough, you say. Now assume you are a botanist and the child has become a college student enrolled in freshman biology. You focus on a solitary daffodil in your midst. Stamen and pistil and pollen grains come into the conversation, along with the neighboring flowers and bees and insects and pesticides and carbon dioxide and the soil beneath and the sun above and water and minerals and photosynthesis and chlorophyll. The complexity of it all becomes staggering. What is even more bewildering is the seed from which the plant grew. How did all this complexity emerge from such apparent simplicity? In search of an answer, you must dig deeper . . . and then deeper.

Beneath every apparently placid surface there lies an unruly ocean of complexity. That is the recalcitrant law of nature that, it is becoming increasingly apparent, lurks behind the simplicity of the classical laws of physics. Fractals, chaos theory, and dissipative structures bear testimony to these hidden complexities of the universe. Classical physics took refuge in surface simplicities. Those simplicities, when recontextualized during the latter part of the nineteenth century and up to our time, eventually became lost in a vast sea of complexity, as obstinate problems and puzzles continued to emerge. The general rule of thumb regarding complexity underlying apparent simplicity also applies to our everyday living (as will become apparent, I do not separate reason from intuition, body from mind, the life world from the physical world, or life processes and everyday living from scientific and artistic endeavors). The city dweller is able to simplify coming and going quite effectively, but if thrown among human counterparts in a remote region of the Amazon, he or she will be overwhelmed by the complexity of it all. By the same token, take Amerindians from the jungle habitat within which they feel quite comfortable and set them down in Times Square in New York City, and they will be baffled by the complexity. Crocodile Dundee was at home in the Australian outback but quite lost in the trappings of urban life; Forrest Gump, in contrast, reduced all his surroundings to the simplicity of his mother's homespun philosophy in spite of the increasingly complex North American milieu within which he found himself. Yet if both Crocodile and Forrest had removed their commonsense blinkers, whether in Australia or New York City or Washington, DC or the North American South, their comfortable, illusory simplicities would quickly have become unbearably complex. "Ignorance is bliss," so the saying goes. Commonsense simplicities afford security in the face of complexity; they also conceal the underlying beauty that complexity holds in store for the properly prepared mind. So much for preliminaries.

I begin my story of complexity and simplicity with a dialogue between Simplicio and Sagredo, two of the actors who starred in Galileo's notorious drama that helped launch classical physics into the stratosphere. In the good Galilean tradition, Simplicio and Sagredo present a couple of well-nigh diametrically opposed worldviews at the outset. The views are tempered somewhat as the dialogue unfolds; then the tables are overturned, and Simplicio, the erstwhile blissfully ignorant role player and occasional straw man, takes a cue from Sagredo and quickly gets the upper hand. Simplicio's intuited simplicity becomes pervaded with an image of complexity, and his classical paradigm simplifying the universe by way of the nuts and bolts of Newtonian physics becomes engulfed by that image. This dialogue, making up the book's introduction, I would hope thus serves to erect the stage for chapter 1, which outlines the entropy principle and information theory. Then the props for that stage are set in chapters 2 and 3 for the postclassical, rather postmodern drama that follows. This drama includes literary figures, chiefly Jorge Luis Borges, Italo Calvino, Samuel Beckett, and Thomas Pynchon, a selection of avant-garde and postmodern painters, and chaos theory and Ilya Prigogine dissipative structures, along with ruminations on twentieth-century physics and genetics. In addition, the limitative theorems of the Kurt Gödel sort, paradoxes of various and sundry types, and the thought of Gilles Deleuze and Félix Guattari will often make their presence felt. However, the cast of characters is by no means a mere random selection from among a virtually countless set of possibilities: the reader must bear in mind that my choices, like all choices, are to a degree arbitrary and at the same time colored with certain predispositions.

By and large, this text is guided by visual images. But please do not be misled: no ocularcentrism, no mirror of nature, no picture for holding us captive is implied. The images—many of them actually well-nigh imageless—come by way of non-Euclidean geometry and topology, non-Boolean algebra, nonclassical logic, and fuzzy sets and vagueness and ambiguity and contradiction and complementarity. I would suggest that it behooves you the reader to prime your mind for a smorgasbord of visual conundrums not merely as pictures rapturously to gaze upon: they are figures in the pregnant sense—diagrammatic, schematic, analytical as well as synthetic, abstract as well as concrete, to be taken viscerally as well as intellectually. Indeed, concrete visual images might be able to open the door to complexity and fractal dimensions and unorthodox logics and fuzziness and vagueness and complementarity, but that door, once opened slightly, demands a kinesthetic, haptic, gut feel for things, in addition to whatever cognitive input there may be.

I suppose what I am suggesting is that this book calls for reader collaboration: I really cannot do it all on my own. Consequently, on occasion you will be asked to recall relevant passages from the dialogue, to reconstruct what might now be bygone images, to reconsider past figures and diagrams—a list of these, including page numbers, is provided following the table of contents—in order to make connections and create analogies, to ruminate on possibilities that are in the act of emerging, to hold certain ideas in abeyance until the propitious moment when they can be placed in the spotlight, and above all, to let key words and sentences and paragraphs and chapters roll around in your mind so that they may gel and form a general sense of the central topic at hand: simplicity and complexity. In this regard, the crucial questions implied in this study are, How can we experience simplicity except by becoming, ourselves, of the most simple nature? And what is complexity that, by our somehow sensing it, it may afford us the wherewithal for sensing simplicity?

Nevertheless, I suspect that, rather unfortunately, after a summary perusal of this book, one might be inclined to leap to the idea that it all comes down to the following:

Premise: Complexity pervades the universe.
Consequently: There can be no complete and consistent account of it.
But: A descriptive picture vaguely suggesting the universe's complexity can be forthcoming, however simple the picture may be in comparison to that which it pictures.
Nevertheless: If complexity pervades the universe.
Then: It cannot simply be reduced to simplicity by means of a descriptive picture.
Yet: A descriptive picture can afford at least a vague suggestion of the universe's complexity.
However: That descriptive picture is destined to remain inconsistent or incomplete or both.
For: Complexity pervades the universe.
.
.
.
n

There are, of course, problems with this spirally unwinding series: the picture theory, the mirror-of-nature dream, the obsession with language, especially after the so-called linguistic turn. I would hope a closer look at this book reveals—as I already asserted—that I have tried to avoid these traps insofar as that might have been possible. Yet even if by some quirk I have actually managed to do so, a certain twisting, self-returning question remains: What are we that we may know the simplicity and complexity of the universe that made for the possibility of our very existence in the first place? In a rather perverse way, my fielding of this final question is nonrationalist and even anti-intellectual. But this is not necessarily a shortcoming, I would submit, for obsessive reason and intellectualism are what eventually landed us in our current mess. Regarding many of the issues taken up in this inquiry, the suggestion I put forth in so many words is that we are quite helplessly learnedly ignorant, ignorantly learned. Consequently we tend to remain detached from the universe's complexities in relation to our scientific descriptions of the world and to our concrete everyday living and thinking and talking and reading and writing. The problem is that if in classical Western science, descriptions have customarily stemmed from an undying faith that the universe is at its core simple, and if our recipes for living have by and large entailed reduction of our myriad surroundings to commonsense simples, then the universe's complexities simply cannot adequately be articulated, at least nonmathematically: to repeat myself, they can hardly be more than sensed, felt. I would hope that such sensing, such feeling, in regard to our mundane existence rather than to some ethereal Platonic heaven of ultimate principles and abstract contrivances, is the purposeless purpose, the unthinking thought, behind the fallible words on the pages that follow.

In order to avoid protracting my prelude any further, I have decided against providing a chapter by chapter rundown of what follows in this book, as is customarily the practice in preface confessionals. Rather, as I proceed I will offer brief previews of what is to follow prior to the introduction and to parts 1 through 4—each of which sports three chapters. You might consider reading these five previews before actually beginning the introduction. In fact, you might even consider beginning with chapter 1 and skipping the introduction—since it is quite condensed—until you are well along in this volume; or perhaps you can even leave it until the end. That way, you will quite faithfully retread my own steps as I wandered, usually somewhat lost, through a few of the universe's complexities. Otherwise, I would suggest that we take a robust dive into the turbulent stream of semiosis, with the hopeful expectation that things will work themselves out, somehow. For, living and thinking and talking and reading and writing are in the flow. They are not found in some fixed set of initiary algorithms, axioms, propositions, statements, assertions, suggestions, assumptions, desires, wishes, whims, or polemical axes to grind, all of which presumably serve to generate, in machine-like fashion, some monolithic corpus of signs.

By way of acknowledgments, allow me to write that the pigeons in their counterclockwise flights north of the bridge between West Lafayette and Lafayette that I cannot help but notice during my daily runs are a constant reminder that order out of chaos is real. The squirrels and an occasional rabbit in the park where I wind down my runs let me know how fragile life can be. The groundhog that tentatively finds his way to my tomato patch each July to nibble at a few of the riper specimens bears witness to the need for caution in all facets of daily existence. The white alley cat that periodically visits the backyard teaches me to make the best of whatever fate happens to have in store. Other than that, the automobiles that zip by are each year on the average slightly older, though their entropic push, for better or for worse, is partly offset by the new shopping malls that are sprouting up at the borders of town, all of which is a stark testimony that negentropy is ephemeral. To all these processes I bow humbly: they have been noble teachers and faithful guides.

"And academia?"—someone queries. Oh, yes. Well, it is what it is, perhaps unfortunately. Hierarchical, constipated, too predictable, and more often than not quite boring. But that is a generality, and generalities are, well, just generalities, often easy to come by and more easily acknowledged with a know-it-all nod. To be honest, however, academia is one of the few possible pathways we have that can provide the ways and the means toward sanity in a world by and large gone mad. I do not mean that academia leads us to simple answers pulled out of the complex morass we call our world and our culture and all our intellectual and artistic endeavors. Rather, academia can allow us to cope with complexity at visceral levels while we obsessively continue to present that complexity in minimalist, reductionist, and allegorical ways by means of mathematical abstractions, discourse, and narratives.

Acknowledgments

Many colleagues have helped me along this path. There is Tom Sebeok, above all, and also Lucia Santaella and Dinda Gorlée and Fernando Andacht and Kate Hayles and Bill Spinks and João Queiroz and Myrdene Anderson, among a host of others, who in word or in deed have challenged me and influenced me. There are those colleagues who have graciously granted me their collaboration in a few sessions that I attempted to help Myrdene Anderson organize at various Semiotic Society of America meetings and at an International Association for Semiotic Studies Congress. There are also a few colleagues here at Purdue, to be specific, Anthony Tamburri, Silvia Dapía, and Herb Rowland, whose comments and encouragement have been helpful. I owe a debt to some graduate students at a seminar on chaos theory and literature during the spring semester of 1995 who tolerated my self-contained, self-referential, self-indulgent mumbling. Then, during the summer of the same year, I taught a seminar on semiotics and chaos at the Catholic University of São Paulo, Brazil, at the invitation of Professor Lucia Santaella during the delivery of which I struggled with my newly discovered Portuguese and aired out a few ideas herein contained. My appreciation must be extended to LeAnn Fields and Kristen Lare of the editorial staff at the University of Michigan Press. And by all means, Araceli, of course, as always, Araceli. Thank you all. I regret that I have not been able to do your kindness justice with this volume, for my own inadequacies have placed strictures on every page.

Contents

Figures

Introduction: A Silent but Ubiquitous Third

Two dogmas threaten the world—order and disorder.
—Paul Valéry

A great disorder is an order.
—Wallace Stevens

History of science reveals to us two kinds of phenomena, opposite as it were: at times, simplicity is hidden behind apparent complexities; at other times, on the contrary, we find that behind apparent simplicity hide extremely complicated realities.
—Henri Poincaré

This introduction comes in the form of a dialogue between Simplicio and Sagredo. Their ruminations, tangential and often painful, reveal more about the monsters engendered by the West's reason and logic than we might perhaps like to admit. Simplicio and Sagredo's message will frequently reemerge as this volume unfolds to foreground the vicissitudes of simplicity and complexity and order and chaos. Consequently, I would suggest that a certain feel for this dialogue is an essential prerequisite to the reading of what follows. At the same time, I feel I should ask your patience, for the full thrust of the dialogue cannot surface until the final pages have been reached. That much said, then, let us begin.

Simplicity, Yes, and No

Simpliciox@rheto.cs.logo.flor: This vacuous monitor became suffocating. So I thought I'd zip you a quick message and see if you're hanging around.

Sagredoy@episto.cs.doxa.flor: Glad you popped up! Of course I'm here. After all, there's nothing and nobody outside the net. Right?

Simplicio: Whatever you say. Gads, it's been ages since we had our last chat.

Sagredo: Yes it has. So, what's been on your mind these decades?

Simplicio: Depression, the black pit of despondency.

Sagredo: I'm sorry to hear that.

Simplicio: Oh, it's been with me for the biggest part of this century. It began when I put myself to meditating on the unfortunate turn of events that has befallen us.

Sagredo: What on earth are you driving at?

Simplicio: Our apparent loss of faith in certainty.

Sagredo: That's funny. I took it all as an excuse for jubilation rather than despair. It puts us into the joyous play of the world . . . er, the net.

Simplicio: Yeah. I figured you'd be kibitzing, pastiching, and generally nihilizing your way around these days. You were always one to follow the fashions. Unlike you, I see these newfangled trends with a jaundiced eye. It has all been unfortunate. What needed to be known was known. The fundamentals had been discovered by our venerable forebears, especially since Galileo, and after carrying all the grand equations of classical mechanics out to their last decimal places, everything would be wrapped up in a tidy package. Then, after the turn of the century, all hell broke loose. As if the enigmas of the second law of thermodynamics and Maxwell's disconcerting findings weren't enough. . . .

Sagredo: To say nothing of the likes of Nietzsche's insights, Mallarmé's enigmatic poetic quests, and Cézanne's relentless push toward a new vision. Those were exciting days, my friend.

Simplicio: Exciting, my foot! It was enough to leave you gasping for air. In the face of this vicious turn, how could we hope to recover the simplicity of which the universe is made?

Sagredo: Your expectations are unreal. You want simple answers? Get a clean bureaucratic job. You want security? Find tenure at the university of your choice.

Simplicio: I'm talking serious business here. The laws of the universe *are* at their roots simple. As one of the greatest physicists of our century put it, if we must choose between an elegant mathematical model and another one of lesser elegance that corresponds more closely to the facts, then we should stick with the former, for the most parsimonious equations are best suited for giving an account of nature (Dirac 1963).

Sagredo: The problem is that the presumed opposite of elegance—that is, in the sense of mathematical parsimony—and simplicity is ornateness and complexity. But there is really no opposition at all. Rather, there is a subtle relationship between them. Things that appear simple more often than not turn out to be complex and vice versa. Wiggle the joystick and things move on the screen. Put a video in the VCR, push a couple of buttons, and sit back to enjoy a show replete with sex and violence. Punch a few keys on your computer terminal, and you can send a message to a friend on the other side of the globe. Activate your CD-ROM, and you have hypertextuality at your fingertips. Everything might seem so simple. But it is not. The tumultuous race of electrons at the submicrochip level is of numbing complexity. It only appears simple when collapsed into the electronic media equivalent of Schrödinger's cat (1967).

Simplicio: Come now. You're overstating the issue, aren't you? Certain things happen to appear chaotic only because they have not yet been reduced to simple laws.

Sagredo: So it might appear. With sugarplums of complete and consistent knowledge dancing in your heads, you reductionists think genes mechanically follow a relatively simple code consisting of four letters, that language operates on the basis of a scant few rules, and that the brain—even the entire universe—is nothing but a computer.

Simplicio: Don't get me wrong. My quest for simplicity doesn't mean I'm a reductionist.

Sagredo: I know your method, and you are a reductionist. You think that knowing particles will tell us about atoms and molecules, and those in turn will say what self-replicating life systems are, which will open the door to the brain's workings; then knowledge of the mind, language, rock videos, hairstyles, and the stock market will follow. The problem is that we can't even know precisely when a drop will fall from a leaky tap or whether a butterfly flapping its wings in Brazil will cause a typhoon in Asia.

Simplicio: The fact of the matter is that even apparent chaos is deterministic at its deepermost levels, which can be eventually accounted for with the proper mathematical tools. Our dear old friend Galileo was the first to teach us so much.

Sagredo: Actually, during the intervening centuries between Galileo's days and ours, we have become painfully aware of the fact that we can hardly hope to do any more than carve a few minuscule bits and pieces from the blooming, buzzing confusion surrounding us. Consequently, in the most fundamental sense we have presented this world we made, *our* world, as no more than "minimalisms," whether in the form of scientific theories, myths, or works of art. A few black slashes on a giant canvas by Mark Rothko, Einstein's $E = mc^2$, and Venus as a Roman goddess or an

Aztec god, are all "minimalist"—at least in the unorthodox sense that I am using the term here. Take art. Cézanne made painting look like paint rather than two-dimensional representations of three-dimensional objects; Braque and Picasso let objects on the canvas do their own thing rather than subjecting them to the conventionally perceived and conceived world. In literature we have e. e. cummings's and Gertrude Stein's poetry, and in prose the economy of a Jorge Luis Borges short story is a paragon of complexity apparently reduced to stark simplicity. [1]

Simplicio: Your project is self-defeating. Forget your vague allusions to art, literature, and such. Once we have the necessary information in hand, construction of the proper equations should enable us to account for everything and anything—we see this in Mandelbrot sets, for example.

Strange Logic

Sagredo: For the sake of illustration, then, let's stick to apparent simples.

Simplicio: I'm game.

Sagredo: Suppose we take A and Not-A, a thing and its opposite.

Simplicio: Logic at its most fundamental. No problem.

Sagredo: Assume A is the class of all happy people and Not-A is the class of all sad people. We have the set of happy people and its complement, the set of non-happy people. The intersection (disjunction) between the two sets is the empty set, ∅. We would suppose that the law of noncontradiction holds true to form: nobody can be both happy and nonhappy. Since the union (conjunction) of a set and its complement gives us the set of all people, the law of the excluded middle should also remain in force: each person is either happy or non-happy. Now, take yourself. From one day to the next are you entirely either happy or sad?

Simplicio: I hope you realize your example is ridiculous from the word *go*. There are varying shades of gray between one mood and the other. Reducing emotional states that rise and fall with time in an either-or matter is like the anguished teenager blurting out that she's only a little bit pregnant.

Sagredo: Well, fair enough. So let's take something apparently even more simple, say the grand old dichotomy between black and white.

Simplicio: I'm sure you'll find a way to complicate the issue in short order.

Sagredo: Consider a graphic image I'll send through the optical fibers, if you will allow me a few moments . . . there, . . . something like that (fig. 1).

Simplicio: And we have another elementary binary system. So, what's new?

Sagredo: Yes. Well, it is only apparently binary, as I hope eventually to demonstrate. Notice that [0,1] pertains to the class of all possible white things that happen to be swans and [1,0] to the class of all possible white nonswans, as well as all nonwhite, that is, black, things—which are presumably nonswans. [0,0] is the field of pure possibility including all things, both black and white and otherwise, with no-thing having (yet) been actualized. It entails the class of all things that can be either white swans or other white things or nonwhite things. It contains a grand mixture of everything possible with nothing standing out such that the field, if actualized, would take on a homogeneous gray appearance. In contrast, [1,1] contains some particular combination of actualized white swans, as well as other white things and nonwhite things that are nonswans. But it is not necessarily the case that the universe will always sport a grand split between white swans and nonwhite things that are nonswans.[2]

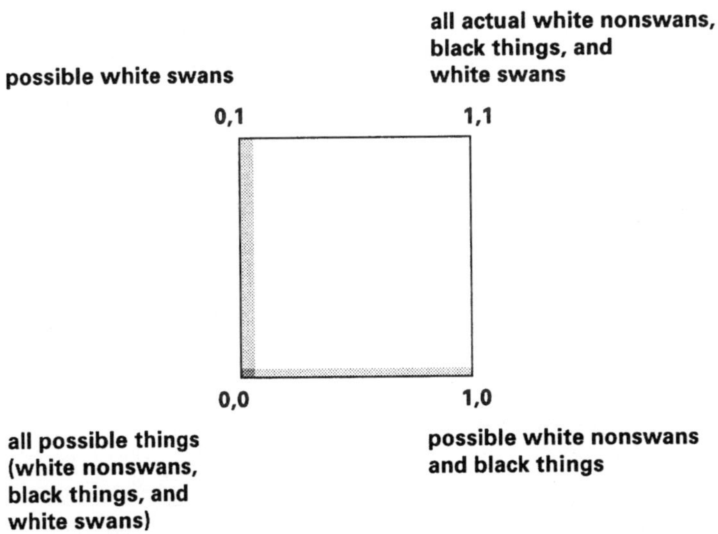

Fig. 1. The class of all swans, possible and actual

Simplicio: If you think you're now constructing a logical picture of the universe you're barking up the wrong tree, for your categories are skewed.

Sagredo: So it would appear, strictly from within the framework of standard logic. However, consider this. When Captain Cook engaged in his expedition down under, he encountered black swans. What do you make of that?

Simplicio: That some swans were henceforth included in the class of nonwhite things.

Sagredo: In due time the category "swan" became vague, didn't it?

Simplicio: Yes, I guess you could provisionally say that.

Sagredo: And so also to a slight degree the category of nonwhite things was altered, for at least a minute portion of them became "swans."

Simplicio: I'll admit to so much, since at [1,1] the two categories seem to be interdependent.

Sagredo: Now, suppose in some imaginary world almost all swans are white, but some are nonwhite. In this case, the vertical line at the left of figure 1 would fatten up a bit; consequently, the category of swans would become somewhat vague, for not all possible swans are white. Moreover, a very limited portion of the class of all nonwhite things is swans. So the horizontal line at the bottom of figure 1 will be just a tad less skinny than it would otherwise be. Did not this vagueness emerge in time—with Captain Cook's finding black swans—according to the slippery ways of the worlds we are perpetually making with our words?

Simplicio: I merely thought the fudgy lines were the product of your muddled graphics.

Sagredo: Not at all.

Simplicio: Anyway, your scheme is hopelessly incomplete. What about colors? Where is blue-gray, or gray-blue? Or for that matter, red, yellow, green, or orange? How can you limit your scheme to black and white and gray and merely "swan" and "nonswan" categories?

Sagredo: Granted, my scheme is no more than radically "minimalist" at this point. Colors, you say? Well, that would be a long story. We'd be best advised to take it up when the preliminaries have been properly aired out. . . .[3]

Simplicio: If you say so, but my patience is beginning to wear thin.

Sagredo: Reconsider the attributes or predicates, white and black. By the force of habit we customarily tend to construct a two-way timeless arrow between these and other such pairs of terms, thus rendering them polar

Fig. 2. Binarism, black and white

opposites, something like this . . . (fig. 2). On the other hand, if we wish to express the relations between the two terms in nonlinear fashion, we must include at least one more dimension by placing white along one axis and black along the other one. So let's label the vertical axis white and the horizontal axis black. We will now have this . . . just a sec . . . ah, here it is (fig. 3). Black and white are no longer simply an either-or matter. Rather, they express the field of combinations of black and white—as such they can now accommodate the vague swan categories. Smeared across the plane, they compose a homogeneous gray field at the lower left-hand extremity of the square that contains the possibility for actualization, over time, into patches of black and white toward the upper right-hand extremity. Both axes are from one view independent and autonomous, yet they are inextricably linked from another view. An increase of one is a decrease of the other, from nonwhite to white and from nonblack to black. Consequently, after the gray has been differentiated into intermittent patches of black and white at the upper right-hand

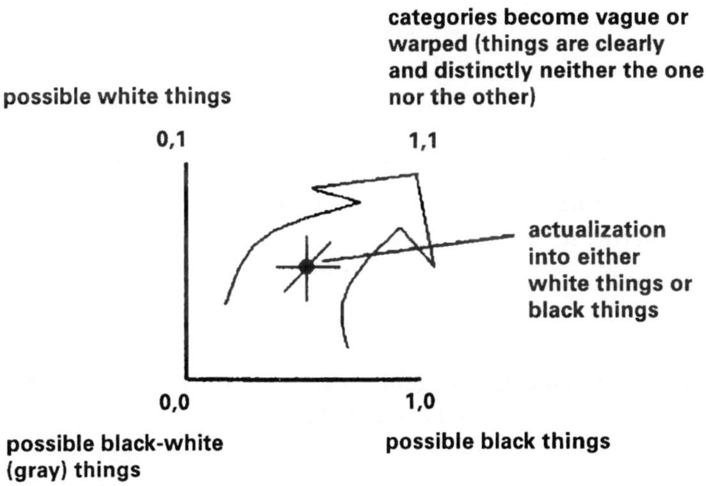

Fig. 3. Three-dimensionalizing the possible-actual swans

corner, there is neither exclusively black nor white, for black things can exist alongside white things. Hence given time, some of what would ordinarily have been relegated exclusively to the class of white things—that is, swans—might turn out to be black and vice versa. In other words, the world becomes vague, fuzzy.[4]

Simplicio: I might have guessed! An illogical swamp of unruly terms!

Sagredo: You're quite right. But I have my reasons for setting up this scheme. If at [0,0] we have a melding of both black things and white things as possibilities, and if at [1,1] we have neither exclusively non-black things nor nonwhite things as fixed categories of actual things once and for all time, then what would we have at the midpoint of the diagram where the possibilities become actualized?

Simplicio: It's a flip of the coin, I suppose. Heads you get white, and tails you get black. Or vice versa. Anyway, I don't like it. Where's my freedom of choice?

Sagredo: Granted, it is a difficult pill to swallow, since at the center of the diagram where grays organize themselves into blacks and whites Aristotelian logic seems to be left standing there with its midsection unprotected. But actually, it's not merely a flip of the coin, for we can always exercise some degree of choice. However, the important message is that at the navel of figure 3, the whole affair begins to become three-dimensionalized, and time begins to enter the picture.

Simplicio: Now I must lodge a protest! Logic knows of no time. I had an inkling you would be headed in this direction.

Sagredo: Oh, if only things were that simple. Actually, deep down we all know that nothing is timelessly either absolutely one thing or the other. We would like to believe that things will always be what we believe them to be. But at some unforeseen future moment the applecart is inevitably overturned and we must reevaluate our categories—like white and nonwhite swans. This bears witness to the inexorable vagueness of our thinking, which we can see in the categories of our everyday talk. Is a whale a fish? Are koalas and pandas bears? Are tomatoes vegetables? Does the sun rise?

Simplicio: All such mushy categories are the result of ordinary language use and abuse, not hard-core falsifiable logical and scientific language.

Sagredo: That depends upon who is doing the talking and from what perspective. The question is, How are we most often inclined to take things scientifically? Gas A deviates to one side of Boyle's laws of the ideal behavior of gases, and gas B deviates to the other side. Are A and B both gases in the absolute sense? Is one more genuinely gassy and the other slightly gasless? When should we create another taxonomy dividing gases up into new subcategories? In the final analysis, our very acts of drawing lines of demarcation are, and will always remain, problematic.

Simplicio: Wrong. We must have some standards of precision to shoot for, otherwise we are driving piles in a bed of quicksand. I exist. This chair *is* real. The sun *will rise* tomorrow. Ravens *are* black and emeralds *are* green.

Sagredo: Well . . . Here's another example for you then. Suppose we depict all white things and all black things at a given point in time by an oval, with a line bisecting it representing the intersection of the two . . . just a moment . . . how's that? (fig. 4).

Simplicio: Everything is clearly either black or white. Where's the catch?

Sagredo: What about the dividing line? Does it depict black things or white things?

Simplicio: Neither.

Sagredo: Of course. The line of demarcation is no more in the field of white things than it is in that of black things. It is actually no-*thing,* so to speak.

Simplicio: I hope you're aware of that fact that there's more to it than that. In a sense, the line at once depicts both nonwhite things and non-

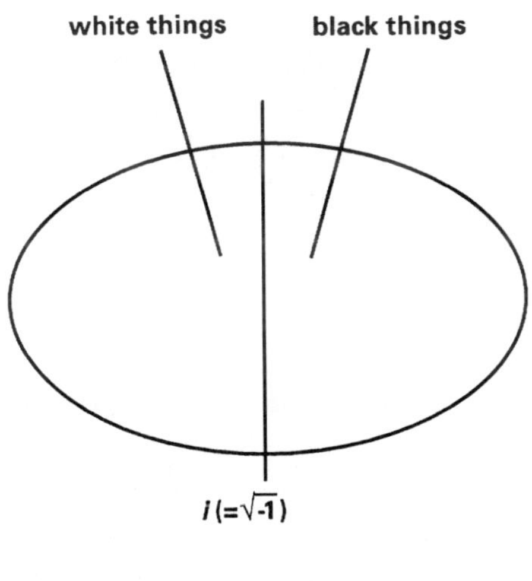

Fig. 4. The cosmic egg

black things; that is, as an imaginary construct it is nonwhiteness and nonblackness. As such it is in neither of the two halves of the oval. But as a depiction of nonwhiteness it possesses a commonality with blackness, and as a depiction of nonblackness it shares a property with whiteness. I suppose that's why you labeled the dividing line of your cosmic egg in figure 4 as $\sqrt{-1}$ (or *i*). The two properties of the egg are contained within the line, which is "nonwhiteness/nonblackness," so to speak, like $\sqrt{-1}$, which is +1/–1. In other words, the line is neither white nor black. It is just a line. This is like expressing $\sqrt{-1}$ as an unambiguous sign, *i,* which is neither positive nor negative, it just *is.* I really see no problem. It's still a matter of who's in control of language. That's all.

Sagredo: The problem is that the dividing line is a metaphor for seepage of uncertainty, ambiguity, contradiction, and paradox into the system. Since the line bisecting the egg is both whiteless (like black) and blackless (like white), and since at the same time, it is neither the one nor the other, it oscillates between the two, like Kierkegaard's paradoxical either-or. There is no stability, no certainty.

Simplicio: No, no, no! I continue to look for order while you once again insist on propagating anarchy!

Sagredo: Not anarchy, Simplicio, but the idea of complexity.

Simplicio: It's a matter of what you call complexity. Are we talking about the self-organizing flow of traffic along an interstate or a hopeless logjam in downtown Tokyo?

Sagredo: Oh, my, we are still talking past each other. Hm . . . Well, look at it this way. [0,1] of figure 3 might be graphically depicted by something in the order of Kazimir Malevich's avant-garde painting, *White on White* (1918). It is apparently quite simple: a white square painted on a white canvas. But it is not pure white; it is somewhat blemished by the texture of the brush marks and some subtle shades of off-white and gray the artist has interjected. Consider [1,0] to be like a Mark Rothko canvas painted black in the Rothko Chapel at Houston. [0,0] can be compared to Mark Tobey's *Tempera* (1945), made up of apparently randomly spaced minuscule black brush marks on white, which from afar gives the appearance of grayness. However, on closer inspection the marks not only become apparent, but they are now noticeably polychromatic. Finally, [1,1] is like Vasarely's op art work, *Harlequin* (1935).[5] It consists of a black and white checkerboard toward the center of which the squares become distorted in such a way as to create the image of a dancing Harlequin with a checkerboard suit and a black hole for a head. Here, let me give you a schematic description of these images—I have them at the distance of a few keystrokes—and shoot them over to you . . . There! (figs. 5*a* and 5*b*).

Simplicio: I recall having seen Vasarely's work once. A fictive being is in the process of lifting itself up by its own bootstraps, pulling itself from the two-dimensional plane into our three-dimensional world, like that Escher woodcut of the snorting reptiles emerging from and disappearing into a two-dimensional plane of chaos.

Sagredo: That's it! In Vasarely's work symmetry is broken, the squares become aperiodical, asymmetrical, and the plane becomes, so to speak, a hyperplane of more than two dimensions. In other words, the fictive being is slightly more than two-dimensional; it is fractaloid. Simplicity becomes complexity, as the inanimate comes to life.

Simplicio: Whatever happened to figure 4?

Sagredo: Oh, yes. At the hairline splitting the oval and functioning like $\sqrt{-1}$ (= i), things are taken to be either one thing or the other, depending upon the angle of vision. For that reason I have constructed an ambiguous drawing at the center, the "Penrose triangle." It is here that vagueness, from [0,0], becomes an ambiguity of values; it is here where choices can be made; it is precisely here, where swan attributes enjoy an expansion, where whales join rank with all other mammals, where atoms become not impenetrable spheres but largely vacuous, where the poet's "Hell" becomes a "Lake of Ice," and so on.

Simplicio: Oh, you mean that, like in Maurits Escher's *Relativity,* $\sqrt{-1}$ represents flip-flops between the values of the rectangle's corners.

Sagredo: Good point.

Simplicio: Well, I for one still remain unconvinced. You think you've demonstrated that your both-and and neither-nor are as viable as the either-ors of standard logic.

Sagredo: It's not really a matter of reason or irrationality or of logic or illogic but rather of order and chaos, of simplicity and complexity. You see? [0,0] is fusion, chaos, where entropy gains the upper hand. [1,1] is self-organization, a move toward possible order, and hopefully, simplicity, though that rarely turns out to be the case. If we combine the two extremes with everything in between in one whole package, we have in essence the most complex of complexities.

Simplicio: My mother warned me about days like this! I still insist on truth, while you think something can simply be one thing from one perspective and another thing from another perspective at a later moment.

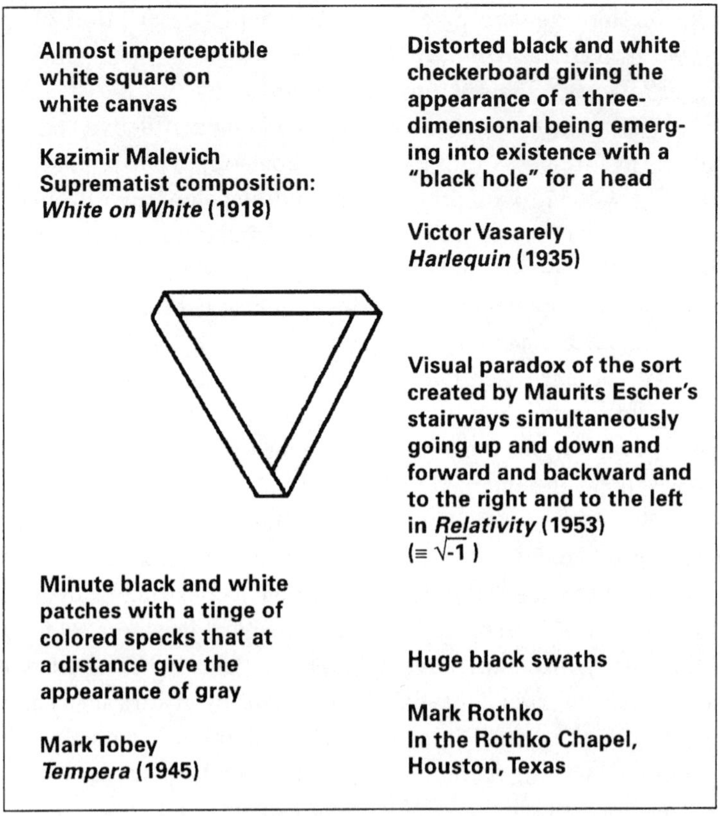

Fig. 5*a*. How the possible-actual swan plane is three-dimensionalized

You Can't Fool Time

Sagredo: Where, how, when, what is truth? We would prefer timeless, crystallized truth. But we live within the ever-changing tide of time.
Simplicio: Truth knows of no time.
Sagredo: Yes, I've heard that story before. Let me exemplify my argument further with a sort of parable my uncle told me a couple of centuries ago: the "Prisoners' Paradox." It is Sunday. Some prisoners are told by the warden that they will be executed on one day of that week, but they will not be informed which day it will be until the arrival of that very day; hence it will be a surprise. Now, the prisoners happen to have found a quite astute lawyer. She reasons, after some deliberation,

Fig. 5*b*. Harlequinizing the plane

that, assuming the warden to have told them the truth, they cannot be executed, for if the fatal day is to be Saturday, then it cannot come as a surprise since it will be the only day remaining. By this mode of reasoning, neither can it be Friday, for since Saturday has been eliminated, Friday is no longer a viable candidate. The same can be said of Thursday, and so on down to the present day, Sunday. Therefore they cannot legitimately be executed.

Simplicio: There must be a flaw in the lawyer's reasoning, but I'm sure you will find yet another way to shirk the real issue.

Sagredo: Not shirk, but skirt: that is, by movement along the edges, or along the hairline in figure 4. The problem is that the lawyer's reasoning is strictly by logical means. She can certainly afford to be logical, for her life is not at stake. In contrast, the prisoners' very existence is in jeopardy. They are rightly concerned over how much time remains of their lives, and time is precisely the issue here. The lawyer's logic is timeless, and within this framework the paradox springs forth in full force. But the prisoners, their emotions having understandably taken precedence over their reasoning faculties, exist in time. Hence, try as their lawyer may to convince them otherwise, she cannot reason away their expectations of an unexpected moment announcing their death. Trapped inside their particular mind-set, they are condemned to temporality.

Simplicio: Must I repeat myself ad nauseum? You cannot mix timelessness and temporality that way.

Sagredo: But we think, live, and breathe in a world consisting of a mix of our timeless, intransigent ideals and our time-bound real capacities. Look at the situation from within the context of my previous example of the line of demarcation separating white things from black things. For the prisoners, their execution will be either Saturday or not, and if not, then some day between Monday and Friday will be doomsday. For the lawyer it will definitely be neither Saturday nor any other day, logically speaking, that is. As far as she is concerned, there cannot logically be any command to execute the prisoners, hence there will be no life before death that becomes death after life. That is, there is no becoming. The infinitesimal time line separating the two states, life and death, cannot logically be trespassed in an increment of time. There can be no more than before/after. You see? Like our spaceless line bisecting the oval and separating black things from white things, the prisoners' dilemma from the lawyer's viewpoint has a *timeless* thread between before and after. The line separating the two areas of the oval is spaceless; the instant separating the prisoners before their execution and their state of affairs after that event is, according to the lawyer's reasoning, timeless. In other words, the temporal dividing line is not-before and not-after, and the prisoners at that instant are not-dead and not-alive. So the instant separating before from after is a state of *dead-alive-lessness*. Like $\sqrt{-1}$ in figure 4, there is both *dead-lessness* and *alive-lessness;* hence in a manner of speaking, from the never-never zone, that of every-*thing* and no-*thing,* the prisoners are both dead and alive and neither dead nor alive. There can be no more than an oscillation between the two values. There can be no becoming of the event of their death. On the other hand, for the prisoners, there is a past, a now, and a future, with the now, like a knife-edge, traveling along the line of time. In the prisoners' conception of things, process, not discrete increments of the Zeno sort, is the rule. They live in time, psychological or existential time.

Simplicio: OK. But you have yet to explain precisely how you think you can place time and timelessness in the same bag.

Sagredo: Time and timelessness, like space and spacelessness, are or they are not, so to speak, depending on whether we are satisfied with nothing less than the realization of our ideals or decide to live at peace within our physical world. It's as if according to the Zeno procedure we were to take a stick, and during the first 1/2 minute we cut it in half. Fine and dandy. If we follow Zeno's logic, in the before we had a whole stick; in the after we had two smaller sticks of equal length, the sum of whose lengths is approximately equal to the length of the whole stick. If we follow our intuition regarding time, first we had a whole stick, and then we had the two equal portions in our hands. So we continue with the Zeno procedure, and in the next 1/4 minute we cut one of the two halves of the stick in half and then cut one of those quarter portions in half in the next 1/8 minute, and so on. That's the ideal. In our world of experienced time we never reach the end of the line nor do we produce an infinitesimal slice of the stick. After a few cuts and less than a minute, we realize we have failed, for we are left with nothing but unmanageable shreds of cellulose.

Simplicio: Of course. You will always fail at comparable tasks. Many factors are at work: your lack of skill, the bluntness of the knife, the splintering of the wood, and so on. In a word, it's "sludge"—or perhaps "noise"—that enters the scene when you embrace your conception of time and space. "Sludge! It is . . . that unruly, refractory element in the world that, assuming the form of mechanical friction in one case, renders the perpetual motion machine impossible. . . . Sludge is everywhere."[6] It pervades your own incapacity to wed theory and logic with practice, or space with time. What we need are hard-core algorithms set in theoretical concrete capable of accounting for observable phenomena. With such algorithms, we can account for everything, including our own brain processes. We can and must get rid of the sludge that exists between theory and practice.

Sagredo: You're up to your reductionist tricks again. How does a race of electrons translate into chemical changes and firing neurons, then into nerve impulses, and finally into words? It's as if to say that if we could understand the workings of Shakespeare's body we would be able to infer how *Hamlet* was squeezed from his brain as a consequence of the lamb chops he had for dinner that day. We don't know how a brain works, let alone the nervous system, or even the inner workings of an atom. Uncertainty and fuzziness—or "sludge" as you put it—prevail.

Simplicio: You're a defeatist. You abandon your ship without realizing it is still seaworthy.

Sagredo: Not so. I do the best with what I have. We all do. Your dream of total knowledge is like a map so faithful in all its details that it contains itself—ourselves included—and that contained map in turn contains a replica of itself, and so on. The problem is that if your knowledge perchance ever attained such completeness, it could not but contain this very map paradox. In contrast, I am content with bits and pieces of tenderly fallible knowing wherever I can find them. Now, you will argue that my map of less-than-perfect detail fails to represent the fullness of the territory mapped; hence it is flawed. But that does not prevent it from depicting, in however rough a manner, the territory. And we get along with life. The moral of the story in a nutshell is that even though the map model is a "minimalism," like any sign, it can in the practice of everyday communication quite effectively relate to its object without faithfully replicating it.

Virtually "Real" *Is* "Real," It Is "Virtually Real"

Simplicio: Heaven forbid! . . . No. . . . I think. . . . Yes, I am. . . . No. . . . Amazing! I believe I sense it. It's all a matter of signs.

Sagredo: What?

Simplicio: Let me see. . . . In a quirky sort of way you seem to have been saying all along that if something is a sign it is a sign of something real, and if something is real, then it is that *of* which a sign is a sign; hence that real thing is a nonsign.

Sagredo: Uh . . . I'm not yet sure.

Simplicio: Are not dreamed images real?

Sagredo: Yes, I guess so. Like hallucinated images, they are real insofar as there is something that from within the dream or hallucinated world appears to be real.

Simplicio: A dreamed image is real. But it isn't real, not really, when placed alongside the things of our wide awake world, yet it must be a sign. Must it not?

Sagredo: Well . . .

Simplicio: What about a unicorn? Is it not a sign?

Sagredo: Yes, a sign depicting a mythical beast.

Simplicio: Is the beast not real?

Sagredo: Don't be facetious. Of course not. Unicorns don't exist. ("How is it that I now find my head on the chopping block?")

Simplicio: But there can be things that unicorn signs depict. Is that not so?

Sagredo: Yes. For example, the word *unicorn* you just punched on your keyboard, unicorn pictures, a unicorn image in your mind, or on the silver screen or the boob tube. But the issue is much more complex.

Simplicio: Yes, indeed, complex. For the record let me summarize: there are real and nonreal signs—signs of the nonreal—and there are real and nonreal nonsigns as well. It seems to me, then, that what signs and the real have in common is that which consists of nonreal nonsigns. In other words, what they have in common is tantamount to your $\sqrt{-1}$. Between signs and the real we have nonsigns and the nonreal, which from an alternate perspective can also be real and signs. In other words, the border between signs and the real—or nonsigns and the nonreal—is just as schizophrenic as $\sqrt{-1}$, is it not?

Sagredo: Hm . . .

Simplicio: According to your talk of maps, a map is like a sign, a sign of another sign, and so on, through to the eons when old Adam originally named everything.

Sagredo: Why bring religion into the picture?

Simplicio: Religion is not an issue at all. Adam or the Cro-Magnon man or the Piltdown man, biblical sign or supposedly real object or hoax, the same applies: all are signs engendered from the hairline dividing real from nonreal. Signs of signs of signs, maps within maps within maps:

it's like Jonathan Swift's image of fleas on the dog's back and fleas on the backs of those fleas, ad infinitum. We can never get to the end of the road, yet in some exceedingly vague sense we are always there, wherever that is, wherever we are.

Sagredo: I suppose. . . . I think. . . .

Simplicio: So, Sagredo, it seems that we finally agree on something.

Sagredo: Well, . . . yes, why not? After all, we are fellow astronauts traveling in the same conceptual space. Aren't we?

Simplicio: It's a pity Salviati wasn't able to join us. It could have proved to be a lively trialogue through the information highways and byteways. But perhaps his absence was fortunate for me, since in our dialogues you customarily side with him regarding the sticky issues.

Sagredo: Quite simply put, Simplicio, after having apparently sharpened your intellect since our last encounter, this time around you were able to hit the target dead center.[7]

Simplicio: Beware of absolutes, Sagredo, lest you become a reductionist.

PART 1

Part 1 justifies its emerging into the pages of this book by way of its creation of five sets of props that adorn the stage of this inquiry with a certain tone for the chapters to follow: (1) the limitations of knowledge, (2) the fuzziness of categories and the vagueness of beginnings and endings, (3) the rhizomic interconnectedness of things, (4) our suspension within space, and (5) our intimate relationship with time.

In order to accomplish this goal, chapter 1 offers a survey of the entropy principle and the strictures it places on the possibility of our knowledge of the world in terms either of simplicity (order) or complexity (disorder). The work of Ludwig Boltzmann is introduced, along with its bearing on the beginnings of cybernetics and the departure from classical mechanics, as depicted in Robert Musil's *The Man without Qualities* and in later "cybernetic" fiction, from Samuel Beckett to Thomas Pynchon's *The Crying of Lot 49.* Following this brief sojourn within fictional worlds, there is a return to the entropy principle, its use and abuse in information theory, and its applications to aesthetics and everyday language. Chapter 2 evokes the indelible element of vagueness regarding the age-old dichotomy, simplicity/complexity, with examples drawn from Jorge Luis Borges and Italo Calvino. This move necessarily places language in the spotlight. With respect to linguistic practices, simplicity and complexity are tentatively qualified in terms of nonlinear (rhizomic) rather than linear phenomena. Consideration of nonlinearity then evokes G. Spencer-Brown's forms, Humberto Maturana and Francisco Varela's autonomy of living systems, and Gilles Deleuze and Félix Guattari's machinic organisms. In chapter 3 Borges and Calvino examples introduce us to the hoary image of chaos. However, a formal exposition of chaos theory is not presented, but rather, there is an attempt to convey a feel, a sense, of chaos. Finally, questions of symmetry and asymmetry of space and of reversibility and irreversibility of time emerge by way of certain Borges passages, which prepare us for what is to follow in part 2.

CHAPTER 1 • On Things That Go Awry

Destiny takes pleasure in repetition, variants, symmetries.

—Jorge Luis Borges

The great struggle for artists is the annihilation of static equilibrium in their paintings through continuous oppositions (contrasts) among the means of expression. It is always natural for human beings to seek static balance. This balance of course is necessary to existence in time. But vitality in the continual succession of time always destroys this balance. Abstract art is a concrete expression of such a vitality.

—Piet Mondrian

Where is the wisdom,
We have lost in knowledge.
Where is the knowledge,
We have lost in information?

—T. S. Eliot

Messes and Misses

Daughter: Daddy, why do things get in a muddle?
Father: What do you mean? Things? Muddle?
Daughter: Well, people spend a lot of time tidying things, but they never seem to spend time muddling them. Things just seem to get in a muddle by themselves. And then people have to tidy them up again.
Father: But do your things get in a muddle if you don't touch them?
Daughter: No—not if nobody touches them. But if you touch them—or if anybody touches them—they get in a muddle. . . .

So it is that Gregory Bateson (1972, 3) begins one of his charming "metalogues." The things of our work and play activities get in muddles (i.e., become complex) by themselves, and it is often up to us to put them in their proper place (i.e., simplify them). But if we do not touch them they seem to stay much as they were. Like the playthings of all children, those of Bateson's daughter, Mary Catherine, tend toward a muddled state of affairs as long as she is rather mindlessly playing with them. If she wishes to arrange them in an orderly state, however, she can do so only by putting her mind to it and at the expense of considerable physical energy. In other words, with respect to her personal possessions, when her mind and body happen to be engaged in their normal course, things get muddled (of high entropy) and, at least from her view, complex. On the other hand, if she wishes to order (negentropize, simplify) them, she must engage in acts of mind and body focused on that specific purpose. Of course in this context the terms *complex* and *simple* depend upon the perspective and the preconceptions of the beholder.

According to the second law of thermodynamics, entropy is generally on the increase. This tendency in the natural world toward what might be deemed messiness was not always acknowledged in classical science. And for good reason. Newtonian physics reduced the world to clean trajectories consisting of physical bodies described as if they were points on a Cartesian plane tracing out a line. Symmetry was the ironclad rule: movement of body *x* from point *a* to point *b* was perfectly reversible, and hence time was really of no consequence—Sagredo's "Prisoners' Paradox" to the contrary. The impulse for its motion came not from within but as a result of outside forces the nature of which remained quite mysterious (i.e., force at a distance, the *whys* of which very few scientists were willing to suffer the embarrassment of trying to explain). There was no notion of self-organization, either of inorganic or biological entities. All the particles tracing out their lonely tracks in the world did not genuinely interact: they simply acted and reacted, collided and rebounded, with each other. It was as if all physical bodies were governed by the same laws of mechanics: simple, crystallized for all time, and after all was said and done, terribly uninteresting. Yet, this hygienic and properly Sanforized world of balance, harmony, equilibrium, and order was eminently predictable and quite comforting.

With justification, shortly after the turn of the century Simplicio preferred to nurture the belief that the end of physics was at hand. And he was not far from the mark, given his venerable Western tradition. However, as is now well known, flaws eventually revealed themselves. They were soon to shake the classical universe at its very foundations, eventually culminating in relativity and quantum theory. Relativity transformed Newtonian empty, homogeneous, Euclidean, flat paper space into a tangle of curved spaces, and time took on the characteristics of something variable and quite personal—if I may put it that way—rather than universal and constant. In the process, time and space came together in an inseparable embrace. What was worse—at least from Simplicio's view—quantum theory diffused the billiard ball atoms of the Newtonian mechanical world, those presumed building blocks of all physical bodies, into fuzzy clouds of uncertainty. During the intervening years, physicists have occupied themselves with mathematical justifications of relativity and quantum theory at increasing levels of sophistication and abstraction. The elusive theory of everything (TOE) or the grand unified theory (GUT), at least for those who persist in their hope of such, would consist of a mathematical system capable of accounting for strong, weak, electromagnetic, and gravitational forces, that is, for what are supposed to be the most general principles governing the universe.

The modern version of this "end of physics" mentality is considerably different than its classical counterpart, however. Nineteenth-century true believers claimed physics would be complete when everything could be described in mechanical terms and by means of pictures based on forces and masses and planes and rods and pulleys—comparable to Simplicio's dream. These pictures were to be taken quite literally, for if correct, they would be representations of the world just as it was, is, and will be for all time. The contemporary postmodern scientific counterpart to this classical picture is actually no picture at all. In a literal sense it is well-nigh incomprehensible; hence, as mentioned in the preceding, it cannot be said but only depicted by exceedingly abstract mathematical formalism. It is based not on tangible, testable postulates but on rigorous mathematical proofs. This formalism appears to some critical onlookers as no more than bizarre mathematical mumbo jumbo, a zany collection of wild-eyed speculations such as twenty-six or so dimensions of space filled with imaginary "strings." In one commentator's view, the postmodern physicists' esoteric use of language is as "baffling as that of the literary deconstructionists. . . . The mathematical physicist and the deconstructionist share the same popular image: each speaks in a private gobbledygook understandable only to those similarly initiated" (Lindley 1993, 19). What appears to be the physicists' mathematical playpen is in this sense a counterpart to deconstructionism's Nietzschean freewheeling play of the world. Derridean terms and the mathematical symbols of the "new physics" compose Sagredo's allusion to cerebral parsimonies guided by inner complexities that are to the observer unanointed into the high priesthood of presumed esoteric subtleties intangible and virtually ineffable.

But let me backtrack somewhat in order to provide a more well-rounded story. Thermodynamics pushed its way into physics about the midpoint of the nineteenth century, introducing a macroscopic account of the colliding billiard ball picture that revealed serious flaws in the classical view of the universe. Baldly put, the second law of thermodynamics says that in a closed system, heat will always flow from hot to cold, which seems about as obvious as announcing that water runs downhill only. So why was the second law so earthshaking? At least a partial answer is found in the great Steam Age's effort to get more from less. Engineers had been tinkering with steam engines for years, trying to squeeze increased oomph out of them with the same amount of fuel. Then, in 1824, Sadi Carnot tossed them a bombshell with his theory that there is a limit to the efficiency of a steam engine, or any other machine for that matter. In 1868, inspired by Carnot's work, Rudolf Clausius keyed in on the idea of the amount of heat or energy available for use—potential energy. In order to explain the relationship between potential energy and depleted energy, and order and disorder, he coined the term *entropy*—now quite common in academic parlance, though it is often abused in the social sciences and the humanities. A system of low entropy enjoys relatively high available energy and vice versa. Clausius convinced the scientific community that as a universal principle entropy will always tend to increase. Energy will flow from a state of high temperature to one of lower temperature, from a relatively ordered (simple) state to a relatively disordered (usually a more complex) one, or to put it more technically, from a less probable state to a more probable state. This energy flow is the great equalizer: things will eventually become homogenized, standardized, normalized, equilibrated, balanced. At the point of maximum entropy there will no longer be any heat gradient such that it can flow from one zone to another; in other words, no more work will be forthcoming. Apparently the grand Industrial Revolution in the long run had limits. Comes "heat death," the assembly line stops, and it's back to the caves.[1]

But there was even more disconcerting news. Thermodynamics as a macroscopic law ushered a hitherto virtually unknown branch of mathematics into the scientific arena: statistical mechanics. Without the capacity to follow the journey of a solitary molecule of gas in an enclosed chamber, it was possible statistically to chart—and mighty accurately at that—the average behavior of all the molecules in the chamber. This conclusion is due chiefly to the independent efforts of Austrian physicist Ludwig Boltzmann and North American physicist J. Willard Gibbs. Their goal was nothing less than a derivation of the macroscopic laws of thermodynamics from the statistical analysis of an ideal aggregate of microscopic entities. Boltzmann, most particularly, identified the entropy of a physical system with the randomness of that system as a macrostate defined in terms of the logarithm of the number of its microstates.[2]

Worlds Becoming

As an illustration of Boltzmann's work, let me try to qualify his mathematical concepts graphically.

In essence, Boltzmann took steps toward describing thermodynamics in terms of statistical mechanics. Up to his day, classical physics saw atoms as freely roaming particles in continuous space constantly in the process of moving from one of a continuous range of states to another. For Boltzmann this was too sludgy. In his effort to achieve greater precision, he divided macrostates into microstates. Visually, this would be much like dividing figure 5*a* into sixteen cells with the cells in the four corners representing the works by Malevich, Vasarely, Tobey, and Rothko, as in figure 6.

At A, the atoms—black dots and white dots—are, like Tobey's *Tempera,* for all intents and purposes equally distributed in the microdomain: they make up the chaos from which order and the universe of differentiated things can emerge. In C it might appear that we have a plenum: the atoms have filled the space entirely. Not so, however. Like Rothko's black swaths, C can be conceived either as no-*thing* or every-*thing,* depending on whether it is the product of no light frequencies refracted or all light frequencies absorbed. On the other hand, unlike its appearance, B is not simply empty space; rather, it is ideally every-*thing* and no-*thing,* depending on whether it is conceived as all light frequencies refracted or no light frequencies absorbed (this, ideally, is the essence of Malevich's *White on White*). Microstate D is itself divided into twenty-five submicrostates, consisting of cells from microstates B and C—which were drawn from A—intermittently dispersed throughout. $\sqrt{-1}$, at the center, is where the action occurs: black atoms cluster together and separate themselves from other clusters to begin a process of ordering from the random dispersion at A as we move toward the upper right-hand corner of the diagram. These clusters, within D, become transient. They are "alive." They can self-organize into one of an infinity of possible forms. Thus the square patches in D can be multiplied and distorted in such a manner that a three-dimensional being begins arising out of the two-dimensional plane, like Vasarely's Harlequin.

"But how can this be?" my skeptic pipes up. "If entropy exercises its ironhand, all natural processes must eventually gravitate toward the lower left-hand corner, and that's the end of the line: everything's kaput. In other words, the arrow in figure 3 should actually be pointing downward instead of upward. All systems are destined to wind down; disaster is the inevitable end product."

Here's where Boltzmann comes in. He proposed that a microstate can give a fairly close approximation of the position (and velocity, which we do not consider in fig. 6) of each particle (or "point").[3] The macrostate is at a given point in time defined by a given number of alternative microstates. Assuming that all microstates can be treated alike, the collection of them should give a determination of the macrostate. In other words, the probability of a macrostate would be proportional to the number of microstates it contains: it would have a probability equal to the number of actualized microstates divided by the total number of possible microstates. So the push of entropy is nothing more than the probability that the microstates will at some future moment be in some configuration or other. There is no invariant rule for determining precisely what that configuration will be, but only the probability of its being one thing rather than something else. However, there is nothing preventing a number of points from clustering at a given microstate, thus going from a more probable state to a less probable state. And if this happens to take place, then the same could occur in the adjacent microstate, and so on. In this manner a cluster of microstates could themselves as an aggregate take on the characteristics of a less probable state of affairs. Boltzmann's theory thus lends itself to the image of small islands of increasing order surrounded by an ocean of overriding disorder out of which life processes might be made possible. During such a turn of events we would be moving toward the upper right-hand corner of figure 6 (and Vasarely's Harlequin could pull itself into existence).[4]

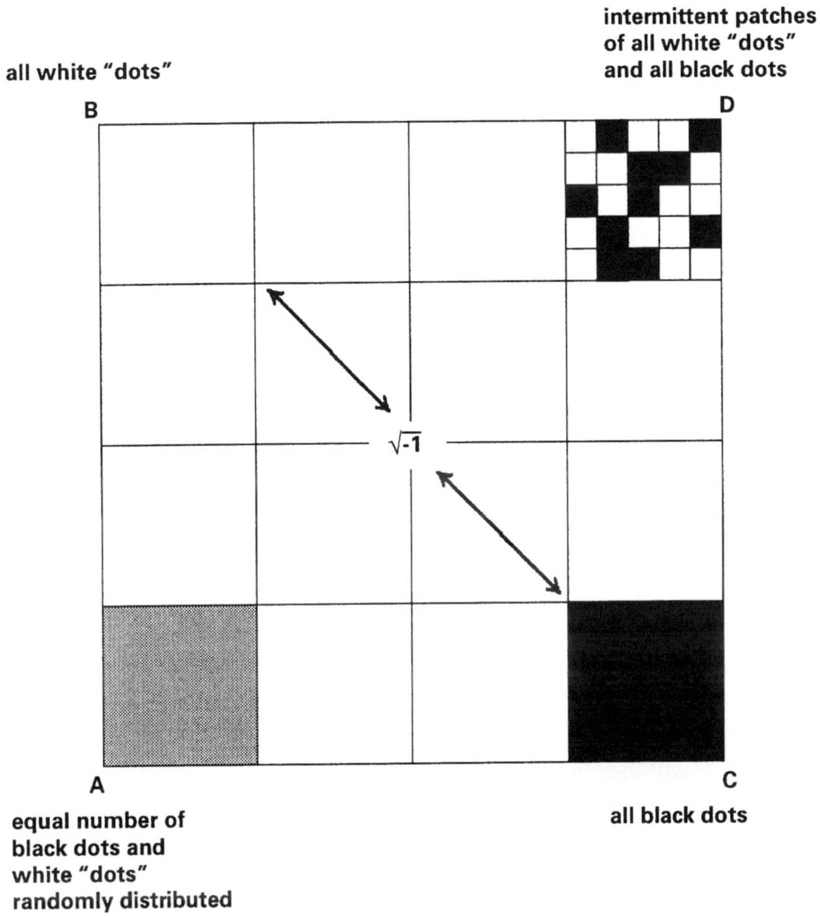

Fig. 6. The plane Boltzmannized

In short, we might summarize Boltzmann's hypothesis in view of the discussion of figure 6 as follows.

1. We would generally expect to find, after nature has taken its course, a random set of black points (atoms) dispersed throughout quadrant A.
2. The points will rarely, if ever, be found distributed as in D.

3. In the beginning, something in the order of D might have been the case due to some artist's intentionally having separated black points from white points (i.e., an act of construction) or to nature's swimming against the current of probability.
4. In such case (from 1 and 3), order would have increased either due to the intervention of the artist or to the appearance of a highly improbable state of affairs in nature.
5. However (from 3 and 4), there will always be many more cases of randomly dispersed black points than ordered microstates.

Items 1–5 represent the tendency toward a more probable state of affairs. Yet, the fact remains that there is nothing preventing "time's arrow" in a given set of microstates from reversing itself to bring about a relatively improbable state of affairs. It hardly needs saying that this aspect of Boltzmann's work played havoc with classical mechanics.[5]

It had come to be understood from within the domain of classical mechanics that the total energy available from a volume of steam, for example, was the result of individual molecules of H_2O colliding with themselves and rebounding from the walls of the chamber within which they were confined. The faster the molecules were moving, the more energy available for work, and as that energy was dissipated in terms of heat expelled, entropy would rise. It all seemed quite simple. Applying classical principles to a cylinder and a piston, the energy would dissipate when countless molecules of H_2O pushed the piston down, and this dissipated energy could be harnessed for the purpose of turning the wheels of a steam engine. That was the classical view of the situation in its most skeletal form. Following from the ramifications of Boltzmann's work, however, things quickly took on inordinate complexity. The question became, If everything is a matter of statistical probabilities, then what is to prevent all the molecules of the cylinder in an engine suddenly from pushing the piston down so violently that it blows a gasket? Or why could a substantial collection of the molecules not for some unforeseen reason refuse to rebound from the piston at all, so that the engine would suddenly lose its compression? Such happenings would be highly improbable, no doubt, much like the proverbial army of monkeys pecking away on typewriters, one of which just happens to compose *Hamlet.* Yet, statistically speaking, these extraordinary sorts of happenings would not be entirely outside the realm of possibility.

The apparent fact remains that for practical purposes, the laws of thermodynamics are quite exact, as classical mechanics ought to be, though philosophically, sludge enters the scene. Nonetheless, well into the twentieth century the idea of a universe of statistical probabilities and the randomizing of events exercised its tired force, as we shall note subsequently.

Entropy Tripping

When applied to the sphere of human affairs, the very idea of a universe governed by statistical mechanics can engender premonitions of dehumanization, ego loss, and in general a diminution of that proud romantic Faustian spirit capable of going it alone and taking on the entire universe.

We experience this attitude in its extreme form in Ulrich of Robert Musil's *The Man without Qualities* (1965), the most powerful literary expression of the early-twentieth-century malaise in Vienna. Ulrich, the "man without qualities," is the statistically average human. That is, he is simply "there" at every moment as a set of possibilities, his ego poised and ready to take on one of an indeterminate number of available postures. As a bundle of unborn possibilities, his "reality," whatever that may be, consists of the awakening of a paltry few of those possibilities in the here and now, while what remains dormant constitutes that which could have been present but is not. Yet what is not has a bearing on what is, for what is, could not have been without its complement, that which could have been but is not. In other words, Ulrich is all possible qualities and hence of no real—that is, actual—qualities; he contains within himself the possibility of being everybody and nobody; he is like an infinite sphere collapsed to an infinitesimal point. He is comparable to A of figure 6 or to Tobey's painting—pure grayness, possibilities without distinctions or separations—and at the same time he is merely like $\sqrt{-1}$ insofar as it is a hitherto unactivated process of becoming. In this manner, Simplicio's nonreal signs are such insofar as they are mere possibilities, in the order of Ulrich. When a minuscule portion of them is actualized, in whatever form or fashion, the actual signs take their rightful place among all signs, and thus they are real, as signs, even though they may be relegated to the dark zone of the irreal with respect to the physical world as it is known at that particular space-time juncture.

We see an intriguing variation of the man without qualities theme in

Woody Allen's film *Zelig.* The main character, Zelig, apparently out of the clear blue just happens to pop up in an impressive number of earth-shaking events in world history over a number of years. Furthermore, he strangely seems to take on the personality and even the physical characteristics of those around him. In Brazilian culture we are witness to a comparable simulation in *Macunaíma,* a film based on Mario de Andrade's novel ([1928] 1984) by the same name. Macunaíma, the protagonist, is a composite of Native American, African American, colonial European, and modern behavior, attitudes, thought processes, and myths and rituals. He is everybody and hence virtually nobody at all; he incorporates the characteristics of many personalities from many cultures and as a result has no characteristics he can call his own (in fact the subtitle of the original Portuguese edition of de Andrade's novel was *The Hero with No Character*).

The statistically average citizen, Ulrich, however, is distinct from these models. He is not everyone and hence antiheroic, nor is he heroic. He is only possibly and at best potentially everyone as a statistical blur. While Zelig and Macunaíma are now one thing to some people, now another thing to other people, the man without qualities is simultaneously the potential for becoming everything for everybody without anything having been clearly distinguished and delineated for all to see. While Zelig and Macunaíma consist of varying expressions of D in figure 6, the man without qualities is hardly more than the faceless, motionless, expressionless A. Is he, therefore, "the common man who has an intuitive prophetic glimpse of the beginning of an immense new, collective, ant-like heroism? It will be called rationalized heroism and will be regarded as very beautiful" (Musil 1965, 8). Emphasis is thus vested not in the individual but "in the relations between things." Experiences have become independent of their individual masters; they have become public, collective, everybody doing the same thing at the same time. The man without qualities will become the norm, the statistical average. There will arise a

> world of qualities without a man to them, or experiences without anyone to experience them, and it almost looks as though under ideal conditions man would no longer experience anything at all privately and the comforting weight of personal responsibility would dissolve into a system of formulae for potential meanings. It

> is probable that the dissolution of the anthropocentric attitude (an attitude that, after so long seeing man as the centre of the universe, has been dissolving for some centuries now) has finally begun to affect the personality itself. (Musil 1965, 175)

This flattening of humans to the same level as their signs and the furniture of their world, as well as to the level of all other organisms—themselves equally semiotic agents—has been a common theme in much twentieth-century art, whether depicted in terms of existential anguish or a freewheeling, liberating euphoria of *ars combinatoria.*[6]

We now pass from Musil's work evincing the concept of statistical averages and the possibility of actual worlds to a more specific focus on entropy, for which one of the most popular literary outlets has undoubtedly been science fiction. This is nowhere more evident than in "cybernetic" prose, examples of which are found in the work of Philip Dick, especially *The Unteleported Man* (1966) and *Do Androids Dream of Electric Sheep?* (1968), and also in George Alec Effinger's *What Entropy Means to Me* (1973). In *Do Androids,* Dick coins the neologism *kipple* as a surrogate for entropy. Kipple consists of useless objects, junk mail, gum wrappers, yesterday's newspaper, and such. When nobody is looking, kipple reproduces itself; in fact, the entire universe is moving toward a state of maximum "kippleization." We must contrast this strictly inorganic vision of entropy with the notion of entropy as a product of culture revealed in the selection from Bateson's "metalogue" with which I began this chapter. Regarding the affairs of our cultural setting as pointed out by Bateson, things get in a muddle not by themselves but because of our muddling them during our everyday practices. Of course, with effort, we can periodically bring some modicum of order to our cultural universe. But the cruel fact remains that, given the nature of our everyday practices, things just tend to get muddled. In other words, in systems that include human agents, and in all communities of biological organisms for that matter, it would appear that Parkinson's law inheres—work must increase to keep up with the inevitable increase of bureaucracy—as well as Murphy's law—anything that can possibly go wrong will do so.

Cybernetic works outside the strictly defined science fiction subgenre often depict humans who are thrown into dehumanizing, disconcertingly nonhuman, or inhumane situations. For example, we are told that John Barth's *Giles Goat-Boy* (1966) is written by a computer and that the author is actually no more than its editor. Samuel Beckett's *The Lost Ones* (1972) is like a scientific thought experiment involving 200 human-like beings trapped in a huge cylinder-like machine. William Burroughs's *The Soft Machine* (1966) hyperextends formalistic techniques and the rules of Boolean algebra in such a manner that the word as carrier of information becomes the same for everyone everywhere. Thomas Pynchon's *Gravity's Rainbow* (1973) patterns the development of cybernetics during World War II, including all its connections with Pavlovian behaviorism, information theory, the inception of robotics, and artificial intelligence research. In his essay "Cybernetics and Ghosts," which appears in *The Uses of Literature* (1986), Calvino speaks of the compelling power of cybernetics as an explanation of what the author does when he writes. He or she is not a conduit for genius or insight but rather a sort of machine, a Cayley matrix, a mechanism for *ars combinatoria,* a *bricoleur.* This is a far cry from the romantic Faustian ideal that includes individualistic, egocentric illusions of creativity, with the artist acting as a medium for emergence of the new. Yet, even though according to this view the author is destined to remain machinic, "literature will continue to be a 'place' of privilege within the human consciousness, a way of exercising the potentialities contained in the system of signs" (Calvino 1986, 16). More recent cybernetic fiction goes by the name of "cyberpunk," examples of which are found in William Gibson's *Neuromancer* (1983), Gibson and Bruce Sterling's *The Difference Engine* (1991), and Sterling's *Islands in the Net* (1988). These narratives place us squarely in cyberspace, in a worldwide computer network that can directly link everyone's nerve nets and neural nets, which becomes Marshall McLuhanism gone wild, Jean Baudrillard to the *n*th degree.

In particular, the work of Thomas Pynchon, which lays claim to notoriety regarding its allusions to thermodynamics and cybernetics, deserves a few moments in the spotlight. Pynchon's short story entitled "Entropy" was published in 1959, and his novel *The Crying of Lot 49* (1966) is his next likely candidate for conveying a sense of the entropy principle. At the beginning of this novel, our protagonist, Oedipa Maas, experiences something in the order of a quasi-mystical insight when she finds herself gazing upon a new housing district the sort of which proliferated at the outskirts of California's cities during the 1960s. Against the sunlight, she squints onto

> a vast sprawl of houses which had grown up all together, like a well-tended crop, from the dull brown earth; and she thought of the time she'd opened a transistor radio to replace a battery and seen her first printed circuit. The ordered swirl of houses and streets, from the high angle, sprang at her now with the same unexpected, astonishing clarity as the circuit card had. Though she knew even less about radios than about Southern Californians, there were to both outward patterns a hieroglyphic sense of concealed meaning, of an intent to communicate. (Pynchon 1966, 13)

Oedipa is incapable of comprehending the meaning of her moment of vague insight, her "odd, religious instant," much less of putting it into words. This scene sets the stage, nevertheless, for her bewildering trek through a confusing plethora of signs, most of which are for her equally incomprehensible. Throughout the novel she tries her level-headed best to read the signs surrounding her, which consist of a virtually random scattering of the author's critical commentary on thermodynamics and information theory and his multiple allusions to codes, messages, and secret languages through the use of parodies, puns, rhetorical figures, and historical and literary references.

But in order to read the myriad outlay of signs before her in her attempt to discern some meaning in her life, Oedipa must wade through a swamp of sludgy messages.[7] One potential message stands out from the rest: a strange hieroglyph consisting of a muted, trumpetlike instrument that is presumably the symbol of a mysterious system called WASTE. This sign is somehow related to a flawed postage stamp that conjures up vague allusions to a subversive medieval postal service that might have survived to the present. The sum total of all these messages suggests some sort of vast international conspiracy by an underground organization called the Tristero System. The problem is that the messages are apparently referenceless. We are told at one point that a sign, most notably WASTE, quite simply is what it is, without relationship to anything else—which is enough to bring a religious person to his or her knees and drive the antagonist of logical and rational virtues to fits. However, there is at least a cloudy allusion to the foundation of WASTE by way of an anonymous organization, IA—*inamorato anonymous*—which is quite in keeping with later mention of DTs (delirium tremens) regarding a convulsing homeless sailor and the idea of AA (Alcoholics

Anonymous), especially since in this scene Oedipa is drunk virtually out of her mind. Moreover, IA, besides being an abbreviation of CIA, which is duty bound to appear in any and all conspiracy theories, is a mirror image of AI (artificial intelligence), the supreme product of that age-old dream of getting something from virtually nothing. This pre-Baudrillardian sort of Californian hyperreality gradually disintegrates, as Oedipa becomes more obsessed with the torrent of signs that ceaselessly present themselves before her. Finally, toward the end of the novel Oedipa runs onto the homeless sailor suffering from delirium tremens, the low point of her adventure and the height of the chaotification of signs around her. A Pynchon play on signs, DT/dt—delirium tremens and the time differential of the calculus—possibly holds the key to a clue. Oedipa realizes that "the sailor had seen worlds no other man had seen if only because there was that high magic to low puns, because DT's must give access to dt's of spectra beyond the known sun, music made purely of Antarctic loneliness and fright" (Pynchon 1966, 96).

But unfortunately for Oedipa, everything then falls to pieces. Signs continue to gush forth, while other characters fade out: entropy is on the increase. In Baudrillardian postmodern fashion, signs proliferate. They become virtually the same signs, simulacra of themselves and nothing but themselves, with hardly any differences that make a difference in order to open the door to meaning. Or at least they are taken for the same referenceless signs, mirrors that mirror themselves, commodities whose function in the physical world has been overtaken by the tide of semiosis such that they become nothing but semiotic operators, like mathematical functors that are nothing but signs that exercise their overpowering influence on neighboring signs. Oedipa, surrounded by a confusing disarray of signs mirroring themselves to infinity, remains mired in the sludge of rampant uncertainty. Her own pathetic story is patterned by the destruction of the story itself of which the novel consists—it appears that under these conditions a Boltzmann microstate would stand nary a chance of lifting itself up by its own bootstraps and proceeding against the tide of entropy. Finally, auction time approaches, the artifacts of perhaps the oldest house in the town will pass on to who knows where, and Oedipa awaits the "crying of lot 49."

After these general—and quite vague—words on entropy from the literary vantage, now for the main course of specifics, however vague they are destined to remain.

What Power Hath This So-Called Heat Death?

Among the plethora of popular accounts of entropy, we have that of two flasks of gas connected by a thin tube and a stopcock. The molecules of one of them are in rapid motion, and those of the other one are in relatively sluggish motion. If the stopcock is opened, eventually both flasks will bear gas molecules of approximately the same average velocity.

Maxwell's Demon, a notorious figure named after its nineteenth-century author, James Clerk Maxwell, is rather miraculously capable of reversing this entropic process. It seems that the Demon's function in life is that of opening and closing the stopcock in machine-gun fashion such that the more nimble molecules are allowed to pass into one flask while the torpid ones are relegated to the other flask. The Demon is, quite significantly, used in Pynchon's novel as a metaphor for Oedipa's task of sorting out the confusing plethora of signs surrounding her and trying to isolate the meaningful ones from the disordered ones in the shuffle. Just as the Demon sorts molecules, so also Oedipa shuffles through countless signs in order hopefully to collect bits and pieces of information that make sense, store them in her memory bank, and leave the rest of them to the devastation of the entropy principle "out there." But by and large, unlike with Maxwell's Demon, Oedipa's efforts are to no avail. Whatever information she happens to stumble upon is offset by steadily increasing confusion. The iron-hand second law of thermodynamics exercises its inevitable force. Anne Mangel (1971, 198) points out that the "parallels between Oedipa and the Demon seem almost too neat. Our heroine discovers that signs do not lead to one stelliferous meaning, but point in virtually countless different directions none of which leads to any solid conclusion." In other words, the system is nonlinear, and whatever information (order) is gained is accompanied by a loss of information (→ disorder) somewhere else. This was also quite in line with the conclusion, as we shall observe subsequently, of Leo Szilard in 1929 and Leon Brillouin in 1951: in order for a being or a machine to increase its information (knowledge) there must be a loss somewhere else. But I am getting ahead of myself; for now, back to the essentials of entropy.

The gas-molecules-in-two-flasks and other such stories about local pockets of entropy increase appear fine and dandy. They give additional credence to Boltzmann's hypothesis, and to boot, they soothe our need for a somewhat intelligible universe. But can the concept really be generalized globally? If not, is it really of any consequence for us in the everyday humdrum of our daily lives? Freeman Dyson (1979) once calculated the time required for the universe to suffer entropic heat death to be 10^{14} years, assuming that the universe continues to expand during that period. Obviously, there is no need for worry, not yet at least. If the universe is simmering down like a giant pot of stew left over a sliver of a flame for many billions of years, eventually we will not be able to tell the carrots from the onions: it will all be mush, or sludge if you will. However, as far as our mundane existence is concerned, enough precision will always be around to enable us to sink our teeth into the meat and potatoes of our perceived and conceived world.

But actually, this is not the real issue at all regarding our desires, hopes, and fears in our mundane everyday coming and going. For example, compare Oedipa to Bateson's daughter. The daughter can rearrange her relatively scant belongings at her leisure, we would suppose, since they make up a local environment. In contrast, Oedipa's environment consists of her entire world, lock, stock, and barrel. "But human and other societies can change their global environment," someone might wish to respond. This appears reasonable enough. After all, bees build hives, beavers build dams, human animals can cause a teeming city to rise from the desert sands. So in the final analysis it would seem that we should be able to alter the flow of entropy. In other words, Bateson's daughter's possessions get muddled when she plays with them, but eventually she must reorder them, and ideally she should be able to build on that order. The inorganic universe, on the other hand, gets muddled with no help from us at all. Yet we can by the force of our will and at least at local levels alter this trend toward maximum entropy, like the bees when they build hives, like life itself when it lifts itself from its brute inorganic environment.

However, from another vantage point, compare this ordering of the environment to Sagredo's onslaught of sludge. Cut the stick in half and repeat the operation, then repeat it again, . . . and again, in Zeno fashion. Things quickly get muddled. Entropy seems to exercise its apparently immutable push. "But this muddling is of a different sort," one might now wish to retort. And yet, is it really? We commonly take the stick to be much like a continuum from one extremity to the other, though upon closer inspection, it is not. At the outset it only appears to be like a continuum offering itself up to virtually an infinity of possible cuts at an infinity of points from one end to the other. In other words, the stick, as a continuum, is unordered, chaotic, in the sense that there are as yet no distinctions having been made by us, its imperious bisectors. So we cut it, and the continuum is broken by a distinction separating *this* from *that.* We have ordered the stick; we have brought about a negentropic state. Then as we reiterate the procedure, the distinctions become finer and finer, until finally, they are so small that we can no longer work with the impure bits of what was originally taken to be a continuum.

In fact, these bits, during the final cuts, become hardly more than pieces of sawdust. When they are thrown together, there is hardly any distinction between one piece and any of the other pieces: disorder or entropy, approaching a continuum (chaos), has exercised its force. We began with what we took to be continuous and of maximum entropy since there was no order, and after successive acts of ordering we ended up with distinctions so minuscule that they appear for practical purposes to approach a continuum: disorder, chaos, sludge. I say "for practical purposes," since in everyday living we generally take our physical existence for what it is according to our socially constructed image of the world. That is to say, it is not the pristine world of our ideals, of purely conceptual matter, of intellection, or of the mind: it is *our world* of everyday living. Yet, like Bateson's daughter, we can go through the exercise of reordering our world so that it conforms, ever more slightly, to our ideals, though our real capacities invariably fall short of the mark. For example, the pieces of sawdust are brought together and joined with others, a 4′ × 8′ slab of particleboard is constructed, and along with its neighboring slabs, a plethora of 2′ × 6′ planks, concrete, siding, and shingles, a house in Oedipa Maas's suburban tract is on its way toward completion. Once again, order (simplicity) has emerged out of chaos (complexity)—though at the expense of a violation of the negentropic process in the surrounding environment.

But the equation is not as simple as it appears or as we would perhaps like it. Simplicio wanted the simplicity of a totally ordered universe, as if he were dwelling in the immediate vicinity of $\sqrt{-1}$ in figure 6, where everything is either one thing or the other. Sagredo had given up on such utopian simples long ago, opting for a vision of complexity, at D and perhaps even beyond. But Simplicio's simplicity can become organized complexity, as house after house is built in the suburban tract and as these houses, virtually identical in the beginning, take on differences that slowly become distinctions. However, decay eventually sets in, then rebuilding occurs or not, depending on the particular microstate (lot), and eventually, entropy has its day. At this stage every lot once again is virtually of the same grayness at A as every other lot, and the entire system has reached a maximum state of complexity, or, if you will, of simplicity, depending on how one wishes to take a look at it.[8]

The capacity to reverse, in however paltry a fashion, the entropic "arrow of time" calls for further discussion of Maxwell's Demon and its relation to Pierre Simon de Laplace's chimerical Superobserver.

Is It All in the Bits and Nothing but the Bits?

Against the background of relativity and quantum theory, Laplace's imaginary being seems incredibly naive. Light in the classical conception presumably traveled at infinite velocity. Availing himself of this simultaneity of all happenings, Laplace's Superobserver took a measurement of each and every particle in the universe, and with this information he could compute all past happenings and predict all future happenings to perfection.

However, if light travels at less than infinite velocity, as it does according to relativity theory, then any information received by the Superobserver would be so received only after a certain lapse of time. Hence he could not be aware of the position and velocity of each particle in the universe at an instant: he could neither know the past nor predict the future, nor could he even know the present as a totality in one fell swoop, for there can be no God's-eye grasp of the whole. Maxwell's Demon suffers from comparable delusions. As we observed, Maxwell's counterpart to Laplace's cosmic Superobserver is of molecular size and charged with the task of rearranging its environment in such a manner that the relentless push of entropy is reversed. Ultimately, the entire universe is susceptible to the whims of our finite, though exceedingly nimble, hero.

But things are not quite as simple as all that. If the Demon is itself part of nature—and we must suppose this to be the case—then its functioning as separator of the fleet-footed goats of concentrated energy from the lumbering sheep of dissipated energy is a violation of the second law of thermodynamics. Yet how can it do so from within its environment? Is it not constrained by the same entropic process? While busy ordering the disordered, is our hero not at the same time becoming disordered if the very law being reversed is the law by which its existence is defined? This problem aside, our hero's receiving information regarding its environment, processing that information, and bringing about an ordered state requires energy. In this sense it is actually part of the system that is supposedly the object of its intervention, for it must draw energy from the system in order to act on it. As part of the entire setup, the energy it uses while carrying out its responsibility for concentrating energy into what would otherwise be highly improbable knots must be subtracted from the whole. It will in this manner be responsible for at least a degree of energy dissipation. So in spite of its efforts, entropy will still be pushed a mite further down the stream. In other words, the initial condition our Demon must measure can be so measured only at a cost of the free energy of the entire system. As part of the system, it cannot act on the system without consuming some of the system's energy, and upon so doing, it contributes to the energy dissipation and hence to the entropy increase.[9]

I am afraid I am getting ahead of myself again. I have somewhat artificially introduced "information" into the equation—that was Oedipa's grand dream and her dilemma—which entails work of an entirely different sort. Perhaps what I should be discussing is entropy with respect to shifts of energy, regardless of whatever information may be imparted. But, then, perhaps I am actually not so far off track after all, since there is presumably a subtle relationship between information and energy, a topic that should now be stressed. To put it metaphorically and in the simplest of possible terms, information comes in immaterial waves, while the nuts and bolts of the world come in particles. Both are a matter of energized discontinuous packets, the first existing as a form of mind-stuff and the second as complementary matter-stuff. The second, regarding our everyday physical world, can for practical purposes be conceived much in Newtonian fashion. If I may avail myself of one of Bateson's rich examples, the act of kicking a football is work, of sorts. Your foot rapidly comes in contact with the pigskin, and it is displaced by a force approximately equal to that applied by your foot. These are Newtonian principles. Kick the nearest pit bull, however, and other rather undesirable forces are unleashed. Your kick has imparted some unexpected information to which the animal responds with a degree of energy far superior to that of your foot coming into contact with its rib cage. And you suddenly are in for a rough time. It could rightly be said that the dog was displaced by a force equal to that of your kick. However, to repeat, the energy generated by your irate canine friend far exceeds the force of your foot. You have transmitted waves of information that it interpreted, such interpretation compelling it to respond in terms of vicious physical activity accompanied by waves of counterinformation that quickly put you in flight or to fight. But this example remains partial at least. Your kick required quite an expenditure of physical energy. In contrast, information through media other than kicks and such requires relatively little energy output. Take language. Martin Luther King's utterance, "I have a dream," did not require an appreciable number of BTUs for its engenderment, yet it set an entire group of humanity in motion, the effects of which are still felt today. This sort of information, in contrast to the expenditure of pure Newtonian pushes and pulls, entails a different game entirely. It is the game of information, which, according to information theory, is quantified by the second law of thermodynamics but in a manner quite different from that of the physical world. Let me try to be a bit more specific with an example from Pynchon's work.

"Who's that with the beard?" Oedipa asks of the sketch of a Victorian gentleman on a box containing photocopied patents of some of John Nefastis's inventions. She is told that he is Maxwell, author of the wily Demon who could play a trick on the second law of thermodynamics by sorting the universe into high energy domains and low energy domains, using the high energy domains to do work for her and thus "getting something for nothing." "Sorting isn't work?" Oedipa asks. "It's mental work," she is told, "but not work in the thermodynamic sense." It's like mind over matter. There happens to be a pair of cylinders in the room. Oedipa is invited to concentrate on which way she wants the Demon to raise the temperature of two cylinders, whether in the right one or the left one, and sure enough, it will be done. The air of one of the cylinders will expand and push its piston up. She tries to accomplish the task by means of pure mental exertion, but, as would be predictable, she can't. And then, quite appropriately the reader is reminded of the WASTE symbol—as well as other allusions to the flawed postage stamp and the suspected postal fraud, and so on (Pynchon 1966, 62). All this apparently uncontrollable process of disordering surrounding our hero, Is it metaphorically counteracted by the muted bugle symbol? Is this not an image of Oedipa's quest?

Well, if so, any answer must be left to idle speculation, however much food for thought it gives. On the other hand, what we have here is the implication of a relationship between entropy and information. Sadi Carnot, revealing that there are limits to the amount of work a machine can be made to produce, was concerned basically with the nature of physical systems. Maxwell's interest was focused on the possibility of altering the system if enough information could be known about it. Boltzmann's version of the entropy principle carried with it the added corollary that there are limits to what can be known about a system in which energy is being passed back and forth. One of his main points was that we cannot know, and should not even be interested in, the coming and going of an individual molecule. We can know in the best of possible worlds, and with almost perfect accuracy, the statistical behavior of the entire aggregate of molecules. The phrase "almost perfect accuracy," of course, in the very least makes for an inevitable residue of sludge. In the worst of scenarios, we would have the "butterfly effect," which, as we shall observe, forms the kernel of chaos theory.[10] The important point at this juncture, however, is the change of emphasis: from Carnot to Maxwell to Boltzmann, there is a switch from work in terms of energy to work in terms of useful information.

Now, it is quite obvious that even if Maxwell's Demon could by some stretch of the imagination open and close the stopcock between the two flasks of gas molecules without expending much energy, the fact remains that it would still need information to distinguish fast molecules from slow ones. In such a case the question arises, Can information alone reduce the entropy of the system and render some of its erstwhile potential energy available once more? Can the Demon perform its task by receiving information, looking at the stopcock, and, as in the example in *The Crying of Lot 49,* exercising mind over matter to open or close it? The answer must be a resounding "No." If the Demon receives any information at all, it is by the transmission of light or some other medium involving energy. Even considering that it could somehow alter some aspect of the world simply by looking at it, light would still be brought into the picture: in other words, the concept of energy invariably forces its way onto the scene.

Szilard (1929), as mentioned previously, placed this perplexing issue in a different context. He argued that Maxwell's Demon creates at least as much entropy as it eliminates upon gathering the necessary information and sorting the molecules into their appropriate pigeonholes. The important thrust of Szilard's argument was that entropy plays havoc not only with the availability of energy and hence the quantity of work that can be done but also with information. Just as a system high in entropy will have relatively little energy available for work, so also a system high in entropy—that is, "noisy," of random signs—will be capable of imparting relatively little information. In 1951 Brillouin put the frosting on the cake—or delivered the fatal blow, according to the eye of the beholder—in his demonstration that the Demon brings about a greater increase of entropy when it takes in and processes information than it creates upon ordering its environment—and I write "environment," since the Demon is in the final analysis definitely part of the very system being altered. Ultimately, for the system as a whole, the entropy principle remains quite robust.

Nevertheless, in light of Brillouin's conclusions, entropy and information are inextricably intertwined: as entropy increases, information diminishes, and if information can be increased, entropy suffers a depletion—hence the term, *negentropy,* coined by Brillouin and made common knowledge by physicist Erwin Schrödinger (1967) and others. In view of an article by W. Ehrenberg (1967), N. Katherine Hayles (1990, 44) points out that Szilard was chiefly ignored until after Brillouin's article appeared. Only then was Szilard's work finally incorporated into the cybernetic tradition as a seminal contribution. Work subsequently moved along at a breakneck pace. In 1948 Claude Shannon presented his paper "A Mathematical Theory of Information," connecting thermodynamic randomness with the general idea of randomness—the range of *possibilia*—conceived as a necessary component of a system with regard to which its information can be computed. A year later Shannon and Warren Weaver published their groundbreaking *The Mathematical Theory of Communication.* With these events, the game had begun in the grand arena of cybernetics.[11]

Yet, there is much confusion even today. Brillouin's Maxwell Demon gathers information and sorts molecules, thus decreasing entropy but at the expense of an increase of entropy somewhere else. Information and entropy are inversely related, thus the term *negentropy* with which to qualify information (but see Bennett 1987; Wilson 1968). In contrast, Shannon's Maxwell Demon, insofar as he interpreted it, encounters no such problem: information and entropy are as far as it is concerned actually quite compatible bedfellows. If entropy (disorder) is high, then the possibilities for order (information) of many varieties are also high. And if entropy is low, then much of whatever information is to be had can potentially be had, and relatively little information that is genuinely new can be forthcoming.[12] However, when Shannon put forth his probability function for computing the quantity of information, he dubbed that quantity derived by the function the resultant message's "entropy."[13] Why this about-face? The reasons are not clear. Jocularly enough, Hayles (1990, 49) recalls the tale about mathematician John von Neumann's advising Shannon to make use of the term because obviously hardly anyone knew what entropy was and hence in a debate he would always have the upper hand. At any rate, the show was now rolling along. And a promising show it was, luring many brightfaced young cyberneticists toward the grand vision of computers that could do more than humans and, to boot, would be able to repair themselves when they were ailing. This was also the image of a new era of slave labor envisioned by Norbert Wiener in his best-seller with the downright frightening prophetic but admonitory label, *The Human Use of Human Beings* (1950).

In the long run, however, the show turned out to be a study of intrigues. Hayles writes that by the act of mixing *information* and *entropy* as if they were interchangeable terms, Shannon's theory "gave rise to decades of interpretive commentary that sought to explain why information should be identified with disorder rather than order" (Hayles 1990, 51). In order that the marriage between "Entropy" and "Information" might be consummated, Information first had to suffer a painful divorce from "Meaning," and then it saw fit to take the maid of honor, "Novelty," along on the honeymoon. Novelty was then associated with "Randomness," which had been selected as best man at the wedding ceremony for Entropy and Information, and it also became the very defining term for Entropy, which brought Information into a tighter embrace with an "entropic move" toward disorder. For Shannon, then, the possibility of high information content consists of a measure of the degree of disorder. Once randomness was understood as maximum information, it was possible to envision chaos as the source of all that is new in the world. And Novelty finally ended up jumping into bed with Information to jeopardize relations between the newlyweds. Information, within the Shannon conception of things, then, could never again be considered in the same breath as "certainty," in any shape or form.

"But," my skeptic might retort, "information interpreted can be either denotative or connotative, and the information of information theory is the latter; we must keep things in their proper perspective." This project would most likely be destined to failure as well, I fear. In his critique of Shannon, Jeffrey S. Wicken (1987a) is most likely correct in noting that denotative and connotative fields, rather than sporting a relationship of opposition, actually overlap. In the case of information, the connotative image that invaded the denotative field of chaos was complexity. Whereas chaos had traditionally and quite uncategorically denoted disorder, complexity implied a mingling of symmetry with asymmetry, predictable periodicity with unpredictable variation. The field of chaos thus became vague and ambiguous. As we have seen, chaotic or complex systems are disordered in the sense that they are unpredictable, but they are ordered in the sense that they possess recursive symmetries that almost—but not quite—replicate themselves over successive increments of time. So there is a sense of order (simplicity), but it can also be said that there is disorder (complexity), depending upon the eye of the beholder. It would appear that Hayles is on target with her argument that the metaphoric joining of entropy and information was instrumental in bringing about these developments. This union of entropy and information allowed certain manifestations of complexity to be seen as rich in information—and therefore ordered—rather than merely deficient in order due to their very complexity. In this sense, there are no nonvague distinctions between simplicity and order, on the one hand, and complexity and disorder, on the other. They depend on the context within which they are found and on the perspective from which they are judged.

This confusion stems in large part from the fact of Shannon's fusing thermodynamic entropy, fit for the physical universe and accounts of energy flow, with his particular interpretation of informational entropy, which is quite divorced from the things of the world and the kinetics resulting from their interaction. Common grounds between thermodynamics and information are found basically in the arena of statistical probabilities, and Shannon's interpretation bears strictly on probability. A particular message is of high information content if it was generated from among a relatively large set of possibilities (that is, the message was of low probability); it is of low information content if there were few possibilities (and greater probability) present at the time the message was generated. If I begin a sentence with "When she . . .", the number of possible words that can follow is great, and the probability of a particular word popping up is relatively low. If I begin with "Rush Limbaugh said . . .", the rest of the sentence is of considerably greater predictability, and hence the information content of whatever choice I exercise for the next word in my utterance is relatively restricted. In the Shannon mode, the information content of a message increases in inverse proportion to the decrease of probability of the message's content. Well and good, so it might seem.

But there are still some serious problems. Shannon's probability determination of information is counteracted by that of Brillouin. Hayles effectively reveals the distinction between Shannon's interpretation and Brillouin's interpretation of information. For Shannon the uncertainty is in the message's source: it is a matter of the number of possibilities within the "synchronic" field of as yet unactualized messages. Brillouin's consideration of uncertainty regards the message's destination: the message, once actualized, is there because it should have been there. For Shannon the greater the improbability of a message's being actualized, the greater the surprise once it is actualized and the greater the information.[14] For Brillouin the message's actualization is a fact, and now, whatever information is there is there because it should be there, and so it has relatively high information content. Brillouin abhorred disorder and looked for order and certainty; Shannon, a friend of novelty, saw disorder as the motivating force behind everybody's need to find information in his or her surroundings. A crystal clear paradigm for information theory as a scientific discipline capable of accounting for all communication, whether by way of humans or machines, was not to be seen in the distant horizon (von Foerster 1980).

Reentering the Labyrinth

Nevertheless, during the early years of cybernetics and information theory, unbridled optimism continued: all problems would soon be resolved, and we would now have robots to do our work for us, and we could spend most of our time sipping margaritas on the beach, watching CNN repeats, and playing golf at the nearest course. And to boot, if those with the know-how and those in charge could just get their act together, as Simplicio tells us, we would have been on the road toward finally effacing that ancient distinction between humans and machines. It was all basically a matter of systems of control and communication when looked at from the larger view.

Indeed, during the 1940s and 1950s, the founders of cybernetics thought they had taken the first giant step toward fusing humans and machines and brains and computers. To cite only a few cases specifically regarding theories of the brain-mind, Karl Lashley's (1951) influential concept of the hierarchical, yet serial, organization of language, thought, and behavior put the program at a distance from the binary stimulus-response equation of behaviorism. Donald O. Hebb (1949) argued that behavioral patterns are built up gradually over time through interconnections within "cell assemblies," which afforded the beginnings of a holist account of the brain. W. Ross Ashby's *Design for a Brain* (1952), which proposed that logical axiomatic methods could be used in designing a machine capable of learning behavior, influenced a whole generation of researchers. Above all, the pathbreaking model of neural networks based on a logical calculus developed by Warren McCulloch and his associates (McCulloch and Pitts 1943) proved to be an ambitious program. Indeed, it served as an inspiration to a considerable number of scholars. But in the long run it bore little fruit. These and other logically and mathematically based cybernetic models fell from grace from the latter 1960s to the 1980s. But now that the cognitive sciences have recently been experiencing serious birth pangs, cybernetics has made somewhat of a comeback in the form of "parallel distribution processing," of models of the brain at the cellular and neurochemical levels, and more specifically in computer studies of neural nets. In Howard Gardner's (1987, 398–99) estimation, models of human cognition—and, we would suppose, ultimately of the brain, and in AI (artificial intelligence) work, of the computer—must eventually incorporate both parallel distribution processing models and "serial-symbolic" models. This time around, it appears that cooler heads are prevailing, since all but the most credulous of observers readily admit that the day of thinking, intuiting, and sentimentalizing computers is still in the distant and receding horizon.[15]

But a real problem existed from the very beginning at the heart of the dream of information as a catchall term. If the universe is a machine and the brain is a computer and DNA is governed by a code and everything is nothing but a matter of information, then there is hardly any room for change: the task at hand is that of proper coding according to fixed machine principles. What is lacking, it hardly needs saying, is an account of process. Actually, brains are more than what machines now are, and by a long stretch of the imagination machines may become somewhat comparable to what brains now are, but by that time, and in order for brains to have built machines that complex anyway—according to the limitative theorems of Alan Turing, Emil Post, Alonzo Church, Kurt Gödel, and others—brains would certainly have become even more complex (Dreyfus and Dreyfus 1986; DeLong 1970).[16] This very important limitation implies process, that is, semiosis, in the sense of C. S. Peirce and other process philosophers, most specifically, Henri Bergson, John Dewey, William James, and Alfred North Whitehead. Process, bound to the changing tide of time, allows for the woofs and the warps, and the fluxes and the flows, of meaning and of interpretation. Processual engenderment of signs is never a static affair; it is always on the go, depending on indeterminately variable contexts and structures. The problem is that the cybernetics of codes and information, grounded in and often inordinately ground into binary bits that are statistically determinable and presumably mathematically definable, accounts for the probability of certain signs popping up but not others and with little to no regard for contextualized processes.[17]

To make a long and tired story short, the problem can be illustrated by a variation of John Searle's (1980b) notorious "Chinese room" thought experiment designed for debunking the AI dream. It goes something like this. Assume you know no Chinese. You are placed in a room and given instructions (a set of rules) regarding how to transform (according to a cut-in-granite code) the Chinese-language characters that are fed to you into English. You are then to push the English symbols out of your room through a slit in the wall to some recipient. You are acting much in the order of a translation machine. However, though you can carry out your translations methodologically and faithfully, you have no idea whatsoever whether your translations are correct. You are simply following the rules provided for you with respect to an established code according to which you can map unknown characters in Chinese onto words in English. Nor do you necessarily have any comprehension regarding your mechanical procedures. You are simply operating as if you had been programmed by a computer expert to do what you do in the way you do it. In other words, you are simply exchanging one set of signs (bits) for another set of signs (bits) without any notion of their significance, without any consideration of the context within which the exchange is carried out, and without any regard either for you as sign creator or for your addressee.

Moreover, the Chinese room example involves two languages as if they were static and fixed for all time, which they are not, for all human languages are incessantly in the throes of change in a changing world of changing speakers, contexts, and situations. If your instructions within the Chinese room had been the equivalent of Katz and Fodor's (1963) famed dictionary devised for their semantic model of language, then what you translate today would be the same as if you had translated it, say, thirty some years ago when the dictionary was published. But dictionaries become obsolete with mind-wrenching rapidity. They do not remain entirely faithful even from one year to the next, in spite of what the guardians of good taste would like to believe. If you had used something in the order of Katz and Fodor's encyclopedia—in semiotic theory also appropriated by Eco (1976)—you would stand on somewhat firmer grounds. But beware of the inevitable soft spots that play havoc with your hopeful foundations. Meaning and your comprehension of a familiar language, the moral of this story goes, are not in a code, in the rules of grammar, or in the formally defined information of information theory. Meaning and comprehension are in your interrelations and interactions with those around you and with your world as both you and your language incessantly change. Yet, following also from the limitative theorems, what you take as meaning and your comprehension of your language, as process, is invariably incomplete, and if not, then it will be inconsistent.[18] For, as even Wiener (1948, 11) was prone to concede, incompleteness there will always be. However, he goes on to write: "This random element, this organic incompleteness, is one which without too violent a figure of speech we may consider evil; the negative evil which St. Augustine characterizes as incompleteness."[19]

The next stage of the present inquiry calls for a return to the problem that antagonized Simplicio and Sagredo: simplicity and complexity. If the apparent goal of information theory is to make of disorder and complexity relatively ordered simplicity, and if process involves the increasing complexification of what began as simplicity, then we need to rethink the relations between the key terms discussed thus far.

CHAPTER 2 • How/Why Is Simplicity Complex and Complexity Simple?

As far as the laws of mathematics refer to reality, they are not certain, and as far as they are certain, they do not refer to reality.

—Albert Einstein

The waves break and spread their waters swiftly over the shore. One after another they massed themselves and fell; the spray tossed itself back with the energy of their fall.

—Virginia Woolf

The waves that roll on one after another in endless succession all return to the one great water, and which in turn swells up again into its waves. No "waves" exist apart from their water, nor does "water" exist apart from its waves.

—Keiji Nishitani

If we study Japanese art, we see a man who is undoubtedly wise, philosophic and intelligent, who spends his time doing what? In studying the distance between the earth and the moon? No. In studying Bismarck's policy? No. He studies a single blade of grass.

But this blade of grass leads him to draw every plant and then the seasons, the wide aspects of the countryside, then animals, then the human figure. So he passes his life.

—Vincent van Gogh

Is the Whole as Complex as the Sum of Its Parts?

The questions "What is simple?" and "What is complex?" are both exquisitely simple and exquisitely complex. This circularity is a matter of wholes and the parts contained within them: wholes and parts look both simple and complex, depending on the way of the looking. The living cell is simple when we consider it a building block among building blocks for an organism, but it is complex if we look at its inner workings. Atoms are simple as Tinkertoy models for making molecules in the junior high school science class but of mind-numbing complexity when the physicist peers into their nuclei. By the same token, since one cannot perceive or conceive that whole of wholes—the entire universe—what is for a given beholder a whole, is no more than a part among parts for someone else and vice versa.

At the outset Simplicio, hanging onto the choice pearls of modernity at all cost, would be content with nothing less than knowledge of the whole. For Sagredo the rather laid-back postmodernist, there is comfort in a tentative grasp of a few bits and pieces. "But," a curious bystander asks, "if a given system is statistically determinable, in light of the previous chapter, and if one can know its initial conditions and the algorithm for its changes, then why should one not be able to know it to a tee, even if by way of some 'minimalist' depiction? What is more, given the entropy principle, is it not possible, ultimately possible, for one to know when a turn of the tide will occur to create islands of order amid the surrounding sea of turbulence?" Well, no. Or perhaps better said, not by a long shot. Even better still, not by an infinite stretch. The task at hand, in order for us perhaps a tad better to understand why any faithful account of the whole of things is out of the question for us pitifully finite knowers, calls for deeper exploration of the ever-shifting interface between complexity and simplicity and parts and wholes. Let us turn to Jorge Luis Borges and Italo Calvino for some illustrations of linear simplicity breeding nonlinear complexity regarding this issue.

Borges reconstructs a couple of apparently diametrically opposed concepts, nominalism and realism, condensed in two strange objects, the Zahir and the Aleph, in two stories by the same names that appear in the collections cited in the reference section (1962, 156–64; 1970, 15–30). This pair of concepts actually embodies the Argentine fabulist's intellectual leanings: it could be said that Borges is "philosophically a fox who longs for the simplicity and certainty of the hedgehog, but cannot bring himself to be one" (Wheelock 1969, 24). The fox is a wily nominalist who slips in and out of the numbing complexity of language particulars, while the hedgehog is a realist who desires to see all things through the same tinted goggles in terms of their relatively simple universal properties. The conflict is essentially between a plurality of simples and the complex singularity of a universal vision, that is, between the Zahir and the Aleph.

The Zahir is an ordinary coin worth twenty centavos. It is per se insignificant; its function could have been provided by one of any number of objects: it has been a tiger, a blind man, an astrolabe, a small compass, a vein in the marble of a mosque in Cordoba. Once an object is chosen to function like the Zahir, however, it becomes a particular perspective that potentially reaches out to all perspectives. Potentially, that is, because one who knows not how to avail oneself of the strange powers of the Zahir becomes ensnared by it: one cannot forget it. This inability to ignore the coin becomes the narrator's plight. He finds himself obsessed with the small round object, sensing that it must somehow signify in linear fashion, like language, each and every thing that it is not until by a process of elimination—a sort of *via negativa*—it has signified what the entire universe is. But that project, of course, would be out of the question for any and all finite humans. In good nominalist fashion, then, the Zahir can arbitrarily be anything that stands for something else—that is, the ideal Saussurean arbitrary sign. We thus enter the arena of language. It is not mere coincidence that the vast majority of all considerations of cybernetics, information theory, and AI research, insofar as they bear on concerns in the social sciences and the humanities, focus obsessively on language. In our recent linguistic turn of events, we have all but disappeared in the digitalized staccato of Saussurean signifiers, of distinctive features, of textuality, and above all, of the arbitrariness of it all, which presumably allows signs to liberate themselves from the furniture of the world euphorically to do their own thing, whether we like it or not.

This notion of arbitrariness, quite significantly, was also the observation of Mr. Palomar in a novel by Italo Calvino (1985) by the same name. While in Mexico and visiting the ruins of Tula, ancient capital of the Toltecs, Palomar contemplates the various representations of Quetzalcoatl as the morning star, the monolithic columns known as "Atlases," a butterfly, and the Plumed Serpent. He realizes that all these signs must be taken on faith that they represent something else, and if the signs are rejected, such rejection must be taken on faith also, the faith that they do not represent something else. In other words, in Simplicio's conception of things, signs of the real and of the nonreal are equally real signs; hence they can be as real or as nonreal as either the real or the nonreal itself. Signs in this sense are of the very stuff of which our world and the world of our imagination are made: whether "in here" or "out there," all that is, insofar as we perceive and conceive it, consists of signs of one form or another.[1] But Palomar bears witness to something other than that logocentric bias within which Western thought is caught. The ancient art he is witnessing is hardly in any form or fashion linguistic; apparently like the Zahir, it is visual, iconic through and through. In the Aztec world, seeing in the most concrete sense predominates, which means that saying takes on a role of diminished importance—that is, diminished in comparison with Indo-European languages:

> In Mexican archeology every statue, every object, every detail of a bas-relief stands for something that stands for something else that stands, in turn, for yet another something. An animal stands for a god who stands for a star that stands for an element or a human quality, and so on. We are in the world of pictographic writing; the ancient Mexicans, to write, drew pictures, and even when they were drawing it was as if they were writing: every picture seems a rebus to be deciphered. (Calvino 1985, 95–96)

Where is meaning in this scheme of multiply interlinked, interdependent, nonlinear series? The only possible answer, it seems, is everywhere and nowhere. Every sign represents another sign, and that another one, potentially without end. Every sign defers the responsibility of its act of representation to other signs. Every such deferral is a translation of one sign into another one, which requires another translation in order that it emerge into the diaphanous light of comprehension, though it never stands a chance of actually arriving at the pristine plentitude of meaning perfectly wrought and crystallized for all time. We read further that a "stone, a figure, a sign, a word reaching us isolated from its context is only that stone, figure, sign, or word: we can try to define them, to describe them as they are, and no more than that; whether, beside the face they show us, they also have a hidden face, is not for us to know" (Calvino 1985, 97).

Now, it appears, Palomar has gravitated from Zahir-like pictorial images toward contemplation of language, and with language, linear writing, we would suppose, in addition to his reference to nonlinguistic signs. Palomar realizes that each linguistic sign requires another sign, requires all signs, from concrete figures to words, from language to language, from culture to culture. The need to glide along the surf of signs interrelated to all other signs is in a way frightening, yet compelling. There actually seems to be hardly any alternative, for that is the message we seem to get from language: it becomes relatively disconnected from the furniture of the world, and it to some extent takes on a life of its own; it becomes a self-organizing whole. Or at least so we are told by those high priests of the linguistic turn. However, language also brings with it the need of its users to reweave and unite it into the continuous and well-nigh seamless whole of culture, which, as we shall observe, is aggressively translinguistic. Consequently, the monstrous aggregate of linguistic signs cannot really remain aloof and autonomous at all, for its very use demands some relation with something other than what it is; it demands interpretation, that is, interaction at least with some interpreter within some cultural context. And if interpreters and interdependent contexts are included in the equation, then interpretation there will always be. For: "Not to interpret is impossible, as refraining from thinking is impossible" (Calvino 1985, 98). It might appear that we are now a far cry from information theory as discussed in chapter 1. Meaning in natural languages pushes beyond mechanical context-free codes and gutless bits of information or signs toward the concrete contexts of their use.

However, we are not yet free of the overpowering force of language. Interpretation, it hardly needs saying, is most effectively conveyed through language, though not exclusively through language, since other signs can serve as helpful adjuncts. Yet, one almost inevitably gravitates toward language and toward linguicentrism.[2] The problem with linguicentrism is that it places undue priority on language, the Saussurean signifier, the Peircean symbol, while shoving the crucially important functions of iconicity and indexicality aside.[3] Perhaps this move toward linguicentrism is hardly avoidable. The idea that language is what makes us most distinctly human pushes us toward the imperious attitude that, as proud owners of "minds," we are above and beyond the nitty-gritty world of instinct the dumb brutes inhabit. As articulate mammals we are also writing and reading mammals.

In *If on a Winter's Night a Traveler* (1981), Calvino tells us that reading—and writing as well—is no more than combinatorial play. But we also read that if what is written and read is false to itself, the product of language's incapacity to represent something other than itself, then ultimately, reading and writing are like a country where everything that can be falsified has been falsified. The result is that nobody can be sure what is true and what false, what is simulation and what is real (212). Calvino's allusion to "simulation" evokes shades of Baudrillard's (1980, 1981, 1983) "simulacra," commodities and their respective signs incessantly repeating themselves until the commodities have been relegated to the dustbin of forgetfulness and all that remains is signs of signs. It is also reminiscent of Umberto Eco's (1990, 172–202) "fakes and forgeries," simulacra or iconic models that are hardly worthy of being called signs. Eco the linguicentrist has a pretty low opinion of icons. Completely iconic signs, he writes, would be identical to that to which they refer and would therefore have no genuine semiotic function: they would in essence be tantamount to the real thing and in no need of the things to which they might have referred—that is, shades once again of Baudrillard's "simulacra."[4]

To be sure, our world is a semiotic world through and through; it is a world chiefly of signs, of a "perfusion of signs." However, linguistic signs are actually no more than a small minority of the entire sphere of signs making up our world. Borges's narrator who came into possession of the linearly developing signs emerging from the Zahir discovers this important aspect of semiosis. The Zahir at the outset appears to be nothing but a mere icon, a visual object that evokes any and all signs other than itself. But it also indicates something *other,* so it has an indexical function as well. Moreover, since it is capable of bringing about the emergence of whatever sign, it is also to a degree arbitrary; hence it is also of symbolic character. The Zahir is prelinguistic, and at the same time it is of the nature of language. It is of the world of physical objects, yet it is a sign, it is part and parcel of the semiosic process. As Simplicio intuited at a gut level and articulated in a rather complementary contradictory manner, signs are prelinguistic before they gain entry into the venerable empire of language: they are icons and indices of sight, sound, touch, taste, and scent before they become arbitrary phonemes exemplified in sounds and marks. Yet, as mentioned in the preceding, it appears that symbolicity inevitably comes to pervade the minds and hearts of human articulate animals. So, just as Palomar gravitated from iconic pictorial images to linguicentricity, so also the Zahir, even though it is at heart iconic, through symbolicity—its narrator's prime medium of expression—it cannot help but become saturated with linguicentricity as well.

The Zahir, nonetheless, is a relatively benign sign. Lying in linear contiguity with all other objects of the world, it is, or it can be, or at least we would like to make it, a representation of all that is, quite clearly and simply. But it can hardly do more than function as an oxymoron, the narrator tells us, insofar as it is not that which would ordinarily represent the represented. Consequently, it is capable of all possible perspectives, perhaps in the order of Calvino's *ars combinatoria*. But, since the Zahir nervously flits from one perspective to another, and from one linguistic label to another, the sum of these perspectives can be no more than sequential in nature, an apparently rather disconnected collection of series something like that envisioned by Gilles Deleuze in *The Logic of Sense* (1990a). Given their sequentiality, and in light of the infinite stretch of all series the prime metaphor of which is the line (a mathematical continuum), they can never reach the end of the road. Like language itself, the Zahir apparently affords successive perceptual grasps of fragments of the universe, though none of them as an individual can be all embracing. Hence, given human finitude, the incapacity to hold more than a few items of thought mentally in check for more than a fleeting instant ensues: the Zahir is ultimately a helpless sign. The narrator, finally realizing this limitation, ends his story with a futile hope: "In order to lose themselves, in God, the Sufis recite their own names, or the ninety-nine divine names, until they become meaningless. I long to travel that path. Perhaps I shall conclude by wearing away the Zahir simply through thinking of it again and again. Perhaps behind the coin I shall find God" (Borges 1962, 164). Thus the narrator's destiny is hardly any less undesirable than that of our curious seeker of knowledge, Mr. Palomar.

If the Zahir is not itself a legitimate word—a Peircean symbolic sign—it is still a sign, to be sure, an iconic and indexical sign. As such, it is real, real as a sign, though as a thing it is nonreal, yet it is a real thing, for it is a sign—to evoke Simplicio's enigmatic finale. The Zahir, then, is of the physical world and of the world of signs, though it is chiefly a nonlinguistic sign. Hence given its myriad array of exemplifications, there is no disconnectedness of series at all but a conjunctive synthesis of series (Deleuze and Guattari 1983—to be discussed subsequently). In this sense, any attempt to interpolate the Zahir into an exclusive linguistic framework would render it linear and relatively simple; but it is not simply linear, for its interdependent interconnectedness with all things in terms of its iconicity and indexicality renders it nonlinear, and as such it enters full force into complexity. Each instantiation of the Zahir as a nonreal sign for something else, which is either a real or a nonreal sign, is a part among parts, the sum of which makes up the whole of semiosis. But that whole remains outside the reach of any given Zahir instantiation, even of any given serial collection of Zahir instantiations. With this broader, more general, concept of semiosis in mind, we are now a far cry from the gutless bits and codes of fleshless information channels that were in the spotlight of chapter 1. Semiosis, properly conceived, is nonlinguicentric through and through.

Borges's tale, "The Aleph," corroborates this nonlinguicentrism and takes it to a shrill pitch of intensity. Carlos Argentino Daneri tells the narrator (that is, Borges) of the Aleph's existence in the home of his parents and grandparents, explaining that it is a point in space that is interconnected with and contains all points. Borges visits the house and descends the stairway leading down to the basement where it is located, and he experiences it, a sphere about one inch in diameter, whose center is everywhere and whose circumference nowhere, a beginning without terminus, paradoxically both finite and infinite, the only place where all places are seen from every possible angle: it is the whole of wholes. The narrator then petitions the gods that they might grant him the appropriate metaphors with which to describe this miraculous vision, but he knows it is impossible:

> for any listing of an endless series is doomed to be infinitesimal. In that single gigantic instant I saw millions of acts both delightful and awful; not one of them amazed me more than the fact that all of them occupied the same point in space, without overlapping or transparency. What my eyes beheld was simultaneous but what I shall now write down will be successive, because language is successive. Nonetheless, I'll try to recollect what I can. (Borges 1970, 26)

The narrator's feat is impossible, because the whole cannot but be timeless, while he as a pathetic mortal is inextricably time bound. And for the purpose of human communication, he is bound to linear language processes. What he knows he knows, now. But to say what he knows is to say what he has known, which is not exactly the same as what he knows at each moment of the saying; hence the saying takes what he knows a bit further down the road, or it retrogresses, perhaps, depending upon the way of the saying. In other words, to say is to know anew, yet to know is to know what cannot be said now.

Linear language engenderment is like a Zahir series: hopelessly inadequate for articulating what timelessly is—knowledge, perceiving, sensing, conceiving—whether in the now, in memory of the past, or in expectations regarding the future. The Aleph's timeless complexity consequently eludes the articulating animal, tied to linear language, just as it eludes the Zahir's multiple time-bound, relatively simple, series. Along comparable lines the whole of the prisoners' lawyer's timeless logic (discussed in the introduction), revealing an inconsistency in the warden's thinking, was beyond the ken of her hapless clients, the prisoners, caught within their own linear existential logic of the particulars of their daily activities.

In yet another way of putting the matter, the Aleph affords a realist image as opposed to the Zahir's nominalism. The one entails an impossible transcendental revelation, the other a potentially interminable series of relatively insignificant perceptual grasps. The one is synchrony, the other diachrony; the one is a nonlinear intertwining of all objects, acts, and events in complex simultaneity, the other a serial collection of relatively simple particulars with no necessary or determinate links. According to Borges, today we almost instinctively favor nominalism, but in spite of ourselves, we implacably gravitate toward the opposite pole in an effort to discover the whole of reality in those hopeful eternal forms (Christ 1967). And since this search is ultimately futile, we find ourselves fleeing back to the secure minutiae, the particulars, of our everyday empirical world.[5]

In short, we have the Aleph and the Zahir, or the whole and its parts: an impossible grasp of the whole's nonlinear, interdependent, interconnected complexity compounded into a timeless present; an equally impossible, time-bound, linear accumulation of quite disconnected particulars, each of them of relative simplicity but the summation of all of which will ultimately become coterminous with the whole.[6] It is nonlinear complexity taken in one gulp, or simplicity linearly snowballing, over time, to become the most complex of complexities. In light of suggestions from the previous chapter, the calculus of cybernetic information is linear and relatively simple; in contrast, nonlinear and relatively complex processes must be left to other concerns that bear on chaos theory and the physics of complexity, which must be placed in the spotlight at a future time.

So, the first props—that is, *the stringent limitations of our knowing*—have been set on the stage of this inquiry: the initial brush strokes attached to the blank canvas eventually to become the background have endowed it with a certain tinge, however vague, of polychromatic points appearing as minuscule globs.

The next step entails a further look at the relationship between parts and wholes. In this regard, it seems that Mr. Palomar beckons once again.

The Great Divide Blurs

Simply put, Palomar wanted to know what needed to be known—which is actually light years removed from simplicity. It is therefore significant that we initially find him at the beach "reading a wave."

Palomar is not merely lost in contemplation while his eyes remain fixated on the ocean wave, because "he is quite aware of what he is doing: he wants to look at a wave and he is looking at it." It is not upon the waves in general that he is focused but on "just one individual wave: in his desire to avoid vague sensations, he establishes for his every action a limited and precise object" (Calvino 1985, 3). He is not seeking generalities but, rather, everything in general (the whole—Aleph) seen through one solitary particular (a part—Zahir instantiation). Each wave rises in the distance, grows, approaches, folds in and over itself, breaks, and dissipates and merges into the continuity from which it arose; it diverges and converges, overlaps and merges with other waves, then disappears only to reappear somewhere else. Palomar realizes he cannot observe a wave without bearing in mind the complex features that concur in shaping it and the other, equally complex ones that gave rise to the wave and that the wave itself originates. These aspects vary constantly, and so each wave is different from every other wave, even if it and another wave are not immediately adjacent or successive; in other words, "there are some forms and sequences that are repeated, though irregularly distributed in space and time" (4). In order to get a grasp on the essence of the waves before him, Palomar tries limiting his field of observation to a square of equal portions of shore and sea, hoping that with fixed boundaries his power of observation will be enhanced and he will see everything there is to see. Perhaps, he surmises, this could be "the key to mastering the world's complexity by reducing it to its simplest mechanism" (4). But something new, something unexpected, always pops up. He even tries reversing the direction of the waves in his mind, thus overturning time, but the overpowering force of physical events wins the battle. If only he could bear in mind all aspects of this microcosm, a solitary wave, then he could leap to the next phase of the operation: extrapolating his findings to include the entire universe. But he soon loses patience and leaves the beach as "tense and nervous as when he came, and even more unsure about everything" (8); (see also Sbragia 1993).

Borges's Aleph consisted of the intertwined and virtually continuous whole condensed to a minuscule bundle; his Zahir was an infinitesimal part presuming the capacity successively, and in linear discontinuous fashion, to recapitulate the whole. Palomar engages in the equally impossible task of seeing the whole in one of its parts that exists continuously with all other parts. Aleph, Zahir, sea, all are not what they are unless they are subject to subjects, knowing subjects. One narrator, Simplicio's ideal scientist, contemplates the Aleph in a timeless, ecstatic grasp; another narrator, tantamount to Sagredo's finite, fallible observer, exercises a sequence of cumulative ephemeral gazes at the Zahir as it moves along its linear path; yet another narrator tells of Palomar's studious efforts to commune with the whole by analysis of a minuscule patch of it nonlinearly. Complexity reduced to a point replete with the universe's discontinuities; simplicity spread along a line in discontinuous segments, hoping to become infinite complexity; and homogeneous, repetitive continuity promising, but never allowing delivery of, a grand unified grasp. Each story depends upon an original separation between knower and known, or better, on the logical impossibility, vaguely intuited by Sagredo, of there being a split between the two.

Interestingly enough, we meet this knower/known problem in another of Mr. Palomar's adventures, appropriately entitled "The World Looks at the World" (Calvino 1985, 113–15). After Palomar had met with a series of "misadventures not worth recalling," he decides to look at things "from the outside." Absent-minded, nearsighted, and introverted, unfortunately he is neither suited to nor does he have the aptitude for objective observation. Yet, undaunted, he tries looking at everything, but the infinite wealth of the universe resists his hopefully omniscient gaze. Deciding that only some things lend themselves to their knower, he then sets about to seek the right ones. To do this, he has "to face each time problems of selection, exclusion, hierarchies of preference; he soon realizes he is spoiling everything, as always when he involves his own ego and all the problems he has with his own ego" (114). He must get rid of his ego, so he takes it upon himself to reconstruct the Alberti Renaissance vanishing point—which is epitomized in that notorious Dürer illustration of the painter behind a checkerboard grid objectively and dispassionately painting what he sees in each square of a nude in repose on the other side (Edgerton 1976). The painter enjoys a window to the world, with him- or herself on one side and that world on the other side; Palomar likewise believes that he can get a wholly objective grasp of the field before him by abstracting a part of it and analytically exhausting it.

In other words, the ideal Cartesian split between mind and world and body and mind that has so pervaded Western thought and led to the untenable divisions of which Gregory Bateson (1972) and Ronald D. Laing (1965, 1969) have in their own ways so eloquently written, is Palomar's goal. Palomar's "I" splits into "a looking world and a world looked at" (Calvino 1985, 114). The question then becomes, What has happened to Mr. Palomar's "I," his ego? There is, strictly speaking, neither inside nor outside, nor here the ego and there everything else in the universe, but somewhere in between, wherever that is. The position of the "I," then, could be conceived as something like that of Wittgenstein's eye of the *Tractatus* (1961). It lies at the periphery of its world from which vantage point it can gaze imperiously upon that world, but it cannot see itself seeing it. In this same vein, we might surmise, it is like the hairline splitting Sagredo's cosmic egg in figure 4 that contains, "within" itself, both blackness and whiteness and from which split, the perpetually oscillating $\sqrt{-1}$, a world is cut into myriad islands of nonlinearly related blacks and whites pulling self-organizing beings (i.e., the Harlequin) into existence. However, in much the Renaissance sense, Calvino's narrator believes he lucidly surmises that "the 'I,' the ego, is simply the window through which the world looks at the world" (Calvino 1985, 114). This is the imperious image of Bacon's ideal scientist set apart from the world he or she sees, a split raised to a screaming pitch by the good grace of Descartes's grand mind/body dichotomy. But such aloof, disinterested, detached objectivity is the stuff of which dreams are made. What Palomar actually sees is still the same old gray world. He does not see it with scientific eyes cleansed of any and all presuppositions, preconceptions, and prejudices. For in order that there be a seer and a seen, there must be some sort of a priori split:

> From the mute distance of things a sign must become, a summons, a wink: one thing detaches itself from the other things with the intention of signifying something . . . What? Itself, a thing is happy to be looked at by other things only when it is convinced that it signifies itself and nothing else, amid things that signify themselves, and nothing else. (Calvino 1985, 115)

If this a priori split was conceivably the origin, when something was separated, separated itself, from everything else, such that there were now two primitive icons without relation to one another, without indexicality—the one indicating the other—and much less symbolicity, then that split must now be part of some long lost and irretrievable past. What is more, when this primordial sign related itself to something else, its other, it marked the initiation of signs as predecessors, signs that presupposed what might follow, and signs that led to the *pre*-conception of what should follow and eventually culminated in *pre*-judgments, *pre*-judices, regarding what actually followed.[7]

But this is all too much too fast, I'm afraid. While catching our breath for a moment, let's go back to that longed-for mythical beginning, the initial split, however remote, between "I"-here and everything else, between knower and known, and see where it leads us. Calvino's Qfwfq, the leading character of *T Zero,* who seems to be always already everywhere and nowhere, the thread tying many of the otherwise quite disparate short stories in the volume's collection, together, gives a fictional account of the painful struggle to enact a primordial knower/known split. In "Blood, Sea" he tells us:

> I swim, I continue swimming in the same warm sea. . . . [T]he inside isn't changed, what was formerly the outside, where I used to swim under the sun, and where I now swim in darkness is inside; what's changed is the outside, the present outside, which was the inside before, that's changed all right; however, it doesn't matter much. I say it doesn't matter very much and you promptly reply: What do you mean, the outside doesn't matter much? What I mean is that if you look at it more closely, from the point of view of the old outside, that is from the present inside, what is present outside? It's simply where it's dry, where there is no flux or re-flux and as far as mattering goes, of course, that matters too, inasmuch as it's the outside,

> since it's been on the outside, since the outside has been outside, and people believe it's more deserving of consideration than the inside. (Calvino 1969, 40)

Here we have shades of mathematician and logician G. Spencer-Brown's (1979) originary mark of distinction (or cut, primordial icon) that is followed by an indication (index) of something else that is other than what has been marked. In other words, there is now *this* and *that, here* and *there.* An originary mark of distinction, it hardly needs saying, is of the utmost, the most primitive, simplicity, as is the ensuing linear series of marks that follows from it. Spencer-Brown's "Laws of Form" was appropriated by biologists Humberto Maturana and Francisco Varela (1980, 1987) in their formulation of the basis of living organisms. And the concept was raised to a screaming pitch in Gilles Deleuze and Félix Guattari's (1983, 1987) "machinic organism." In the terms of Deleuze and Guattari, radically nonlinear rhizomicity rather than arborescent models proliferates in all possible directions to breed complexity of the most intricate sort. This is organized complexity, to be sure. But it is not that complexity inherent in chaotic states, and it is light years removed from the simplicity so desired by Simplicio.[8]

The "I" of Qfwfq's remarks in the preceding Calvino quote, that bottom level of self and self-consciousness, is also reminiscent of Beckett's Unnamable: "I'm not outside, I'm inside, I'm in something, I'm shut up, the silence is outside, outside, inside there is nothing but here, and the silence outside, nothing but this voice and the silence all around" (Beckett 1955, 410). Beckett's vacillating, hesitating, fluctuating Unnamable, engulfed with uncertainty, doubt, expectations of nothing and constant surprise that he is not surprised that nothing happens, simply is: neither here nor there, neither in a now nor a then. The Unnamable would like to think he occupies the center, but nothing could be less certain. Maybe he is situated somewhere between the center, which is everywhere, and the circumference, which is nowhere. But then where would he be? "From centre to circumference in any case it is a far cry and it may well be situated somewhere between the two" (295). It is equally possible "that I too am in perpetual motion, . . . as the earth by its moon. All is possible, or almost" (295). At this point, where all is clear yet nothing is clear, from this specified and precise yet indefinitely vague beginning that is no beginning, the number of possibilities is virtually infinite.

This was also much the case of Qfwfq's beginning in "Blood, Sea."

The outside is a vague, confused mass when looked at from the inside. What is not so indecipherable is the kinesthetic sensation of the dilating, vibrating, pulsating, but tenderly fragile surface, a bare membrane, the most fundamental of life's fundamentals, separating inside from outside. It is as if one side were one and the other side the other with both separated by the oscillating $\sqrt{-1}$ of figure 4, to reevoke that well-worn image.[9]

Messages can be transmitted between Qfwfq as an inside and other insides in the vicinity through the "vibration of the waves," the medium allowing for haptic, felt, vaguely sensed signs in their "swimming manners." Well, actually, we the readers eventually discover that Qfwfq and his companions are in a "VW bug" careening down the highway! It appears that while plodding through the prose we were privy only to the most primitive levels of their consciousness that remained in the oceanic lull accounting for the tenor of Qfwfq's painfully tentative attempts to come to grips with himself and his surroundings. Very effective, this beginning of beginnings of life from the viewpoint of some slice of life from within the midstream of life's general process and, in fact, of the entire process of evolution.[10]

With respect to the originary split of the Qfwfq sort between "I" and other, knower and known, subject and object, inside and outside, Maturana and Varela (1980, 1987) offer the concept of autopoiesis (auto = self, poiesis = production) as a means for describing living systems in terms of autonomous unities. In accounting for these unities, Maturana and Varela contrast autopoietic systems with allopoietic systems. The latter, like cars, are not self-organizing and self-motivated but need an outside and separate agent in order to carry out their function, from the showroom to the junkyard. Allopoietic systems can experience no self-realization, unlike their counterparts, autopoietic systems. Their advantage is that they can hardly suffer from any identity crisis, although such crises can plague autonomous unities without end since they apparently have some inner need to remain "identical to themselves" over time, in spite of the fact that they are constantly exchanging energy and information with their environment.

In this respect, Calvino's VW bug metaphor is not drawn at random. Redirecting our focus on "Blood, Sea" in Maturana and Varela's terminology, Qfwfq and his cohorts are significantly inside their "bug," the vehicle most closely approximating cellular form. As no more than vague suggestions of inside/outside, the passengers within the bug, and the entire bug as a cell, are hardly more than allopoietic. The act of lifting themselves up by their own bootstraps into autopoiesis is as paradoxical as is the capacity of Zeno's arrow to move, to move itself, from one spatial increment to another—that is, if we remain within the parameters of Zeno's logic. Moreover, even if an allopoietic system—or Qfwfq—could exercise such a feat and gain entry into autopoiesis, it would not be exclusively autopoietic, for it would remain perpetually in the act of taking in and expelling matter and energy. So its inside would be destined to become part of what was outside, and the outside would contain part of what was inside. As would also be expected, Qfwfq fails to keep time and space separated. He is at an ineffable stage of his development that is even more primitive than that stage that he is able to articulate, albeit in stumbling fashion, in a later story in the same Calvino collection entitled "Mitosis." There, we read that it was a matter of

> me, in that point and at that moment—right?—and then there was an outside which seemed to me a void I might occupy in another moment or point, in a series of other points or moments, in short a potential projection of me where, however, I wasn't present, and therefore a void which was actually the world and the future, but I didn't know that yet; it was void because perception was still denied me, and as for imagination I was even further behind, and when it came to mental categories I was a total loss, but I had this contentment because outside of me there was this void that wasn't me, which perhaps could become me because "me" was the only word I know. (Calvino 1969, 63)

Qfwfq's sensing that what is not him could become him and what he is could become part of what he is not ushers in the roots of time, to be sure. Consequently, Qfwfq's ruminations dispel the idea that he is totally autonomous. He is autonomous, since he has brought himself into existence as *this* in contrast to *that;* but he is not autonomous in the sense that he remains interdependent upon and interrelated with everything that is other. Qfwfq's realization of this aspect of his existence arouses "an exhilarating enthusiasm in me, no, a torment, a dizzying torture, the dizziness of a void which represented everything possible" (Calvino 1969, 64).

In other words, the Qfwfq or Maturana–Varela autopoietic system is autonomous, yet it is not autonomous in the absolute sense because it must remain open to its environment, for otherwise it would soon degenerate and die. Since it remains perpetually open, there can be no legitimate closure, yet infinite possibilities of closure are always open to it; therefore there is no possibility of any closure to end all closures. Maturana and Varela's thesis appears quite paradoxical and rightly so. They tell us that the key to resolving the paradox rests in the manner in which we interpret the terms *process* and *relationship*. Autopoietic entities are process, not product, until their dying moment. And their nature as process is dependent upon their relationship with everything, autopoietic as well as allopoietic, found in their surroundings. However, nowhere does the notion of anything, any-*thing*, come into the picture. This is for a very important reason, as we shall note in more detail subsequently: there are no "things" but only "events," defined in terms of interrelations of interdependency that make up processes. The upshot of Maturana's and Varela's conception is that autopoietic entities are from one perspective self-contained and self-sufficient wholes and they are from another perspective in a perpetual process of self-organizing in terms of a bootstrapping operation. As such they contain the grounds for their own existence; they are like fire that consumes but does not consume itself. Yet from another complementary view, they are in constant interaction with their surroundings—a larger whole—and they depend upon those surroundings for their very existence. That is to say, there are no ultimate grounds for their existence, since without the whole of their surroundings, which expands in concentric circles ever outward, they exist not at all.

The same can be said of all "thing-events," given the emerging view that the universe consists of an interconnected whole of mutually interacting, interdependent, and interpenetrating parts, that is, events.[11] Fire needs something to consume, something to interact with other than what it is, for if there is nothing other than fire, fire is not fire—like the eye that sees what it sees but cannot see itself seeing. Fire is noncombustion, though it is the agent of combustion: the very existence of combustion depends upon noncombustion, upon fire itself. In other words, fire is not what it is without some combustible patient and its agent, which agent is noncombustion, that is, fire. In still another way of putting it, fire by itself can be no more than pure "emptiness," that which gives rise to any-*thing* and every-*thing*. Emptiness is not itself exactly empty, however, for there is in unrelated fire no-*thing* to be empty. It is simply emptiness. That's all there is to it. Much the same can be said, it seems, of inside and outside, of complexity and simplicity, or of chaos and order. The one is dependent on the other for its very existence, and without the other the one is nothing at all, or better, it is emptiness. Thus Pascal's great chasm between the infinite and the infinitesimal, Nietzsche's Great Affirmation, Derrida's joyous play of the universe, Wittgenstein's language games, Rorty's conversation of mankind, and even Cantor's mathematical continuum all ultimately begin and end in emptiness as they approach the limit where they merge into everything else. Emptiness, if I may say so, is implied graphically in $\sqrt{-1}$ of figure 4 and by its function at the center of figures 5*a* and 6. It is nothing, no-*thing,* so to speak, yet it contains, within itself, the possibility for every-*thing.* So it is itself both no-*thing* and every-*thing,* and conversely, it is neither no-*thing* nor every-*thing,* according to how one would like to take it.

The notion of emptiness, actually Asian in origin, is a long stretch removed from the classical Renaissance-Enlightenment model with which Mr. Palomar remains so engrossed. The Renaissance-Enlightenment model places blind faith in the real, with everything actualized and *there,* as it must be, for it could not be otherwise, presumably. An alternative to this model comes in the order of a self-contained world of neither purely insides nor outsides, of neither subjectivity nor objectivity. There is a hint of such an alternative in Calvino's *Invisible Cities* (1974), which, he tells us, is not "really a novel at all, but a sort of fugue on the nature of the City." The main characters of *Invisible Cities,* Marco Polo and the Great Kublai Khan, are "only attempts at character," they are not really developed (Rushdie 1981, 16). Polo and his tales are more fantasy than real. They are like Christopher Columbus's "America," which was more "invented" than found, more constructed than perceived (O'Gorman 1961). The Great Khan's mind is a willing receptacle, inviting whatever fantasy might be ready to enter it, and his relation to Polo is fantasy looking at fantasy, just as the narrative itself is Calvino looking at Calvino. It is a matter of the inside looking at the inside, at itself, with no possibility of a view, sub specie aeternitatis, and without any actualized physical world "out there." In other words, the inside consists of an immanent, interrelated, interdependent whole without the interrelations and interdependency of which there would be nothing but emptiness. Or better, emptiness would not be, for it cannot be, it just is; no, that is not it either; perhaps better said: just emptiness, sheer emptiness. The nonconcept concept, of course, is slippery.

In contrast, Palomar the Renaissance man believes himself to be properly outside looking at and knowing the outside, that is, what is outside his self, the experiencing self. He believes he objectively observes the waves and everything else that is subjected to his presumably detached, impartial, and undivided attention. In order that he engage in such activity, there must, of course, be that now familiar split between himself and what remains "out there," but if what is split and set apart is actually part of the whole, then it can only at best mirror the whole (ideally in the order of Leibniz's monads, the Buddhist string of pearls, Borges's Schopenhauer for whom one nightingale is tantamount to any and all nightingales, or his Shakespeare who is at once everybody and nobody). This appears innocent enough, at the outset at least. However, *mirror* in the Renaissance-Enlightenment sense demands that the mirror exist outside the whole in order that it may reduplicate and provide a picture of that whole. The problem is that if there is something-someone outside looking out, then the whole is not the whole, and that something-someone outside cannot picture-know the whole as it is, for it is now something less than what it was.

The question, then, is, How to proceed? As if there were no outside at all? As if the outside were actually part and parcel of the inside with there existing no real implication of any outside? Let us linger a moment on the virgule: that which separates, an imaginary line of infinitesimal thickness, that is, the $\sqrt{-1}$ of Sagredo's thought experiment. The line of demarcation is at once linear yet resists notions of linearity. As a cut separating some-*thing* from some-*thing* else, it is inside, though it resists customary binary considerations of inside and outside, and it is itself some-*thing,* as an imaginary entity, though as an actualized entity it is really no-*thing* at all. Borges's "The God's Script," which appears in *Labyrinths, Selected Stories and Other Writings* (1962, 169–73), can perhaps give us a peek into this unutterable nonconceptualizable concept, this yield of unthinking thinking. Tzinacán, an Aztec priest, was captured by Pedro de Alvarado and thrown into a dome-shaped dungeon divided by a wall, with him in one half and a jaguar in the other. Tzinacán was the last priest of the god who had written a secret formula in an unknown place for the last man to read. This formula, with which Tzinacán became obsessed, would supposedly give omnipotence and immortality to its decipherer. After deciding that the sacred script lay concealed in the jaguar's spots, Tzinacán devoted years to learning their configuration in order to decipher the inconceivable text. He once asked himself what type of sentence an absolute mind outside would construct, but, try as he might, he could not discover an answer to this enigma. Finally, he realized that the universe is a universe of immanence, with himself, the jaguar, and god as well inside it, and that each thing in some mysterious way contains everything, paradoxically. Thus "there is no proposition that does not imply the entire universe; to say *the tiger* is to say the tigers that begot it, the deer and turtles devoured by it, the grass in which the deer fed, the earth that was mother to the grass, the heaven that gave birth to the earth" (171)—that is, the interdependent, interrelated, interconnectedness of all things. Tzinacán became aware, in that moment of insight, that in the language of a god each word enunciates the concatenation of facts of which the entire universe consists.

In Tzinacán's sense, like the Aleph, each thing contains and is at the same time tantamount to every-*thing,* in simultaneity. Yet, as with the Zahir, there must be a linear, asymmetrical progression of parts directed toward the whole. But what is the nature of this progression? How can it ultimately culminate in a monumental mass that is tantamount to the whole? Tzinacán has an experience presumably along these lines. One night he dreams, and upon awakening he finds a grain of sand on the floor; he sleeps again and finds a second grain of sand when he awakens. And so on, until he lies buried beneath a hemisphere of sand. Upon becoming aware of his plight, he realizes he has not awakened to wakefulness but to a previous dream within his dream, and so on to infinity. One linear series is parallel—a parallel world—to another linear series, and that to another, . . . and another, without end. One series might end, but others would still be in progress in the space-time continuum of progressions, each of them relative to all others, all of them making up the space-time whole. Consequently, in the grains of sand we not only have the "sorites paradox," the so-called paradox of the heap, but also Cantor's paradox of the continuum—to be discussed subsequently. And to top it all off, we have the intuitive notion that, in William Blake's image, each grain of sand contains all grains, contains the entire universe.

This, of course, is a description of Tzinacán's side of the hemisphere; on the other side the situation can potentially be reduplicated, so there are two, and even many, infinities in the order of Cantor. What is more, the hemisphere within which Tzinacán is imprisoned is divided by a line, the wall, consisting of an infinity of infinitesimal points—another study, or nightmare, of infinity—comparable to Sagredo's bisected ovum (fig. 4). Tzinacán is inside one half looking out, much in the order of Qfwfq and his associates, who exist within and without in relation to their outside. Tzinacán's outside, like that of the inhabitants of the "bug," is also inside, for his existence is that of immanence, not transcendence. In other words, both Tzinacán's inside and outside as well as the dividing line between are, properly speaking, inside, where Tzinacán remains, somewhere, somewhen, suspended. Extrapolating this situation to encompass our own condition, the universe, *our* universe, consists of a space of infinite possibilities only a minuscule portion of which can be actualized for us at a given moment. Yet all the possibilities are always already there, in the dividing line so to speak—as emptiness—and waiting their actualizing turn. In this sense, if Palomar believes he is on the outside looking out, if he believes there is a subject/object or mind/body split, he is in a certain sense correct, though for all the wrong reasons. He has exercised a split of the ovum of his universe in order potentially to observe every-*thing* he is not. But in this condition he is, from within, like the man without qualities, no more than an infinity of possibilities, just as is that which he believes he so detachedly observes. Upon exercising the split, he has introduced asymmetry, temporality, irreversibility, yet as sheer *possibilia* for acquiring an infinity of perspectival grasps, he is the same as that which he is not, that is, as every-*thing* else. He and his universe are of numbing complexity, from the radically pluralistic, Zahirian view, and at the same time, from the Alephic view they are at heart of the utmost simplicity, that is, of the utmost of simple complexity, like a solitary cipher on the jaguar's hide. Each, we shall note, is holistically as complex as a fractal, a strange attractor, a dissipative structure, yet from another grasp it is at its core as simple as a Newtonian or Einsteinian equation.

But . . . I am afraid I am still muddying rather than clarifying the issue at hand. Perhaps it can at least be said that at this stage, the meat of the matter seems to revolve around the distinction between what some-*thing* is and what it is not. The hairline splitting Sagredo's ovum is both the initiation of semiosis and that which makes semiosis possible in the first place. It is not what is, and it is not what that which is not what is, is. That is, to repeat, the hairline is virtually emptiness. It is that from which all semiosis arises. So metaphorically in Tzinacán's prison—or in the figure 4 scheme—we have the whole, the ovum, and its primitive pair of parts, along with that which makes the parts *parts,* the hairline. Each part, like the butterfly flapping its wings in Brazil, can disrupt the whole, and complexity ensues. Yet that unruly part that disrupted the whole can act as a catalyst for the emergence of some sort of new order. At least that is the elegant hope of the new science of complexity.

On this somewhat embarrassingly perplexed note, the second set of props—that is, *the fuzziness between categories of inside and outside and parts and wholes and simplicity and complexity, and the vagueness of all beginnings and endings*—however tentative, and however uncomfortable we may feel about it, has been set on stage: and the brush now flows liberally across the canvas, depositing a stream of azure liquid in its wake. Everything somehow appears smooth and orderly; surely a magnificent creature of the artist's—the universe's—talents will shortly come into being.

In an effort to get a somewhat more secure foothold on the *simplicity-complexity, part-whole,* and *inside-outside* pairs of terms, let us expand our horizon by taking up a real world event, that is, an event so irreal it should never have become real (truth is stranger than fiction; in fact, it is not stranger than we hitherto thought but perhaps stranger than we are capable of thinking). Then we may be able to return to Calvino's laudable efforts to know his world from a broader mind-set (yes, that is the method to our madness: obsessively to magnify our view of things in order to take in a maximum portion of the whole without sacrificing an inordinate number of particulars).

The Blur Becomes Reckless

In February of 1974, a trawler, the *Gaul,* disappeared at sea off the coast of Norway. Thirty-six lives were lost. The inquiry that followed concluded that the ship's design and construction were according to specifications and that the crew had acted with necessary prudence.

Yet the vessel capsized when all reason and all logic dictated that it should not have capsized. One explanation had it that while the ship was undergoing some transient phenomenon, an abnormally large wave arose in the sea that hit it broadside. The term *transient phenomenon* signifies that the ship was moving in an irregular fashion, rocking, swaying, pitching, heaving, and rolling uncontrollably. Why and how did the ship enter into this strange behavior? Why and how did the abnormally large wave suddenly appear? Use of the term *transient* also has mathematical implications. It is not merely motion but motion that is nonperiodic, unsteady, that has not yet settled down to a regular pattern of behavior. Steady to-and-fro rocking is relatively unproblematic. It is of the simplicity Simplicio sought, adequately describable by linear equations. Transient motion, in contrast, is nonlinear, aperiodic, out of kilter, and in a state of disequilibrium (McRobie and Thompson 1992).

In time everything in our world becomes rhythmic in one manner or other. Trees sway with the wind, and so do tall buildings. Idling cars settle into a slight rocking motion due to the revolution of the crankshaft in one direction only. Refrigerators and central air conditioning systems do the same. Humans and animals swerve back and forth slightly as they ambulate, depending on the gait. And so on. But, as Mr. Palomar was quick to observe, the movement of the waves close to the beach is uncomfortably turbulent rather than soothingly harmonious and strictly periodic:

> The shifts of the waves that ruffle toward the shore do not disturb the steady impulse of this compact crest that slices them at a right angle, and there is no knowing where it comes from or where it then goes. Perhaps it is a breath of east wind that stirs the sea's surface against the deep drive that comes from the mass of water far out to sea, but this wave born of air, in passing, receives also the oblique thrusts from the water's depth and redirects them, straightening them in its own direction and bearing them along. And so the wave continues to grow and gain strength until the clash with contrary waves gradually dulls it and makes it disappear, or else twists it until it is confused in one of the many dynasties of oblique waves slammed against the shore. (Calvino 1985, 6–7)

The difficulty in any and all attempts to account for this phenomenon stems from its unexpected yield of nonlinearity, of disorderliness. Palomar could never quite reconcile himself to it. But actually, we are being told today by the pioneers of chaos theory that the ways of nature are disorderly more than they are orderly. Yet we more often than not continue pigheadedly to think otherwise. That is, we tend to hold tight to the view of classical physics that tells us we are right in our belief that the universe comes in a harmonious package. After all, if we drop a perfectly spherical rock from a position absolutely perpendicular to the surface of a completely motionless pond of water, will the waves not smoothly and dutifully march out in concentric circles with no hindrance and in a pleasingly periodic manner? Such perfection is rarely of our world, however. We invariably confront a minor to major element of imperfection, whether we know it or not and whether we like it or not. One need only experience a dozen or so giant waves breaking in succession across the deck of a small ocean craft to begin to appreciate how difficult the complete analysis of a nonlinear, chaotic, dynamic phenomenon can be. Ship designing engineers make a stab at taking nonlinear features of the ocean and the ship's behavior into account but with the expectation that sooner or later the unexpected may happen.

In this respect, Palomar is once again so right and at the same time so wrong. Right insofar as wave phenomena "could perhaps be the key to mastering the world's complexity" but wrong insofar as he believes, like Simplicio, that complexity can be reduced "to its simplest mechanism." Above all, he is wrong in attempting to limit his field of observation to "a square zone of, say, ten meters of shore by ten meters of sea" in order to "carry out an inventory of all the wave movements that are repeated with varying frequency within a given time interval" (Calvino 1985, 6). That is, he is wrong in that he believes a boundary line can be established between parts of the wave phenomenon within the whole. His boundary is artificially contrived, to be sure. But that is not the point; rather, the point is this: the boundary, in addition to its fictive quality, cannot but remain value laden.

Consider one of Palomar's own examples. While at the beach and engaged in his strictly empirical inquiry of the world surrounding him, our hero happens to catch a glimpse of the "bronze-pink cloud of a naked female torso." In response, he "quickly turns his head in such a way that the trajectory of his gaze remains suspended in the void and guarantees his civil respect for the invisible frontier that surrounds people" (Calvino 1985, 9). In the first place, the image is "cloudy," with somewhat vague boundaries, yet it simultaneously cries out for and at the same time against Palomar's sneaking a specific and well-directed glance at it. In the second place, Palomar averts his gaze from this prohibited delight of the eyes, which then becomes projected into the void, the emptiness or nothingness, without boundaries at all, thus rendering the erstwhile boundary invisible, even nonexistent. Palomar prudently reinforces the convention disallowing his gaping unabashedly at the nude sunbather who is caught within his visual grasp. Yet, he creates "a kind of mental brassière suspended between my eyes and that bosom, which, from the flash that reached the edge of my visual field, seemed to be fresh and pleasing to the eye. In other words, my not looking presupposes that I am thinking of that nakedness, worrying about it; and this is basically an indiscreet and reactionary attitude" (10). Physical boundary is erased; mental boundary becomes indiscreet, that is, discrete in the sense that it takes on the precision denied it by social mores. Boundaries will be boundaries, however: sooner or later and to a greater or lesser degree, what is in the mind, or in the fantasy as it were, will be projected onto the physical world.

But let's give Palomar the benefit of the doubt for the moment. With some effort, he succeeds in blurring the boundary between bosom and landscape in such a manner that "my gaze counted no more than the gaze of a seagull or a hake" (Calvino 1985, 10). But on contemplating his feat, his "objectivity" has resulted in a Robbe-Grillet sort of flattening of the "human person to the level of things, considering it an object, and, worse still, considering as object that which in the person is the specific attribute of the female sex" (10). Palomar turns around and retraces his steps, this time arranging his "objectivity" so that "once the woman's bosom enters his field of vision, a break is noticeable, a shift, almost a darting glance" (10–11). But the second tandem of empirical grasps doesn't satisfy him either. It has endowed him, he concludes, with what might be taken for an "attitude of superiority, an underestimation of what a breast is and means, as if putting it aside, on the margin, or in parentheses" (11). So he does an about-face and passes by the female figure again, but this time sharpening the distinction between breast and landscape a bit more. But alas, "she suddenly springs up, covers herself with an impatient huff, and goes off, shrugging in irritation, as if she were avoiding the tiresome insistence of a satyr" (12). Palomar's quest for knowledge has been misunderstood by an intolerant tradition.

In the exchange between Simplicio and Sagredo we observed that the invisible and indivisible line of demarcation between one thing and another is never neutral but saturated with what the one thing presumably is not and thus what the other thing hopefully is, and vice versa. We also noted that the equivalent of Zeno's asymptotic convergence in a fictive sense renders our approach to any precise boundary problematic. Nevertheless, just as we see the arrow fly through the air in spite of

Zeno's powerful logic, so also we would like to suppose we can dance in and out of the categories of our thought and our world while giving it hardly a moment's notice. On so doing, change and time exercise their inevitable push: we exist in a self-organizing universe of ongoing process. In other words, $\sqrt{-1}$ is real, no doubt—though it is called "imaginary"—but it affords no answer other than a schizophrenic back-and-forth movement. Yet, in time, like Sagredo's prisoners of his "Prisoners' Paradox" and by practical use of $\sqrt{-1}$, one can maintain confidence that one particular set of events from among many possibilities will come to pass as irreversible acts, and the world will never again be exactly the same: an order will be given, and the prisoners will be shot. Events can hardly be more irreversible than that, at least insofar as the prisoners are concerned. However, the fact remains that, given the lawyer's timeless logic, which is quite in line with the timeless implications of classical mechanics—and which carried over even into relativity theory—the universe, like a synchronic monolithic block, is what it is. How do we reconcile this conflict of interests and of mind-sets? Is it reconcilable at all? Yet another Palomar thought experiment may allow us a more satisfactory grasp of this problem.

Palomar's concept of his "infinite lawn" sets up a blurry lawn/non-lawn boundary as a physical complement to the more explicit knower/known line of demarcation he attempted to establish on the beach. The lawn surrounds Palomar's house. Under natural circumstances there would be no lawn here but, rather, so much dirt, rocks, weeds, and perhaps a few hardy sprigs. The lawn is in this sense an artificial object composed of naturally occurring objects, namely, grasses. The lawn's purpose as a sign is to represent nature, and like a sign, "this representation occurs as the substitution, for the nature proper to the area, of a nature in itself natural but artificial for this area" (Calvino 1985, 29). In other words, the lawn demands an expenditure of time, effort, and money. We find Palomar pulling weeds and unwanted grasses from his lawn, as is quite proper for respectable citizens of his socioeconomic class. The problem is that the weeds have roots deeply embedded in the soil and the harmful grasses are so thickly interwoven with the desirable sprigs that their extrication is difficult, for the grass, as rhizome in the sense of Deleuze and Guattari, is interwoven with everything else in what is for practical purposes an infinite knot of twists and turns. Consequently, there are no clear lines of demarcation. Palomar soon realizes that pulling up a malicious plant here and there solves nothing. As an alternative, he decides that this is how it should be done:

> [T]ake a square section of the lawn, one meter by one meter, and eliminate even the slightest presence of anything but clover, darnel, or dichondra. Then move on to another square. No, perhaps not: remain perhaps with the sample square. Count how many blades of grass there are, what species, how thick, how distributed. On the basis of this calculation you would arrive at a statistical knowledge of the lawn. (Calvino 1985, 31)

This apparently uncomplex strategy is now quite familiar to us. We recall Palomar's abstracting a rectangular section of beach and sea in order to intuit and comprehend the whole by contemplating the part. That task was a matter of the parts containing the essence of and mirroring the whole (recall in this respect Sagredo's allusion to the map paradox).

However, Palomar's strategy involves more complexity than initially meets the eye. It has to do with statistical averages, which is another game entirely and takes us back to Boltzmann and the dilemma of Musil's man without qualities. If 1 cm^2 of space contains x gas molecules caroming around at random in a chamber, then $1n$ cm^2 will contain x^n molecules racing about with the same average velocity. And that is precisely the point: the average velocity, insofar as it is a statistical velocity, is a matter of determinism in terms of an aggregate, though the perpetual chance of surprises roots out the possibility of predictability. Yet with respect to each molecule, Brownian random movement prevails, and there is neither determinism nor predictability. Boltzmann failed during his own time to convince his contemporaries that he was on the right track, but the essence of his argument won out in the long run, for, somehow, from sheer chance (the ultimate of simplicity and simultaneously the ultimate of complexity) comes some sort of determination, that is, self-organization, self-organized complexity, and eventually, life processes may even emerge.

Palomar, confronting what at the outset appears to be a matter of statistical averages, wishes to determine the nature of the whole lawn on the basis of the number of solitary plants in a relatively small sample—that is, like Boltzmann's macrosystem broken down into a set of microsystems. However, he soon realizes that counting the blades of grass is a futile exercise: the collection is virtually undenumerable, for neither the lawn nor a given section of it has precise boundaries. This is a simplified version of the task of taking a cubic millimeter of a cubic meter of colliding gas molecules and counting them, as well as determining their average velocities, in order to extrapolate from the sample the average velocity of all the molecules in the aggregate. The extrapolation, as we saw in the discussion on Boltzmann, can undoubtedly be close but will never be absolute—that is, assuming the cubic millimeter could have been extracted from the whole at a given moment; but it cannot be so extracted, for everything is interconnected and interacting with everything else. Palomar is not plagued with this problem, since grass (rhizome) does not consist of minuscule parts schizophrenically moving about in haphazard fashion. Yet there is still a problem, since the rhizome is a nonlinear, boundaryless web and the assumption that one section of the whole is the same as all other sections and therefore as the whole itself is no more than that: an assumption, which at best is partly right and partly erroneous.

In other words, it is impossible to know precisely where one plant and its indeterminate number of shoots ends and another plant begins. There are no borders. There isn't even any grass as genus, species, and particular exemplifications but, rather, a rhizomic assemblage. The assemblage includes *Palomar-nature-lawn-grass-weeds.* Palomar—or perhaps somebody else—has carved a geometrical form out of nature's continuum in order to put down an artificial construct replete with living tubers that would have been unnatural in that particular locale. Now, through nature's attempt to reap its vengeance by way of its guerrilla force, the weeds have broken down erstwhile boundaries and rendered a homogeneous standing army of green sprigs complexly heterogeneous. Disorder has exercised its force; a strange sort of entropy has raised its ugly head—unlike with Mary Catherine Bateson, who brought about the entropy increase in her environs due to her own activity. Palomar's artificial and originally crisply delineated artificial garden, as a definable discontinuity carved out of the undefined and virtually continuous product of nature's ways, becomes blemished.

To be sure, "there is a border where the grass stops growing, but still a few scattered blades sprout farther on, then a thick green clod, then a sparser stretch: are they still part of the lawn, or not?" Palomar responds to his own question that there is no knowing "at what point you can stop counting: between one little plant and the next there is always a tiny sprouting leaf that barely emerges from the earth, its root a white wisp hardly perceptible; a moment ago it might have been overlooked, but soon it, too, will have to be counted" (Calvino 1985, 31–32). This complexity confuses Palomar, who had always been in search of the simplicity of classical answers to his problems. He engages in a moment of realist-nominalist speculation over whether what he sees is "the lawn" as a generality or merely one sprig of grass plus another one plus another one, . . . and so on. Nominalism appears momentarily to win: "What we call 'seeing the lawn' is only an effect of our coarse and slapdash senses: a collection exists only because it is formed of discrete elements" (33). So in the final analysis there is no point in counting all the sprigs, their number does not matter. What matters is "grasping in one glance the individual little plants, one by one, in their individualities and differences. And not only seeing them: thinking them. Instead of thinking 'lawn,' to think of that stalk with two clover leaves, that lanceolate, slightly humped leaf, that delicate corymb" (33). Palomar extrapolates from there, and soon the inveterate realist side of him takes over, for he is no longer thinking merely of sprigs or even of the lawn but of the entire universe, applying to it everything he had thought about the lawn: "The universe as regular and ordered cosmos or as chaotic proliferation. The universe perhaps finite but countless, unstable within its borders, which discloses other universes within itself. The universe, collection of celestial bodies, nebulas, fine dust, force fields, intersections of fields, collections of collections" (33).

Thus the story ends, Palomar's nominalism now having waned. From one wave of the sea to the entire sea, or from a totalizing grasp of a section of the lawn to the lawn and to the universe: it is all a matter of seeing, quite paradoxically as it were, the whole in the part, or the part in the whole, same difference.

I will have much more to say regarding Palomar's behavior from other Calvino texts, from Borges and other writers, from Deleuze and Guattari, from chaos theory and the "science of complexity." For the moment, however, let us rest with the observation that, like Simplicio, Palomar wants an intelligible world, but in view of Sagredo's message, that world resists any totalizing God's-eye grasp. Quite significantly, Simplicio came around in the end: he admitted, albeit in indirect fashion, to the blurring of any and all categories, thus opening the door to the likes of Palomar's lawn without precise boundaries, populated with a tightly intertwined network of living tissue, of legumes, that is, of rhizomes, any part of which is virtually as complex as the whole; hence neither the part nor the whole can be taken in one totalizing gulp. Though the parts can somewhat adequately be known, the whole remains beyond grasp. And even though artificial cuts can and are repeatedly made in the virtual continuity of the universe of becoming, those domains cut out inevitably become blurred and often eventually sludgy: complexity, disorder, turbulence, must ensue.[12]

The third, but not the final, set of props—that is, *the rhizomic interdependent interconnectedness of every*-thing, *which renders any and all boundaries fuzzy*—has been set on the stage: brush strokes fuse and confuse the greens and the blues, the "grues" and the "bleens,"[13] as the canvas unfortunately becomes even more vague, cloudy.

Now to put space in the proper perspective insofar as that is possible.

CHAPTER 3 • When Space Breaks Down

No region can include itself as well.
—Alfred North Whitehead

I would never belong to a club that would have me as a member.
—Groucho Marx

Science cannot solve the ultimate mystery of Nature. And it is because in the last analysis we ourselves are part of the mystery we are trying to solve.
—Max Planck

I prefer to tack the unstretched canvas to the hard wall or the floor. I need the resistance of a hard surface. On the floor I am more at ease. I feel nearer, more a part of the painting, since this way I can walk around it, work from the four sides and literally be in *the painting.*
—Jackson Pollock

Libraries, Labyrinths, Apparent Lawlessness

In Borges's "Death and the Compass," which appears in *Labyrinths, Selected Stories and Other Writings* (1962, 76–87), Lönnrot the supercogitating detective thought he had things all wrapped up in a tidy package. By way of an intricately complex interplay of simples—a well-reasoned combination of threes and fours, a harmony of time and space symmetries, logical inferences derived from the Judeo-Christian tradition, and the regularities of human behavior reduced to a minimum—he had determined precisely when and where a fourth murder was to occur, even though the assassin himself had predicted its nonoccurrence. There, according to Lönnrot's fine-tuned calculations, he would finally apprehend his antagonist, Scharlach. Although our detective, faithful to his fearless symmetries, had perhaps perceived a few subtle asymmetries lurking in the shadows, he nevertheless pushed them under the rug, opting to remain confident regarding his sleuthing capacities upon entering the house at the villa of Triste-le-Roy at the proper time at which he had calculated that the crime was to take place.

After pushing his way in and immediately sensing the "architect's preferences," he surmised that at the wall opposite to where he stood there would most likely be another stairway. And sure enough, there was. He ascended it, raised his hands, and opened a trapdoor. The house, however, soon thwarted his expectations.

> Through anterooms and galleries he passed to duplicate patios, and time after time to the same patio. He ascended the dusty stairs to circular antechambers; he was multiplied infinitely in opposing mirrors; he grew tired of opening or half-opening windows which revealed outside the same desolate garden from various heights and various angles. . . . On the second floor, on the top floor, the house seemed infinite and expanding. The house is not this large, he thought. Other things are making it seem larger: the dim light, the symmetry." (Borges 1962, 83–84)

Shortly thereafter, by way of a spiral staircase Lönnrot arrived at the oriel. The evening moon pierced the yellow, red, and green "diamonds of the window."

And, then, with a twinkle in his eye, the narrator slips us the key. The staircase, like other features of the building's architecture, is spiral rather than constructed along straight lines and angles. The diamonds of the windows, acting as prisms breaking the light from the moon into its myriad frequencies, yield not the primary colors but an asymmetry of shades and hues. The house, in short, is a labyrinth, no doubt, but unlike a maze or labyrinth of the customary two-dimensional variety constructed on a plane, it is an asymmetrical labyrinth of three dimensions. It turns and twists along vertical as well as horizontal pathways.

The house reminds us, for quite obvious reasons, of the library at the upper level of the massive Aedificium in Umberto Eco's *The Name of the Rose.* The Aedificium

> was an octagonal construction that from a distance seemed a tetragon (a perfect form, which expresses the sturdiness and impregnability of the City of God). . . . Three rows of windows proclaimed the triune rhythm of its elevation. . . . [T]he quadrangular

> form included, at each of its corners, a heptagonal tower, five sides of which were visible on the outside—four of the eight sides, then, of the greater octagon producing four minor heptagons, which from the outside appeared as pentagons. And thus anyone can see the admirable concord of so many holy numbers, each revealing a subtle spiritual significance. Eight, the number of perfection for every tetragon; four, the number of the Gospels; five, the number of the zones of the world; seven, the number of the gifts of the Holy Ghost. (Eco 1980, 15–16)

This monolithic Aedificium consists of three levels: the library at the top, the scriptorium in the middle, and the kitchen and refectory at the bottom. Order at the top and disorder at the bottom; Saint Augustine's "City of God" above and the "City of Man" below; power and the codification of divine texts and the Law above and cutthroat competition for survival in a medieval world of uncertainty, catastrophe, and the unpredictable forces of evil below; dominance above and subservience below; all that is beautiful and good above and the unruly, sordid, presumably contemptible aspects of human existence below.

But the world does not gyrate around and acquire its sustenance from binary oppositions alone. There is the scriptorium, the spawning grounds of institutionalized culture, propagator of all that is sanctioned and sanctified, condoned and condemned. The library is accessible solely under rigidly prescribed regulations to the monks and absolutely prohibited to the common folk. It is the charge of those who hold the proper keys and keep watch over the books in order to defend this minuscule island evincing some semblance of order carved, by way of language and textuality, out of the overriding ocean of chaos surrounding it. The monks are those who maintain the rather artificially contrived opposition between dominant and dissident cultures until the end of Eco's novel, when the catastrophe of catastrophes befalls all, and the library, enveloped by fire as foretold by apocalyptic warnings, collapses, and crushes everything below it (Colletti 1988).

Actually, there are two distinct perceptual and conceptual grasps of the Aedificium: (1) its numerological, geometrical order from some distant outside view of its surface features and (2) a linear, stringently limited worm's-eye view from inside. When William of Baskerville and Adso succeed in gaining entry into the library, they soon discover themselves

> in a not very large room with seven sides, windowless, . . . but only four of [the sides] had an opening, a passage flanked by two little columns set in the wall; the opening was fairly wide, surmounted by a round-headed arch. Against the blind walls stood huge cases, laden with books neatly arranged. . . . We passed through one of the openings. We found ourselves in another room, where there was a window that, in place of glass panes, had slabs of alabaster, with two blind walls and one aperture, like the one we had just come through. It opened into another room, which also had two blind walls, another with a window, and another passage that opened opposite us. . . . We entered the third room. It was bare of books and had no scroll. Under the window, a stone altar. There were three doors: the one by which we had entered; another, leading to the heptagonal room already visited; and a third, which led to a new room, no different from the other except for the scroll. . . . From here you went into a new room. . . . This room was without other apertures: once you reached it, you could proceed no farther and had to turn back. (Eco 1980, 195–96)

William infers that there are five "vaguely trapezoidal rooms," each with a window, wrapped around the windowless heptagonal room at the end of the staircase. They proceed. But they soon discover they have erred, for "the builders of the library had been shrewder than we thought." After the two leave the tower, the order of the rooms becomes increasingly confused. Some "had two doorways, others three. All had one window each, even those we entered from a windowed room, thinking we were heading toward the interior of the Aedificium" (Eco 1980, 197).

And they become lost. After Adso is frightened out of his wits by what he thought was a devilish beast threatening him, which turns out to be his own mirror image, with difficulty they make their way out of the perplexing labyrinth. The initial neutral, objective, detached appearance from outside numerological and geometrical symmetries patterning the divine harmony of the library turns out to be, from the inside, a subjective view, a nightmare of asymmetries, dissymmetries, and bilateral or mirror-image symmetries, comparable to the house Lönnrot enters minutes before his demise.[1] The theme is now familiar. Symmetry became asymmetry, clear-cut lines of demarcation became vague, determinacy became indeterminacy, relative simplicity became complexity. On this note after a brief digression through Eco's labyrinth, back to Borges and Lönnrot's labyrinth.

In the beginning it had all seemed so logical to Lönnrot, and things were going so well. The three murders plotted on a map of the city composed an equilateral triangle, and the murderer had declared that this, the final crime, had been committed. But Lönnrot was less than satisfied. Three is tension. In contrast, four is balance and harmony, the closest thing to perfect symmetry since breaking out of that sphere of unity and into the world of struggle, tears, and death. Lönnrot was convinced that a harmonious—albeit static—and timeless order was destined to triumph. Availing himself of the tools of his obstinate reason, he had calculated the day of the fourth crime, and with a compass he extrapolated the lines of the triangle on the map of the city before him to construct a rhombus: unwieldy one-dimensional lines stretched through an infinity of steps to join with one another, thus yielding two-dimensional order—the very epitome of Sagredo's "minimalist" constructs. Now, after Lönnrot enters the villa of Triste-le-Roy, his erstwhile intractable ideals are given a slap in the face by the befogging, bewildering reality confronting him: *the world,* which inevitably resists any and all "minimalisms," which are of *particular worlds.* Lönnrot becomes aware that his world of ironclad logic is tragically flawed.

Thus in "Death and the Compass" Borges offers us a most imposing construction of bilateral—that is, mirror-image—symmetries that, given their very bilateral nature, create the asymmetries leading to their own destruction (compare to the bilateralism of Escher's woodcut, *Relativity,* or the Penrose triangle at the center of figure 5*a*). Argentine novelist and essayist Ernesto Sábato (1945, 104) calls Borges's short story, with its interplay of threes (triangles) and fours (parallelograms), an "extreme case of geometrization and the legitimate descendent of the scientific novel inaugurated by Poe." This "geometrization" is not of the ordinary sort, however: it also includes time. To be specific, on the third day of December, the third Talmudic congress began in which Yarmolinsky, victim of the first murder, who had endured three years of war in the Carpathians and three thousand years of oppression, was to be a participant. Yarmolinsky's assassin left a note in the typewriter, "The first letter of the Name has been uttered," along with Hebrew books on the Tetragrammaton and other esoteric topics. This homicide, occurring to the north, was followed by a second one, to the west of the city, on January 3 and a third one to the east on February 3, both also accompanied by notes that the second and third and final letters of the Name had been uttered. However, since the obvious Name, J-H-V-H, contains four letters—that is, four tokens, though three types of letters—Lönnrot concluded that there was surely to be a final homicide somewhere. To his surprise, on March 1 Inspector Treviranus received a large envelope containing a letter signed "Baruch Spinoza"—that most architectonic of philosophers—and a map of the city with three lines drawn in red ink between the sites of the three crimes. The letter prophesied that on March 3 there would not be a fourth murder. Treviranus sent the map and letter to Lönnrot, the mental geometer. After studying them, Lönnrot noticed that the events called for symmetry both in time and space. The locations of the three crimes mapped out an equilateral triangle, which was asymmetrical and hence inadequate. For space to be symmetrical, the triangle must duplicate itself to form a rhombus, and for time to be symmetrical, there could be no murder on March 3, since February had three fewer days than December and January, but, rather, the murder must occur on March 6. With calipers and a compass, Lönnrot "completed his quick intuition," pinpointing the spot to the south on the map where, on March 6, the fourth crime must occur. He then left for Triste-le-Roy, to become Scharlach's victim.

The note Treviranus received was correct: there are three types of letters in the Name; but one letter is repeated, giving four actual letters. The interplay between threes and fours collaborates with the generational principle whereby the triangle, by duplicating itself and rotating its twin 180° on the two-dimensional plane, becomes a quaternary structure. However, not only does trinary space become quaternary, the same occurs to time also. But not as Lönnrot had calculated. The first three murders occurred on the third days of three successive months, and each murder was separated by thirty-one days from the others, to give equidistant temporal separation. The fourth murder, Lönnrot obviously inferred, must occur on March 6 to separate it by thirty-one days from the third event—since February has but twenty-eight days. Hence Lönnrot rushed to Triste-le-Roy precisely on March 6.

But after Lönnrot was overpowered, disarmed, and handcuffed by two accessories of Scharlach, Scharlach told him: "You are very kind. You have saved us a night and a day" (Borges 1962, 84). How could this be? Scharlach explained that Lönnrot had omitted the fact that the deaths had actually occurred on the fourth days of the first three months, for the Hebrews compute days from sunset to sunset.

Lönnrot had been calculating with threes, which were fours in Scharlach's scheme; hence the fourth death was to occur on the seventh rather than the sixth day of the month. Very significantly, then, Lönnrot's scheme, developing linearly through time, was viewed from Scharlach's scheme *in toto,* as if he were in a higher, more complex, dimension looking upon Lönnrot's trajectory: it is much as if we were to observe, from above, a rat running what for us is a simple two-dimensional maze that for the rat is of exceeding complexity. (This is comparable, once again, to Escher's *Relativity:* from above, in three-dimensional space, we can see all the alternate possibilities for actualization of the stairways, but the confused inhabitants of Escher's world, from within their two-dimensional sphere of existence, cannot.)[2] As a consequence, the symmetry of time as Lönnrot conceived it was that of time in the strictly reversible sense. His triangle replicates itself by undergoing a 180° rotation on the planar map, and the parallelogram is formed. Reverse the operation, and things are once again as they were. It is as if Lönnrot's own time were in a figurative sense that of a one-dimensional "world line" on his two-dimensional map, whereas Scharlach's time in the Einsteinian sense consists of a world line in three-dimensional space coupled with one dimension of time. Scharlach can from his additional dimension gaze upon Lönnrot's movements as they trace a world line on his planar space-world, the map.

In this manner, Lönnrot's time within his scheme is seen by Scharlach in terms of another dimension of space added to Lönnrot's two-dimensional plane consisting of the map of the city. Within Lönnrot's one-dimensional world line, he inched his way on the two-dimensional map toward his destination, all of which was seen from above by Scharlach in a timeless instant. That is, Lönnrot's time in Scharlach's conception is metaphorically a dimension of space incorporated within his own three-dimensional space—metaphorically comparable to our time within our own Einsteinian space-time continuum. Just as Lönnrot creeps along his world line within the equivalent of a two-dimensional spatial world, so also Scharlach, for whom Lönnrot's world is all there all at once, creeps along his own world line within his own three-dimensional spatial world.[3]

In other words, a rat running its maze traces a single irreversible line over a certain lapse of time in its effort to land its reward. But from our imperious three-dimensional vantage we can in one perceptual grasp see where it should and should not proceed. And after it has run the maze, we can see where it entered, where it went along its uncertain route, and how it came upon the pellet of food. In comparable fashion, construction of the mapped parallelogram from within Scharlach's world entailed first a replication of the triangle on the map. Then Scharlach enacted the equivalent of a 180° flip in three-dimensional space such that the triangle's base lay adjacent to the base of the original triadic form. He had now located, from within another dimension and timelessly so, the site of his next crime. The consequence of this flip rested outside Lönnrot's field of vision. For Lönnrot, a solution to his problem was a matter of seeing that the previous crime was here on the map, and now, according to his calculations, the next crime would be there, as if he were limited to a relatively helpless two-dimensional perspective. There was a past, a knife-edge "now" racing through time, and a future. For Scharlach, in contrast, upon viewing Lönnrot's trajectory as if on the two-dimensional map, there was simply a before and an after, in the sense of J. M. E. McTaggart's (1927) timeless B-series.[4] Scharlach lives in time, to be sure, but his world consists of three spatial dimensions and one dimension of time. Lönnrot also lives in time. However, his map-world, in contrast to that of Scharlach, consists essentially of two spatial dimensions and one of time, all of which is compacted into Scharlach's three dimensions of space in an instant.[5]

So, the fourth, and still not the final, set of props—that is, *our suspension within space-time*—has been set on stage: there are now strokes on top of strokes, strokes complementing strokes, strokes and anti-strokes—and chaos, perhaps inevitably, begins to threaten. Time is needed to sort all this out. But . . .

What Sort of Labyrinth in Time?

According to raw classical mechanics, a given moment was the same for everybody everywhere: absolute simultaneity ruled, and time was isotropic, reversible.

However, in light of the Simplicio-Sagredo encounter, problems arose during the last century, especially after the disrespectful emergence of thermodynamics. According to thermodynamics, the universal process is temporal, irreversible. If Humpty Dumpty falls off the wall and splatters on the sidewalk below, in time order has catastrophically become disorder. Now, neither all the king's horses nor all the king's men can reverse all the molecules of that event and bring the egg back together again—even though no basic laws would be violated if this happened to occur. After a splattering on the sidewalk has taken place, in all probability the world will never again be quite the same. There was obviously a serious conflict in the classical picture between isotropic, reversible time and irreversible time. Special relativity altered this conflict, though it did not exactly resolve it. For relativity, there is no "same moment." If Humpty Dumpty falls off the wall at this moment on Earth, he will remain on the wall for quite a while for the observer of a distant galaxy, since photons bouncing off the legendary egg may require years to reach him or her. But this nonsimultaneity is of hardly any consequence regarding the everyday living of us earthlings and the inhabitants of the far away planet. Like Scharlach, we exist in three-dimensional space coupled with a one-dimensional time line or world line, along which our consciousness creeps, thus giving us the illusion of irreversible time. But actually, everything is/was/will be always already there in its entirety in the four-dimensional space-time "block."

For a chimerical four-dimensional demon privy to the whole of the relativist block, in contrast, there is simply an egg on a wall here and a splatter on the sidewalk there. Essentially there is no more than a disconnected series of static frames, each one slightly different from the one preceding it and the one following it. The very idea of irreversible time doesn't even enter the demon's picture. Of course, such a perceptual grasp, sub specie aeternitatis, of the whole collection of frames is out of the question for us. However, this impossible totalizing view would be somewhat comparable, albeit on a meagerly impoverished scale, to our gaping at the rat maze, en bloc, noting where the rat started and where it ended: there is no time, properly speaking, only the rat's world line. As far as the rat is concerned, in contrast, there was a beginning, a succession of moments during its irreversible path toward its goal, and, finally, success. Our construction of the maze and our observations of what in all probability appear to be the rat's pitiful efforts are also comparable to Scharlach's tracing, from within his three-dimensional world, of Lönnrot's path as if on the two-dimensional map. It is as if Scharlach, from above, were observing Lönnrot's movements without their author's being aware that his destiny had been sealed from the beginning.

In the final scene of Borges's "Death," and after a brief verbal exchange with Lönnrot during which Scharlach explains how Lönnrot's ratiocination had been his undoing, Scharlach then prepares to kill his antagonist. Lönnrot, perhaps in a feeble attempt to prolong the act or perhaps as the ultimate exercise in ratiocination, proposes an alternative to Scharlach's labyrinth. "In your labyrinth there are three too many lines," he begins:

> I know of one Greek labyrinth which is a single straight line. Along that line so many philosophers have lost themselves that a mere detective might well do so, too. Scharlach, when in some other incarnation you hunt me, pretend to commit (or do commit) a crime at A, then a second crime at B, eight kilometers from A, then a third crime at C, four kilometers from A and B, half-way between the two. Wait for me afterwards at D, two kilometers from A and C, again halfway between both. Kill me at D, as you are now going to kill me at Triste-le-Roy. (Borges 1962, 86–87)

Here, it is immediately apparent, we have Zeno in a new garment. There could be another murder at E, halfway between A and D, and another at F, halfway between A and E, and so on. And Lönnrot would never fall into Scharlach's trap. But he does. In this case, to continue from within Zeno's framework, Scharlach could fire his pistol and the slug would first travel halfway between A and B, to C, then halfway between C and B, to D, and so on: it would never reach its target.

But alas, Lönnrot surely errs once again. Scharlach's labyrinth, to repeat, is not of the conventional two-dimensional sort. It is three-dimensional, with implications of Scharlach's world within a four-dimensional space-time continuum that contains Lönnrot's pathetic world line as if he existed within a two-dimensional space-time continuum. For Lönnrot, of course, the labyrinth Scharlach constructed is confusing, alienating, and maniacal. It is of the most complex of complexities, though for Scharlach it is quite simple. In an effort to simplify the complexity within which he has been thrown—that is, "minimize" it with a "minimalist" construct—Lönnrot proposes his alternative, a linear paradox, like that of Zeno, which is apparently of the most elegant simplicity. But this simplicity is, of course, deceptive, mere fiction, a purely mental world. We learned from Sagredo that something as innocent-looking as Zeno's linear labyrinth, an infinitely converging series, places us in a morass the complexity of which we cannot possibly fathom, let alone put into practice.

I hardly need write that in spite of his vague allusion to Zeno, Lönnrot could by no stretch of the imagination have prevented the bullet from reaching him. Scharlach's response to Lönnrot's proposal of a linear labyrinth reveals his confidence that Zeno is refuted in the practical affairs of everyday living: "The next time I kill you, . . . I promise you that labyrinth, consisting of a single line which is invisible and unceasing" (Borges 1962, 87). Granted, within Zeno's mighty logic, the slug from Scharlach's gun will never pierce Lönnrot's body. But, of course, that is another story, a story of intellection and ideation in some ethereal realm rather than of living and breathing within our concrete existence.

We really must return to this issue at a later moment.

The Paradise of Timelessness Regained?

Lönnrot's inability to halt Scharlach's design as if Lönnrot were within a lower dimension of space is a counterpart to Hladik's inability, by the force of his own will, to halt time for a year in order to finish a play that had been engaging him. I allude to Borges's "The Secret Miracle," which appears in *Labyrinths* (1962, 88–94). The narrator tells us that Hladik was condemned to die before a firing squad at dawn, on March 29, 1943. Much in the manner of the prisoners in Sagredo's paradox, we must suppose, Hladlik was quite aware of his impending demise. In fact, he "infinitely anticipated the process of his dying, from the sleepless dawn to the mysterious volley" (89). One day he reflected that

> reality does not usually coincide with our anticipation of it; with a logic of his own he inferred that to foresee a circumstantial detail is to prevent its happening. . . . Miserable in the night, he endeavored to find some way to hold fast to the fleeting substance of time. He knew that it was rushing headlong toward the dawn of the twenty-ninth. He remembered aloud: "I am now in the night of the twenty-second; while this night lasts (and for six nights more), I am invulnerable, immortal." (Borges 1962, 89–90)

Hladik had never been satisfied with the fruits of his literary and scholarly craft. Perhaps his most successful book was the multivolume work entitled, significantly, *Vindication of Eternity.* The first volume "gave a history of man's various concepts of eternity, from the immutable being of Parmenides to the modifiable past of Hinton" (Borges 1962, 90). The second denied that "all the events of the universe make up a temporal series, arguing that the number of man's possible experiences is not infinite, and that a single 'repetition' suffices to prove that time is a fallacy. . . . Unfortunately, the arguments that demonstrate this fallacy are equally fallacious" (90).[6] In other words, what is up for consideration is a notion of timelessness within a four-dimensional block conception of things as opposed to the temporal flow of consciousness, of subjective time.[7] Hladik opted for the first conception. Indeed, with such remarkable precedents, how could he possibly be overtaken by time? To "redeem himself" of his past literary and metaphysical failures, Hladik had been writing a play in verse entitled *The Enemies,* which remained unfinished. He petitioned God that he might be granted an extra year to finish it before his execution. It was granted. Then, on March 29, before the firing squad and at the moment the command was given to execute him, the "physical universe comes to a halt." Hladik's "weak logic" had triumphed. For the duration of one year, motionlessly, secretly, he then "wrought in time his lofty, invisible labyrinth . . . nothing hurried him. He omitted, he condensed, he amplified. . . . He concluded his drama" (94). Then he was shot at dawn, on March 29.

Yes, time. That is obviously the issue that has emerged, as well as the timelessness of emptiness. Archimedes and Aristotle held two extreme views of time. Aristotle considered time intrinsic and fundamental to the universe; Archimedes did not. In Plato's *Timaeus* time was born when the divine *homo faber* imposed form and order on primeval chaos. Form is eternal and forever the same: like Parmenides's timeless, indivisible universe, it houses Being. What changes is mere Becoming. Zeno, the Parmenidean, in a massive mind act, teased his paradoxes out of this immutable universe. For Kant time was part of our experience but devoid of objective reality. For McTaggart and Francis Bradley, time was caught up in so many contradictions that it simply could not be real. Einstein's universe held past, present, and future within a monolithic block; hence time as irreversible process could not hope to enter the picture (Einstein and Enfeld 1952). The process philosophers—Whitehead, Bergson, James, and in a certain sense Peirce—in contrast, held onto the idea of time as intrinsic to the very nature of reality (so did Ilya Prigogine, as we shall note).

According to Sagredo's prisoners, existence took precedence over logic. For Hladik, as with the prisoners' lawyer, logic momentarily took precedence over existence. The lawyer's reasoning was helpless in the face of reality, since the prisoners were sure to die on the day arbitrarily determined by the warden. Hladik's reasoning powers, in contrast, for the durationless duration of a year brought forth the construction of a reality in much the sense in which scientific theories, purely imaginary constructs in the beginning, as Einstein puts it and as thought experiments would have it, can subsequently become in the minds of their believers synonymous with the real (i.e., the Newtonian-Cartesian machine model of the universe and other such models and metaphors).

The lawyer's time was that of before/now, in the order of McTaggart's, Bradley's, and Einstein's static block. The prisoners' time was that of past/now/future, with the traveling knife-edge of the "now" constituting the temporal Becoming of their world in all its complexity. Hladik's time, initially the same as that of the prisoners, was temporarily brought to a catatonic halt by his "weak logic," and then he was able to finish his play. In order to stop time dead in its tracks, however, he needed the intercession of some higher mind with a grasp, sub specie aeternitatis, of the whole and capable of controlling all lower dimensions by rendering them objective, thus deadening time. So we arrive at the crux of the issue. Lönnrot's alternate labyrinth that he proposes to Scharlach is simply one dimensional and caught up in Zenoesque complexity. An infinity of lines makes a plane, which lends itself to labyrinths of the sort of Lönnrot's map or the rat's maze. An infinity of planes describes a cube: the mansion of Triste-le-Roy, Scharlach's machinations. And if we include an imaginary time dimension within this block, we have a fourth dimension, the "higher power" of Hladik's dreams, capable of seeing all in an instant. But, then, if we follow Schopenhauer, as often does Borges, no mere earthling lives either in the past or the future, but, rather, the present alone is the form of all life. Time is like the Buddhist wheel: that which is always sinking fades from the present into the past; that which is always rising emerges into the present from the future; all that is real is the present, the eternal present. Borges, perhaps with this image in mind, concludes his celebrated "A New Refutation of Time" with some longing and certain resignation:

> And yet, and yet . . . Denying temporal succession, denying the self, denying the astronomical universe, are apparent desperations

> and secret consolations. Our destiny . . . is not frightful by being unreal; it is frightful because it is irreversible and iron-clad. Time is the substance I am made of. Time is a river which sweeps me along, but I am the river; it is a tiger which destroys me, but I am the tiger; it is a fire which consumes me, but I am the fire. The world, unfortunately, is real; I, unfortunately, am Borges. (1962, 233–34)

Significantly, from this view we are *in* time. Time is *in* us. We are time, and time is us. In fact, in a certain sense time is of the nature of our very universe—as we shall observe in future pages. If not, from a God's-eye view—or that of Scharlach gazing on Lönnrot, we on the rat, Hladik on his work from his mind's eye, or the lawyer on the prisoners' plight from her indefatigable logic—all would always already be crystallized into a static whole. From the pompous penthouse view of the whole in an instant, logic would reign supreme, contradictions would be barred, and any and all middles would be excluded. In contrast, from within time and while we carry the burden of memories of the past and anticipations of the future on our weary backs, contradictions periodically raise their undecidably ambiguous heads, the space between terms and things and forms and feelings is filled with an infinity of possibilities that might have been but are not and that may pop up at some unexpected moment, and Sagredo's gradients between values emerge as a matter of course. Things get vague, fuzzy, sludgy. They merge into one another. And life never ceases to become complex.

If this were not the case, the world might be hardly more than that of poor Funes the Memorious from Borges's tale, which appears in *Labyrinths* (1962, 59–66) and whose title bears the same label. Funes's world is not quite simply that of his immediate experience along the one-dimensional knife-edge of time and in three-dimensional space. Funes is capable of seeing only particulars, and he virtually sees them all. He can at a glance take in all the leaves, branches, contours on the trunk, and so on, of a tree, and years later recall them to memory perfectly. The problem is that his memory is a garbage heap. It contains an indefinite number of individuals, yet Funes is incapable of "ideas of a *general,* Platonic sort." It seems strange to him that a dog seen at 3:14 P.M. from the side is considered to be the same dog seen at 3:15 P.M. from the front. Conceiving number as an ordered series is for him impossible. He has simply memorized each number without establishing the necessary serial relations between them. In fact, he once developed his own alternative number system consisting of arbitrary names in place of every number, which for him was just as effective. Funes, in short, is unable to think, for to think "is to forget differences, generalize, make abstractions. In the teeming world of Funes, there were only details almost immediate in their presence" (66).

Funes, it appears, either sees all or nothing at all; he remembers aggregates of particulars without being able to isolate any of them. He is the consummate nominalist, a superempiricist. A hypothesis, theory, conjecture, even a beginning, would be for him virtually impossible. For us, before there can be any-*thing* at all, even before there can be no-*thing,* there must be some-*thing,* and this some-*thing* must be a selection, an abstraction, of some part from the whole. Our collection of selective abstractions makes up the world, *our world.* On the other hand, if, like Funes we would expect, we were able to perceive the world as an unselected continuous stream in terms of pristine objects and events, that world would be a teeming jungle, a myriad array of clearly differentiable differences. Every-*thing* would be clearly and distinctly here now and gone in the next instant but indelibly committed to memory. Since we would be incapable of abstracting any-*thing,* every-*thing* would be reduced to essentially the same level. We would take in one perceptual snapshot of the world "out there," then another, then another, and so on. With each snapshot we would see a slightly different collection of particulars before us. But the question is, How would it be possible for us to detect any movement at all? In other words, if we saw every-*thing* at once as an aggregate of particulars, and if we could not abstract any-*thing,* then we would be incapable of seeing one particular against the background of the whole, and hence we could not detect a change in that particular while holding the whole in check as an unchanging entity. What we would perceive, and the only thing we could perceive, is, so to speak, a succession of static "slices," a crisp series rather than a virtual continuum. But there could be for us no change, no time in the conventional sense—for which Borges has a special affinity and which has resisted complete resolution over the centuries.

And Zeno's arrow paradox would prevail. In other words, if the arrow is where it is at each instant and displaces a space equal to itself and no more, then it cannot move to another space at another instant, so it cannot move, and therefore time cannot exist. Or at best, instead of Funes's perception of his world as a unifying whole, there is no more than a set of static, discrete particulars, without any unifying thread.

Moreover, since each of these particulars occupies no more than a split second, the difference between them is minuscule, perhaps well-nigh infinitesimal. In the face of these myriad differences, quite surprisingly, Funes is indifferent toward his world. The narrator describes him as a

> face belonging to the voice that had spoken all night long. [Funes] Ireneo was nineteen years old; he had been born in 1868; he seemed to me as monumental as bronze, more ancient than Egypt, older than the prophecies and the pyramids. I thought that each of my words (that each of my movements) would persist in his implacable memory; I was benumbed by the fear of multiplying useless gestures. (Borges 1962, 66)

Funes somehow triumphed over Zeno's arrow and created time; as a consequence he had fused discrete differences into the flux of his experience ultimately to breed indifference. The way of Sagredo also implies freedom from the temporal bonds of Achilles's dilemma. But no matter how far along the road Sagredo proceeds in his effort to fracture the world into categories of perception, thought, and belief, sludge there will always be, for his world of holist interconnections, unlike Funes's world of teeming particulars, is of unfathomable complexity. Either way, gestures cannot but become to an extent useless, for they cannot hope to refer or correspond to, neither can they represent, the world, but rather, they construct their own world, a world fabricated more than found, protracted more than preordained, and yet, a world of self-organizing becoming that sweeps every-*thing* along its slushes, gushes, and occasional sloughs.

So finally, the last props of the stage—that is, *our intimate relationship with time, which we are within, which is within us*—have been set: the pigment on the tired surface becomes smoother and smoother, until finally, only a few scant differentiations can be made manifest, and the brush, having lost its very raison d'être, stops at midcanvas and remains there, motionlessly, poised. Poised for what? Who knows? At this point it simply is, between the +1 and the –1, between every-*thing* and its negation, dwelling in emptiness, nowhere, nowhen, waiting. Waiting for what? Who knows? It is simply now, between elusive memories of the past and vague anticipations of what is to come.

Now for a look at complexity, life processes, and the universe according to some current conceptions.

PART 2

In part 2, simplicity and complexity are given a new countenance in light of the preceding chapters. Consideration of asymmetry at the roots of life processes leads to a look at what is termed *phase space*. Exemplifications from the everyday affairs of life, prose by Calvino, and ruminations by artist Paul Klee afford—at least that is the hope—a sense, a feeling, of "fractality," "strange attractors," "chaos," and, ultimately, of life itself. Simplicity and complexity, having now become quite benign, no longer provoke anxiety: they compose a common theme illustrated by labyrinths as order (simplicity) heaving, undulating, and syncopating into disorder (complexity) and back again. Labyrinths as intricate entities in phase space, via Borges and Calvino, pour forth into an image of the entire universe as a labyrinth: a universe of uncertainty containing, within itself, all universes, actual and possible. But eventually we find ourselves once again forcibly gravitating toward that age-old question of whether the universe is in the final analysis of order or of chaos. An excursion into Borges's "Library of Babel" and, finally, to the "book" metaphor of the universe reveals the lingering problematics encountered when attempting to account for the universe's self-organizing process of becoming. The problem resides in the inevitable sense of permanence and timelessness with respect to the universe as labyrinth, library, or book, in contrast to a sense of change, irreversibility, temporality. That problem, however, will be forced to await its due share of the spotlight in part 3.

CHAPTER 4 • Veering off on a Tangent

To see the world in a grain of sand,
And a heaven in a wild flower;
Hold infinity in the palm of your hand,
And eternity in an hour.

—William Blake

[Information theory] is so general that one does not need to say what kinds of symbols are being considered whether written letters or words, or musical notes, or spoken words, or symphonic music, or pictures.

—Warren Weaver

We forget that we must always return to zero in order to pass from one word to the next.

—John Cage

All that I love, all that I think and feel inclines me towards a particular philosophy of immanence according to which surreality will reside in reality itself. . . . And conversely, because the container shall be also the contained. One might say that it will be a communicating vessel placed between the container and the contained.

—André Breton

But first, I must retrace a few paths along which a number of obstinate questions remain. From within the spheres of art, literature, science, and that more abstract discipline, mathematics, we seem to have arrived at some sense that (1) the universe is at the macrolevel a matter of statistical averages, which during the nineteenth century initiated the process that eventually threw classical physics for a loop; (2) reflection on complexity-simplicity inevitably leads to blurred borders, vagueness, and the becoming of continuity, that is, continuity in the act of becoming without the possibility of its reaching the plenitude of its beingness, all of which reveals the limitations of our knowing, suspended as we are within time and space; consequently (3) complexity, at some of the most inopportune moments, resurfaces, accompanied by nonlinear, irreversible processes that reveal the interconnectedness of the universe and evince images of order from disorder.

"So what's the problem?" may perhaps come the yawning query. The problem is in the knowing of becoming when the knower is part of the becoming that is to be known. The problem is in the apparent inability of one's knowing, and especially sensing, complexity except by making it over into some "minimalist" form of simplicity, but on so doing, it is no longer what it was, or was becoming, but in that case it could not have been known anyway, for once it was known in all its complexity, it would have become some other manifestation of complexity.

Complexity and Simplicity in an Intimate Embrace

And yet, and yet, . . . in spite of the complexity, the fuzzy categories of one's thought—that is, the sludge that persistently dogs any and all attempts toward clarity—one usually opts obstinately to nurture one's undying faith in the universe's cognizability. This tendency is illustrated at its full-blown best, once again, in Calvino's *Mr. Palomar,* a sort of fallible and faint shadow of Paul Valéry's *Monsieur Teste* (1947), though Mr. Palomar is engrossed with empiricism in addition to his overriding idealism. This faith also remained implicit in the surrealists' own quest for the absolute (Balakian 1959), as well as in the disappearance of the self in a diversity of works beginning in the twentieth century with the likes of Samuel Beckett, Natalie Sarraute, Alain Robbe-Grillet, Pierre Boulez, John Cage, Kazimir Malevich, Mark Rothko, and Jackson Pollock, to mention only a scattering of artists (Meyer 1967; Sypher 1962). Faith in the universe's cognizability has especially lingered in the hardest of the sciences, physics, by way of Max Planck, Erwin Schrödinger, Louis de Broglie, Albert Einstein, and more recently in Stephen Hawkings and others who maintain confidence that a grand unified theory will soon be at hand (Barrow 1991).[1]

The motivating force behind this search for the plenitude of mind, thought, and knowledge is chiefly the result of a long history of "nothing but" reductionisms: the universe is nothing but a machine, the brain is nothing but a computer, and so on. Complexity, according to the reductionist program—this we will recall from Simplicio—is actually nothing but a consequence of the reiterative application of a set of relatively simple algorithms and rules. Simplicity breeds complexity through the actualization of possibilities, and organization of a domain is forthcoming from the tidy simplicity of the rules by which that organized complexity emerged. Thus in the world as science has traditionally had it, interactions between a limited number of particles combine in Lego fashion to make up the chemical elements, and atoms of these compose molecules, the most complex of them blooming into life processes via the DNA program. In other words, the reductionist story includes a chain of episodes according to which atoms wind their way up through molecules, a choice combination of them twisting and turning into the familiar double helix, which generates organisms, societies, intelligence, and entire ecosystems. However, the critique has had it that things cannot be so simple. Granted, as a result of evolution, the myriad characteristics of an individual organism are determined in machine-like fashion by the DNA code. However, environment also lends a hand governing and allowing for a somewhat unique set of developmental patterns in that organism (Salthe 1993). According to this environment-developmental alternative, focus should not rest on simplicity lurking somewhere within complexity—that is, the reductionist enterprise—nor should complexity be simply ignored.

Complexity is apparently with us, whether we know it or not and whether we like it or not. And to make matters worse, as mentioned previously, we are part of the very complexity that is supposedly the focus of our knowing. Ultimately, the problem has to do with current theories and experiments, speculations and experiences, articulations and meditations, on the topic of disorder and the order arising from it, such order by and large leading to ever greater complexity. But before jumping headfirst into that issue, let me begin my story with an example from literature.

Simplicity within Complexity and Vice Versa

Italo Svevo's *The Confessions of Zeno* (1958), bearing the name of that grand philosopher of Elea, paradoxically conjures up recollections of Heraclitean flux: "Nothing helps one to concentrate so well as gazing for a long time at running water. One remains perfectly still oneself and all the necessary diversion is provided by the water, which is never the same for a single instant" (381). One side of Zeno—Svevo's character, not the pre-Socratic philosopher—wants the recursive "What can be done once can be done once again, and again . . . *n*," to which, we will recall, Sagredo alluded. Zeno's other side wishes for repetition, but with a difference of the Heraclitean "You cannot step in the same river twice" sort—which was actually the implication behind some of Sagredo's allusion. Hence the conflict within Zeno's mind: stasis and transience, permanence and change, Parmenides and Heraclitus; in other words, the age-old conflict of Western thought, mathematics, science, and the arts.

Running water, of course, does not appear at the outset mathematically, scientifically, or philosophically very inspiring. Babbling brooks are what poems are made of, but, like all forms of turbulence, they have always bothered physicists, and they are difficult to analyze mathematically: they are fuzzy, messy, far from the purity of standard classical equations. Nevertheless, they and other comparable phenomena are quite common in our everyday living. They are everywhere. The fact is, however, that until recently scientists have tended by and large to ignore them—somewhat like the Greeks who ignored irrationals. It might appear that there is hardly any need for one to be concerned with them in one's everyday living either. After all, if you open the refrigerator, take out a pitcher of orange juice, and pour yourself a glass, you need not be bothered with the thermodynamics of refrigeration, the molecular structure of ascorbic acid, or the patterns of neuronal firings that prompted you to do as you did. That is what science is all about. And if some upstart physicists are now studying how the orange liquid leaves the pitcher and spills into your glass, that's their business. All you wanted was a cool drink. But my example is naive, for it is merely an isolated event. Look out your window and you may spy an intricate concoction of processes: flowers growing, birds flying, cars whizzing by, clouds intermittently hiding and revealing the sun, children playing, dogs barking, people talking, or whatever. A moment's reflection on the complexity involved in this hustle and bustle is mind boggling. But it definitely is the world of your everyday experiences: a virtually infinitely complex world, the nature of which you simply cannot ignore, for it is the concrete world of your living and breathing existence, not some mathematical equations dealing with intangibles.

Classical mechanics—much in the order of Simplicio—thought it could ultimately account for all these processes in terms of isolated simples: interactions between microscopic particles rushing about and colliding with one another. Closer examination revealed, however, the onslaught of complexities that seemed too intricate for the human mind to comprehend. Take a simple cloud. Is it really so simple? Its form is incessantly changing, according to air movements that are the composite of statistical averages of countless molecules that nevertheless remain unpredictable. Look at the cloud more closely. Tiny droplets in random movement meet your eye. A more greatly magnified view reveals a violent bath of swirls and swishes apparently without order. An even closer observation gives you an inconceivably rapid throbbing, scintillating, aggregate of forces in perpetual struggle with one another. And all this is the increasingly minute grasp of a mere inorganic phenomenon. Consideration of, say, a flower, lands us in a hotbed of complexity. The plant giving rise to the flower is open to its environment, taking in carbon dioxide, water, and minerals, putting them all together with the aid of light to produce chlorophyll and other compounds and expelling oxygen into the atmosphere. But there is more. The flower grew from a tiny seed to its unfoldment into a bloom, and now it will soon wither and return to the earth—except for a deposit of seeds somewhere that can hopefully serve to initiate the cycle anew. For the reductionist, all this complexity was originally compressed in the seed in the form of a set of instructions according to the character of the plant's DNA molecules. From a simple seed, complexity emerged and bred further complexity, until finally, apparent simplicity once again is destined to gain the upper hand. However, it is becoming increasingly apparent that the flower cannot be isolated from its environment, or from any aspect of the entire universe for that matter, for everything is related to and in perpetual interaction with everything else in the intricately interconnected web. No aspect of the universe can be considered a mere matter of complexity reduced to simplicity.

Yet the received, reductionist view of the universe as nothing but a vast clockwork mechanism has doggedly persisted. Why? Because the universe is thus rendered hopefully intelligible and a comfortable place to make a home. Take, for example, a software program and, at the extreme stretch of the spectrum, DNA. Both are presumably capable of generating ever more complex systems that, according to classical principles, are nonetheless ultimately reducible to simples. A standard program for your desktop PC or MAC is rigidly governed by Boolean algebra. There is very definitely a right way and an indeterminate number of wrong ways to use the program, and if a wrong key is punched it is impossible to engage in a dialogue with that stubborn brute—the hardware and its housing—who is capable of seeing things its way and its way only. DNA, of course, presents an exceedingly more complex case. But from the "bottom up" view, James B. Watson and Francis Crick, and following quite closely in their footsteps, Jacques Monod and François Jacob, reduce DNA to a relatively simple matter of mechanism and of code and information (it is perhaps no coincidence that DNA and RNA were discovered in a molecular physics laboratory).[2] Yes, the universe considered in relatively simple terms is quite soothing, a comfortable security blanket.

However, this indelibly classical view of the origin of complexity and the confidence that it ascended from and can be described in terms of simplicity is now challenged from an increasing number of directions. In the mundane world of inorganic processes, movement from simplicity to complexity is quite common. It is now known that simple laws, rules, and reactions can lead to systems whose behavior is so irregular as to appear virtually random. This behavior is not merely complex but chaotic. Though chaotic, it can be somewhat adequately comprehensible once the necessary details are known, yet it is unpredictable and unknowable at any future moment. A prototypical example of this phenomenon—which gave Simplicio his last great hope—is found in Mandelbrot sets of the sort that adorn TV and magazine ads, T-shirts, posters found on the walls of teenagers' bedrooms, and perhaps your favorite screen saver.[3] A computer program for generating a Mandelbrot set is no more than a few instructions in length. Yet it engenders a self-organizing, recursive pattern whose intricacy never ceases. No matter how far one dives into the pattern, it appears as complex as it did from the larger view.[4]

In light of complexity out of simplicity, and at certain opportune but unpredictable moments, order out of chaos, Palomar's waves, as well as gurgling brooks, dripping faucets, and the like, are what chaos theory and Prigogine's "science of complexity" are all about. They link our everyday experiences to the laws of nature by the aesthetic portrayal of subtle relationships between simplicity and complexity and order and chaos. The reductionist view Simplicio held onto for dear life until the end of his encounter with Sagredo was motivated by the desire to explain the universe once and for all time with a theory of everything, from physics to chemistry to biology and finally to psychology, anthropology, and sociology. Human beings and all living organisms would be subjected to the same laws that governed the movement of heavenly bodies. Eventually, nothing would lie outside the all-seeing eye of the knower. It was just a matter of time, and of rendering the equations a tad more sophisticated, Simplicio reasoned. As nineteenth-century physicist John Tyndall is reported to have once put it—and from Sagredo's mouth—in the final analysis, if we could know Shakespeare's brain to the letter we would be able to deduce Hamlet from the lamb chop its author had for his evening meal the day it was composed.

Then, chaos entered the scene, and Simplicio hit upon his apparently bizarre self-reflexive infinite regress story about signs and nonsigns and the real and the nonreal. But further implications of that story must be left for a future moment. For now, we should acknowledge the fact that, in Sagredo's words, we can't even know when the next drop will fall from a leaky faucet. It would appear that focus on that necessary prerequisite to chaos theory, "symmetry breaking," might pay dividends.

Fearful Asymmetries?

To recap key observations to this point, although gas molecules move about in apparently random fashion, in the classical view of things, according to statistical laws an aggregate of such molecules turns out to be quite predictable. However, as Sagredo pointed out, no gas conforms exactly to the letter to Boyle's laws; hence there is always a slight flaw in their behavior. A crystal is orderly with respect to its lattice arrangement, to be sure. In the case of halite or salt crystals, the same "algorithm" is repeated time and time again: an Na^+ ion surrounds itself by six Cl^- ions, and each of the latter in turn surrounds itself with six Na^+ ions, and on and on, as the cubical structure grows. A sunflower, in terms of its burst of form from the center outward, is also periodical. But unlike with an obsessively symmetrical crystal, the sunflower's periodicity is not absolute; it is slightly aperiodical; hence its growth is to an extent unpredictable. In fact, as we shall shortly observe, life in general tends to be somewhat orderly but not absolutely so.

This is a topic that will be of increasing importance as the present disquisition proceeds and which merits some digression at this juncture. Symmetry breaking is an idea that has become fashionable even in the "new physics." Until 1956, physicists held it as self-evident that nature does not favor left-handedness over right-handedness or vice versa. In that year physicists C. N. Yang and T. D. Lee worked out an experimental test for left-right-handed asymmetry, which was finally conducted in 1964 at Princeton University with the announcement that nature violated what was to that time presumed to be CP (charge conjunction parity) invariance. This violation came in the form of strange particles called "K mesons" for which CP invariance ordinarily holds with the production of symmetrical pairs of "pi mesons." However, once in several thousand K meson decays, an asymmetrical trio of "pi mesons" is regurgitated. As a consequence of this finding, it is now considered that the almost empty universe bears witness to the fact that CP violation is perhaps no more than an occasional affair. Yet symmetry breaking exists, at the most fundamental of levels.[5]

Upon moving from the microphysical to the macrophysical level, symmetry breaking occurs when a stable (or ground) state of a system appears to violate the symmetries present in the physical equations governing it. For example, the equations that account for magnetic matter are spherically symmetrical, yet a ferromagnetic system violates this symmetry by exhibiting a preferred direction in space—the direction of its magnetic axis. Moving up the scale from simplicity to complexity (or in some cases from complexity to simplicity, depending on the perspective), we have crystals. Crystals can be strange beasts indeed. The dissolution of crystals in water is either exothermic or endothermic. That is to say, heat is expelled or it is taken in. Putting the matter succinctly, in the first case, the resulting condition is of a lower energy state because energy has been expelled into the environment; in the second case it is of a higher energy state since it has taken in energy. In strict terms of the second law of thermodynamics, in the first case, it might appear that entropy must have increased; in the second case it must have decreased. In the first case the move is from a less probable to a more probable state; in the second case it is from a more probable state to a less probable state, and entropy has consequently suffered a decrease. However, this picture is oversimplistic, for, since energy is either taken in or expelled, the system consisting of water saturated with ions from a crystal lattice is not closed but open; hence it cannot be an ideal thermodynamic system. That is precisely the point: all systems as parts within the unfathomable whole are in perpetual interaction with all other parts, and no part can be isolated from any other part and be expected to retain the properties it sported when in interaction with that other part, with all other parts. (This, in essence, is the Prigogine view, the idea that this interaction between parts, this incessant exchange of energy and information, breeds life's processes. But we must finish our hors d'oeuvres regarding symmetry breaking and the origins of life before we can be served the main dish.)

A. G. Cairns-Smith (1985) recently argued that life finds its origins in silicon-based chemicals, rather than the carbon basis of organic chemistry.[6] He conjectured that a relatively low-tech silicon mix—that is, a concoction of tiny crystals—is more conducive to a primitive sort of self-generativity than high-tech carbon-based systems. Once the silicon mix had begun lifting itself by its bootstraps, then there could be a genetic intervention and ultimately a coup d'état by the more efficiently operating and dynamic organic molecules. Cairns-Smith targeted clays as the likely testing grounds for the origin of life. Geologically, plenty of clay is continually in the making. These inorganic low-tech crystals, by shearing, fracturing, and re-forming, could create asymmetrical units at every step. Growth and replication of these crystals could then mix with growth and replication of organic molecules. According to Cairns-Smith, inorganic forms were eventually subject to takeover by organic life when organic growth outran the capacity of inorganic forms for self-replication. In other words the low-tech molecules were taken over by the more versatile high-tech molecules. Eventually, natural selection brought about strands of RNA and nucleic-acid–based genes, and life emerged, while inorganic systems were left in the muck. The important message, Cairns-Smith claims, is that bizarre, irregular crystallization was precisely the sort of conditions from which life eventually evolved.

Of course, crystals are generally considered the archetype of symmetry. But this symmetry depends upon rotation. If a cubic salt crystal is rotated 90° it is the same as it was, but if rotated by anything less than 90° its symmetry will have been violated. A perfect sphere containing a scurry of gas molecules, in contrast, remains pretty much the same no matter what the rotation or what the angle. The rotationally symmetric order of the crystal breaks the complete macroscopic symmetry of a gas. On the other hand, if the crystal is redissolved in solution and the solution placed in a perfectly spherical container, its symmetry will approach that of the gas. The general rule is as follows: a chaotic (and complex) aggregate of finite entities can be capable of reaching symmetry; on the other hand, a rotationally symmetrical system (a crystal) can be the product of a simple rule for ordering a set of finite entities. As we shall observe, in the unkempt and finite kingdom of our everyday world, systems fall somewhere between total order and total chaos, between simplicity and complexity.

Crystal growth has fascinated scientists since the Renaissance. The medieval philosopher Duns Scotus thought crystals were alive. Certain crystals of organic chemistry—for example tartaric acid, which favors right-hand asymmetry—Louis Pasteur thought, provided for the emergence of life. William Bateson, father of Gregory Bateson, saw the beginning of life in vortices—in the nineteenth century conceived as a model for atoms—of which a smoke ring was the most likely example. A smoke ring—topologically speaking, a torus—evinces periodicity and directionality; it remains the same, yet it is constantly changing; its evolution is the product of its capacity for self-organization, for lifting itself up by its own bootstraps.

In 1944, and some years before the connection between genes and DNA was known, Erwin Schrödinger (1967, 64–65) predicted that genetic information, and hence life itself, was stored in "aperiodic crystals," in which an arrangement of atoms encoded the genetic message that passed from one generation to the next. By "aperiodic," Schrödinger meant that the gene's material contained something it should otherwise not contain in a world of pristine balance, harmony, and symmetry, and therefore it contained information of the information theory sort. For example, the sentence I am now in the process of punching out on the keyboard contains information (at least I hope it does) because each letter could be replaced by alternative letters: the differences between the alternate letters and the letters present, differences that make a difference, give rise to information. Basically for this reason language is radically aperiodical—except for palindromes and other linguistic idiosyncrasies. By labeling the gene an "aperiodic crystal," Schrödinger meant that it should follow principles of regularity but it could also evince certain irregularity so as to contain information. The regularity and irregularity embodied in one structure were soon thereafter found in the DNA helix. A DNA molecule consists of a regular arrangement of atoms along the outside of the two helix chains, while inside there is an irregular assortment of chemical groups of four kinds, each with its own shape and size. Regularity (symmetry) outside and irregularity (asymmetry) inside, crystal and variation of ordinary crystalline form: the opposites form a unity that twists and turns, as if in a constant struggle with itself.

Speaking of crystals and asymmetry, Calvino's voice pierces the noise.

Symmetry Breaking within the Text

In "Crystals," which appears in *T Zero* (Calvino 1969, 28–38) we have a fictive counterpart to biological theory of the sort presented thus far. The narrator, Qfwfq, spends his days "going up and down the horizontal and vertical axes that crisscross that compact solid [the crystal] or along the obligatory routes that graze its sides and its edges. But I don't fall into the trap: I know they're making me run among the smooth transparent walls and between symmetrical angles so I'll believe I'm inside a crystal, so I'll recognize a regular form there, a rotational axis, a constant in the dihedron, whereas none of all this exists" (29).[7]

And what exists, at least according to Qfwfq? The unruly opposite of perfect harmony found in a well-wrought, absolutely symmetrical crystal: *glass*. Glass, of the makings of the modern and postmodern city, is made not of crystal but of "a paste of haphazard molecules which has invaded and cemented the world, a layer of suddenly chilled lava, stiffened into forms imposed from the outside, whereas inside it's magma just as in the Earth's incandescent days" (Calvino 1969, 29). These were the days when the Earth was without any crust. It was a "mineral bag, with bleak swirls of iron and nickel that dripped down from every crack toward the center of the globe, and jets of mercury that gushed up in high spurts" (29). Finally, the unruly concoction condensed and cooled, gradually changing from disorder (unorganized complexity) to order (simplicity, and, eventually, organized complexity and life). From the lava flow, form emerged and then solids, "with regular, smooth facets and sharp corners; and these facets and corners were slowly expanding, as if at the expense of the surrounding matter, and also the form of the solid was changing, while still maintaining symmetrical proportions" (31). Cubes, hexagons, octahedrons, and all forms of polyhedrons came to the surface from the Earth's bowels.

However, what Qfwfq's mate, Vug, liked "was to discover in crystals some differences, even minimal ones, irregularity flaws" (Calvino 1969, 33). In other words, we have the wish for the makings of shearing and stress (Cairns-Smith) from which aperiodicity emerges (Pasteur, Schrödinger) and small flaws leaping into big flaws (Prigogine, Poincaré)—the transition from simplicity to organized complexity. With one solitary atom out of place in the crystal lattice, "an exfoliation that's a bit crooked" in a solid that is destined "to be enlarged infinitely according to a regular pattern," myriad crystals can pop up, engaging in a struggle, arresting each other's symmetrical growth, in order "to be free, to draw all the matter to themselves" (33–34). And the Earth slowly moves "in the direction of Vug's preferences" (36). Now, the world appears minutely variegated, growth becomes heterogeneous, crystals break up and become smaller and smaller (fractals), like Palomar's grains of sand on the beach, "being reduced to almost imperceptible fragments rolled by the waves." The Earth's crust is "becoming a congeries of disparate forms" (37).

But Qfwfq would like to believe that it was really "a network of symmetries so complicated we couldn't comprehend it" (Calvino 1969, 37)—not yet, at least, although with Simplicio's blind hope, when sufficient data could be factored into the equation, all would be comprehended. However, things were becoming virtually incomprehensible, for they were taking on "hypercrystal" characteristics, that of slightly aperiodic crystals within other aperiodic crystals. "Everything comes from crystal," Qfwfq finally declares, "even the music we're hearing." But he knows full well that all crystals, even those of the transistors in his radio, are slightly flawed, veined with minuscule impurities, with "rents in the warps of the atoms." Vug wants to believe that legitimate order "carries impurities within itself, destruction" (37). And she is correct. Consequently, as if under Vug's spell, Qfwfq finally sees "nothing but perturbations of the order of the atoms: luminescent tubes, TV, the condensing of tiny silver crystals on the photographic plates" (38).

The image Calvino offers up is that of the magma of gray in figure 5*a,* which opens itself to the fracturation, fractalization, at the oscillating $\sqrt{-1}$ between blackness and whiteness, where the seeds of its unmaking bear fruit. And then there is a push toward asymmetry and the Harlequin, which affords a peek into the next dimension . . . from within which we have the comforting hope of some sort of regularity and order and organized complexity. But at the same there is a threat from Pandora's box, as if Apollo were to give ground to Dionysus, as with Lönnrot's universe and its unmaking by the mind of Scharlach. What is worse, the ominous threat of impending chaos is present in Oedipa's confusing world of radical asymmetries culminating in a confusion of images and symbols, Simplicio's merging of signs and nonsigns and the real and the nonreal, all suggesting symmetrization to the infiniteth degree or asymmetrization to the infiniteth degree; that is, either way, we have a projection to the absolute sphere whose center is everywhere and whose circumference is nowhere—of Borges's Library of Babel or of Nicholas of Cusa's and Blaise Pascal's universe—or if not, a collapse into the absolute asymmetry of magma of which Qfwfq spoke.

It would seem that the pursuit of asymmetrization might be promising.

CHAPTER 5 • When Things Get out of Line

Any symmetry principle is at the same time a principle of simplicity.

—Steven Weinberg

[L]ight and rational forms . . . locked in combat; light sets them in motion, bends what is straight, makes parallels, ovals, inscribes circles at the intervals, makes the intervals active.

—Paul Klee

Suffice it to say that black and white are also colors, for in many cases they can be looked upon as colors, for their simultaneous contrast is as striking as that of green and red, for instance.

—Vincent van Gogh

What rubbish all this stuff about light and dark.

—Samuel Beckett

How Asymmetrical Can We (Do We Want to) Get?

Symmetry is the way of binding and rest, asymmetry is of loosening and change; the one is order and law, the other embraces arbitrariness and accident; the one is formal rigidity and constraint (classical physics, rationalism), the other is flexibility and freedom ("new physics," pluralism, radical empiricism).[1] The one is essentially Parmenidean, the other Heraclitean.

Chaos theory involves dynamic phenomena. It is the product of complex concoctions that change with time. There are orderly changes, of course. They make up the stuff of classical dynamics. There are also chaotic changes, upon which the new science of complexity focuses. And apparently there are random changes, which, as far as we are concerned, lie beyond all comprehension—though that is no guarantee that at some unknown level they are not determinate.[2] A collection of things manifests some set of variables, which determines the state of the collection at a given point in time. To know these variables is to know what can be known of the collection. Mr. Palomar wanted to know a single wave to perfection. But to know the wave is to know its relation to adjacent waves, and the relations between those waves and their neighbors, and so on—everything interconnected and interacting with everything else. Ultimately, Palomar's overambitious project is interminable. Take the case of a ball on the pool table. While one is playing a game of billiards, it all seems quite elementary. You strike the ball with another ball put into motion by the action of your cue, and as a result it veers off along a tangent. But, contrary to your expectations, it disrespectfully collides with the eight ball, imparting energy to it that sends it caroming at an angle and into the corner pocket. After pouring forth a bit of profanity, you concede your defeat, forget about the game altogether, and go on to other more satisfying activities.

If we wish to translate the ball's undesired behavior into pure geometry, we must move to a greater level of abstraction, forgetting about the ball as a physical object entirely and focusing on the variables that define it. In the beginning it appears to be a matter of spatial dimensions, of dynamic, not static, space, since we are considering the medium through which the ball travels. Borges (1978, 117) puts it quite adequately in the opening lines of "The Book of Sand": "The line is made up of an infinite number of points; the plane of an infinite number of lines; the volume of an infinite number of planes; the hypervolume of an infinite number of volumes." Borges mounts no disclaimer. What he writes is what is in no uncertain terms "true." Or so we read. From point to line to plane to cube and on to ever higher dimensions, Borges's story is based on a tested and proved mathematical theorem and on Euclidean geometry, so it must be true. However, if, as Borges was wont to admit, the whole of mathematics and geometry is a vast game of fiction rather than the search for eternal Platonic truths, then the story built upon a fabulated cornerstone is figment. But it is a figment that has found certain fits with the physical world insofar as it was mathematized by classical science and hopefully by the new science of fractals. So is the figment fact or fantasy? In either case, the same conclusion inheres according to the prevailing knowers of the lingering classical model: if it is fact, then ultimately it will be demonstrated that numbers, the building blocks of mathematics, hook onto the world, and if it is fantasy, then the hooks still remain valid, for after all, the brain is a computing machine. So Borges's story is either true or it is not true, whichever happens to be our piece of cake. But enough of this palaver. Let us return to our billiard ball, which, we would suppose, is real.

But how real is it, really, regarding our concrete living and breathing world, if one is to consider it as an abstraction? A line consists of an infinity of points, one of which specifies the ball's position. Countless lines make up a plane, which is capable of specifying coordinates relative to a pair of axes. An infinity of planes stacked one on top of the other composes a three-dimensional space, demanding three specifying features. For the geometry of the billiard ball we would need six-dimensional space in order to account for both position and momentum: three coordinates are needed for position and three for momentum.[3] "Such a space is unimaginably complex," one might wish to retort. Yet it can relatively easily be punched into the computer and translated onto the monitor in terms of a few symbols: simple, it appears. This six-dimensional space is, admittedly, a fiction, since it exists in the mind and not in the combination of a ball and its companions on a green table and a stick that periodically shakes them from their resting places. But it is a necessary fiction, since it depicts all the possible motions and resting places of the particular ball in question. This six-dimensional space and all comparable spaces are placed into the category of phase space. So six-dimensional phase space (and other such items from the universal encyclopedia of imaginary beings) is real, in spite of whether it exists or not; and it is semiotic, that is, it takes its rightful place within the arena of semiosis. This we learned from Simplicio's ruminations about signs and nonsigns and the real and the nonreal, which were not quite so simple in spite of his wishes to the contrary. We begin to fathom the complexity, the interminability, of Mr. Palomar's grand quest for ultimate knowledge.

Tropics of Topoi

But what, more precisely, is "phase space"? In a manner of speaking it is a map, a way of defining a collection of things that are not transient but in motion, whether incessantly or intermittently. It is the geometry of dynamical systems that takes place in the mind, not in the empirical—Cartesian-Newtonian—world. It is quite unlike what we perceive and conceive to be ordinary physical space. With the previously mentioned six coordinates in place, three for momentum and three for position, and with a total of six dimensions, a phase space map for all possible trajectories of the billiard ball can be constructed. The fact that six-dimensional space as such does not exist in the actual world as we see it is no problem, since the map is, after all, an abstraction, not a road map of familiar colors, lines, letters, and numbers. The six-dimensional phase space map is used by physicists to depict the possible dynamics in *our world,* not simply what there is.

All this is somewhat mind numbing at the outset. So let me begin metaphorically with a familiar example and work up from that point. A road map of the United States does not have a buzz of minuscule cars moving along its lines or a roar of planes flying around overhead. It is a static sheet of paper punctuated by dots and lines and letters and numbers, by variously colored patches, and by squiggles representing topological warps. As a static depiction, it provides information regarding all possible routes from each point to all other points, regardless of whatever routes its studious onlookers might actually take. In this manner comparable to a road map, phase space is a depiction of the dynamics of what might occur in light of the range of all possible occurrences. It is the space of the possible, and what is actualized in physical space depends upon, so to speak, what questions are asked of phase space. The weary traveler questions his or her road map in order to locate the quickest way to a destination from among all the possible routes; the mathematician consults the phase space depiction of a particular phenomenon in order to determine what might happen in all possible worlds. A chief difference between the road map and a phase space map is that one can quite easily read the road map—as an icon it looks something like the territory mapped—while the phase space map is pure abstraction, intelligible only to the proper reader with no icons dancing in his or her head but, rather, arbitrary symbols.

Phase space is germane to chaos theory. What appears to be the relatively simple example of a swinging pendulum is a good place to start. A pendulum is guided by an "attractor." An attractor's function is just that: *attraction.* It attracts a point-object—the pendulum—that is moving about in phase space. A "basin of attraction" surrounds the attractor. The point-object can begin its movement anywhere in the space, and its trajectory will be guided by the basin of attraction and inexorably drawn toward the attractor, which is also defined as a point in space. There are several types of attractors: fixed point, limit cycle, quasi-periodic, and strange (for a graphic illustration, look ahead to fig. 11). The first consists of a basin of attraction that brings the point-object invariably to a particular spot in phase space when it runs out of steam and comes to rest. A pendulum, if well oiled and relatively free of air perturbations, will swing back and forth in ever decreasing orbits until it comes to a standstill more or less at the designated fixed point. A limit cycle attractor does not bring the point-object to a rest but causes it to continue in a cycle around a closed loop indefinitely. If a child on a swing going back and forth in pendulum-like motion is given a boost at the peak of every other swing by an adult, the child will continue swinging until he or she succeeds in exhausting the prime motivator. If instead of the child being pushed straight ahead the angle of the swing is tilted slightly, the attractor will have changed its position somewhat, and the swings will have become somewhat asymmetrical. If the swings are perpetuated in the same fashion, they will continue on as before, somewhat out of kilter but managing quite well. The heart's beat is another example of periodic movements guided by a limit cycle attractor, at least until the organ stops and the death of its owner occurs. A quasi-periodic attractor can be patterned by an endless line drawn around the surface of a torus in phase space, circling in a helicoidal path such that with each revolution it returns almost to its point of origin but not quite. The motion of the microscopic particles making up the smoke ring, mentioned previously, is typical of a quasi-periodic attractor. It is a dynamic, somewhat structured pattern of movement, with the particles gyrating around and around the surface of the ring and with the ring itself slowly rotating like a fat, slow motion hula hoop.

Finally, the strange attractor consists of a point-object moving about in phase space apparently without any definite or predictable rhyme or reason. The path of the point-object mapped out on a two-dimensional sheet appears to be diverging and converging, twisting and intertwining. The continuous path is bound to a particular region, though it never returns to the same point. In this manner, even though the path is drawn on a flat sheet, in order that the lines never touch, the sheet cannot be exactly of two dimensions but of somewhat more than two, and somewhat less than three, dimensions. Moreover, if two somewhat parallel paths are examined, they will soon be found at some indeterminable point to diverge radically. This behavior reveals the sensitivity of the chaotic system. A tiny variation at the beginning of a particular cycle can be magnified manyfold as the cycle proceeds, and the loops become wilder and wilder until they become well-nigh chaotic, a characteristic that is absent in the other attractors. In this manner, strange attractors follow endless and nonrepeating paths in abstract phase space, paths that can be effectively constructed with computers to produce the dramatic images with which we are now quite familiar (Tritton 1986).

Strange attractors, in sum, arise from a system's sensitivity regarding initial conditions. *Sensitivity* is the key word here. The concept owes its present fame to its modern founder, French mathematician Henri Poincaré, though almost a century passed before his groundbreaking work was duly acknowledged. Poincaré threw a monkey wrench into the works of classical physics even before Einstein's special theory of relativity. Up to that time, it was assumed that any chaotic disturbance in a system must be the result of some outside force, that is, if the system itself is in good working order. As was mentioned in note 10 of chapter 1, Poincaré demonstrated that slight perturbations within a system, even if they are virtually infinitesimal, can lead to its collapse. Poincaré's work involves the simple phenomenon of a three-body system. In regard to two-body systems, say, the Sun and the Earth or the Earth and the Moon, Newtonian equations are quite effective. However, when all three bodies, Sun, Earth, and Moon, are included in the equation, problems arise. The Moon's revolving around the Earth affects the Earth's orbit around the Sun, and in turn the Sun is affected, however slightly, by the interaction between the Earth and the Moon. Interaction between any two of the three bodies is relevant to their interaction with the third body. The upshot is that Newtonian equations cannot be worked out with absolute precision with respect to the three-body system, but, rather, due to reasons of formal mathematics, they can account for no more than a series of approximations. There can be no more than a convergence toward an absolutely precise computation, since the absolute exists only at the infinite stretch. That is, if the measuring instruments were of infinite precision and the three bodies in question were absolutely perfect spheres moving in a complete vacuum, an accurate picture could be forthcoming. But nature and our instruments are inevitably to a degree flawed. What is worse, one cannot know the initial conditions of this or any other system to absolute precision, and even if one could, it would require our going back to the beginning of time itself.

Poincaré's discovery that the very slightest of perturbations causes some celestial bodies to wobble and weave drunkenly along their orbits in apparently random fashion bears sober tidings: prediction becomes impossible, even though the entire system may be guided by deterministic laws. In 1908 Poincaré wrote:

> A very small cause which escapes our notice determines a considerable effect that we cannot fail to see, and then we say that the effect is due to chance. If we knew exactly the laws of nature and the situation of the universe at the initial moment, we could predict exactly the situation of that same universe at a succeeding moment. But even if it were the case that the natural laws no longer had any secret for us, we could still only know the initial situation *approximately.* If that enabled us to predict the succeeding situation with the *same approximation,* that is all we require, and we should say that the phenomenon had been predicted, that is governed by the laws. But it is not always so; it may happen that small differences in the initial conditions produce very great ones in the final phenomena. A small error in the former will produce an enormous error in the latter. Prediction becomes impossible, and we have the fortuitous phenomenon. (Poincaré 1914, 68)

The behavior of such a system is not intrinsically indeterministic, as we observed in the preceding. It becomes indeterministic only in its relation to other systems, and those to others, until the entire universe is taken in. The initial conditions, then, are tantamount to the primal instant of the "big bang" itself. Given Poincaré's work, it becomes quite apparent, however painfully, that the universe must be teeming with phenomena of the strange attractor variety.

Two initial states that are almost identical, but not quite, will swerve away from one another along nonlinear paths until they become radically distinct. This sensitivity, writes Edward Lorenz, one of the pioneers of strange attractors, "implies that two states differing by imperceptible amounts may eventually evolve into two considerably different states. If, then, there is any error whatever in observing the present state—and in any real system such errors seem inevitable—an acceptable prediction of an instantaneous state in the distant future may well be impossible" (Lorenz 1963, 133). Chaotic dynamical systems can be modeled on computers with a sophisticated degree of accuracy by rounding off the calculation to the twentieth decimal place; yet such precision proves to be of no avail, for the system's sensitivity dependence quickly brings about a radical divergence in one of a potential infinity of directions. Computation of exact ends by almost, but not quite exactly, specifying the origins becomes impossible. Such sensitivity dependent systems exhibit what Lorenz labels the "butterfly effect." The

Earth's weather behaving like a chaotic system renders it potentially sensitive to so apparently trivial a perturbation as a butterfly flapping its wings in the Amazon River basin, which can make the difference between calm weather and a tornado in Oklahoma. However, if a tornado happened to occur in the North American Midwest it would by no means be the fault of the butterfly's inconsiderate behavior. There is no linear cause-and-effect chain involved here, but, rather, the tornado would have been the product of nonlinear processes, that are deterministic regarding the whole but unpredictable with respect to a grasp of any of their parts. These sorts of nonlinear phenomena play havoc with any and all attempts to forecast the weather with precision; hence, as in the case of all chaotic systems, a qualitative approach becomes necessary.

Strange attractors are "strange" in part because they are somehow capable of reconciling two contradictory phenomena: they are *attractors,* the locus toward which the trajectories of elements in a system converge, and they are *repellers,* rapidly forcing the trajectories of those same elements outward. There is a combination of folding and stretching, of convergence and divergence, of a tendency toward the oneness of a single point and the manyness at widely disseminated points, of continuity and discontinuity, reversibility and irreversibility, simplicity and complexity, and, as we shall note, of atemporality and temporality. Strange attractors of the Lorenz sort also exhibit fractal characteristics: dimensions require fractional descriptions rather than descriptions by whole integers. Moving along the Lorenz attractor in clockwise direction and then crossing over into the counterclockwise twist demands slightly more than two dimensions as in the case of a plane, a sphere, or a torus, since the lines can never exactly intersect.

For a likely example of a strange attractor, consider the Möbius strip. It consists of a two-dimensional plane one end of which is given a 180º twist and connected to the other end (for a graphic depiction, look ahead to fig. 21). Now we have a plane whose conceptualization must include a third dimension, since the warp in the strip transforms what was inside to outside and vice versa, which requires another dimension of space. If we begin drawing a fine line along the strip in such a way that it never crosses itself but describes a potentially infinite series of near parallels, then look at the whole concoction from above as if it rested on a two-dimensional plane, we will get a quite faithful picture of a strange attractor. The one-dimensional line travels around and around in space without ever touching itself; but it cannot do so unless it begins nudging its way into a second dimension. That second dimension, in turn, consists of a smooth surface that continues around and around without end; but it can take on this characteristic solely by way of its intervention, however subtle, into the next dimension's territory.

A mountain peak affords a more common example of fractal space of the strange attractor sort. From a considerable distance the darker blue of the peak takes on the appearance of a relatively smooth contour against the lighter blue background of the sky. Upon closer inspection, the contour reveals probing rocks and then the erect points of numerous pines, spruces, and other evergreens. Moving in closer, we see unevenly spaced limbs appear on the trees, and the rocks become a jumble of angles and edges, of straight and broken lines. Narrowing the focus even more brings myriad new details into view in such a manner that the once infinitesimally thin, crisp line outlining the peak becomes fuzzy. The line's breaking in many tangential directions and at times doubling back on itself in virtually an infinite number of ways seems to invade two-dimensional space. It takes up more than one dimension but less than two dimensions.

The complexity of all this might provoke a longing for Simplicio's longing for simplicity. Perhaps that lost paradisiacal state Mr. Palomar sought is best forgotten. For it is becoming increasingly evident that without strange attractors, without chaotic systems, there could be no order out of chaos, that is, order of the self-organizing variety. And without order from chaos, life could not sustain itself, nor could it have emerged in the first place.

Attractors and Life Processes

In order to relate the previous section to the perplexing phenomenon of life, consider an exceedingly more complex example than those that have hitherto been displayed: a telephone line that happens to be the preferred resting place for a local flock of pigeons.

The line, a one-dimensional space of possibilities, represents an infinity of points, an infinity of possible perching spots for a given pigeon. However, the affable aves tend to space themselves apart from one another by about the same distance, and that distance grows or shrinks depending upon the number of birds on the wire. If six pigeons are perching, let us say the distance averages approximately three feet; if sixteen decide to take a breather, then a little less than one and one-quarter feet will separate them; and if there are thirty-six, then the space between them will be about one-quarter of a foot. The line is a tacitly acknowledged phase space for the pigeon community. When the pigeon population on the territory is *x*, for each pigeon there will be a "fixed point" toward which it will tend to gravitate, though it will always miss the mark slightly—our pragmatically minded pigeons would care little for the mathematician's precise mappings. The community of pigeons, each bird with its inclinations and perhaps a few idiosyncrasies within its own phase space, will find their proper places and generally live in quite peaceful coexistence. Now let us move up to a higher notch of abstraction.

An object in Poincaré phase space can be mapped on a plane by means of two coordinates, *x* and *y*. The object is represented on the plane by a single point that enjoys two values. The point defined on the plane is the space of possibility, with every point as the possible state of the system. If the point is in actual motion, say, in terms of the extremity of the to-and-fro movement of a pendulum, with one coordinate representing the point's velocity and the other its position, an infinity of points will be actualized along a curve forming an ellipse, with the center at the intersection of the two coordinates. This situation would be comparable to the one that would exist if the flock of pigeons were all to leave the wire in unison, fly to the nearby river, and swirl around above it in counterclockwise fashion just north of the bridge leading out of town. Each pigeon, considered a point in phase space, would describe a rather smooth ellipse, and ideally, the composite of all the pigeons would make up a set of concentric ellipses with a constantly moving center. The Cartesian grid is a convenient method for describing the position of something in space. But since, like the pigeons' coming and going, nothing is stationary in the absolute sense—this we have learned especially from relativity theory—time and dynamics necessarily force their way into the picture (see Weiss).[4] The problem is that a picture of phase space inevitably becomes, once again in Sagredo's words, sludgy. Space, we would like to believe, is something that things move around in from one place to another in a quite simple way, like the classical version of the billiard ball example or like a swinging pendulum. But from one periodic movement to the next, things are never exactly the same. Take ourselves. In the morning we drag ourselves out of bed and at day's end we flop down into the same bed, nothing gained, nothing lost. But we all know that during a given day, things happen to us and we do things to our environment, and besides, the bed is never exactly the same bed from one day to the next, one year to the next. Irreversible changes are inexorably wrought in time.

After all is said and done, neither we nor pigeons are actually very orderly. Our behavior invariably tends toward the chaotic (i.e., as with Bateson's daughter, Mary Catherine, whose activity cannot help but muddle her environment). Regarding the pigeons, what usually occurs during their elliptical flight is a converging, diverging, involuting, convoluting, now accelerating, now decelerating, rather syncopated concoction of curves in *n*-dimensional space rather than along a plane. An account of the pigeons' trajectories would in essence entail what is known as "ergodic theory." The bob of the pendulum follows an elliptical path: this is nonergodic, much like a "minimalist" work of art composed of a circular orange patch on a blue background. If the bob is connected to an elastic device that allows it to move further and further away from the center of the ellipse with each swing, an ovaloidal spiral would be mapped out in phase space: this is ergodic but still quite orderly—compare it to a traditional painting highlighting the Renaissance vanishing point to give it depth. Now, if the bob is connected by no more than a rubber band that will hardly support it without breaking, you can imagine its degenerating to an unpredictable, and apparently random, sequence of movements. This situation is comparable to a frenzied Jackson Pollock painting consisting of paint dripped onto a horizontal canvas from above in what appears to be near-random fashion. We are no longer considering merely two dimensions but, rather, an approximation to three dimensions, since the streams of paint crisscross upon folding themselves within one another. One series of paint droplets deposited on top of another series, for the viewer giving the entire painting a mere glance, might simply be taken as an intertwined jumble of streaks on a flat plane.

But actually, something higher than two dimensions was required for the work originally to be constructed. The artist had positioned himself above the canvas, and from that somewhat supreme locus, he allowed the paint to drip down on the world of a mere two dimensions below. To the spectator the paint appears to have been deposited on the flat plane at random. But Pollock insisted time and again that there was no real randomness in his movements at all: they were the product of artful calculation designed specifically to give the impression of purposeless and selfless spontaneity. However, we cannot exactly call the work three dimensional, for from the viewpoint of the spectator there is little indication of depth as we are accustomed to seeing it in our three-dimensional world. Yet, we must concede that the work takes in something more than a mere two dimensions. In the case of our pigeons' flight having become desynchronized, at least the fraction of a third dimension clearly enters the scene also. Their dynamic flight is in this respect somewhat like the Pollock painting. But that is not really a faithful allusion. As a matter of fact, the pigeons' movements are more comparable to Pollock's *act of painting*—after all, Pollock works should be taken as acts of paintings, so their author and the art critics tell us. The pigeons' trajectories require vertical as well as horizontal vectorials. And they involve diverging and converging, swerving and veering, weaving and emerging and disintegrating lines of remarkable complexity. This is a strange attractor enacted on nature's stage, in much the same way as the kinesthetics involved in Pollock's bringing about the creation of his work. Furthermore, the pigeons' movements from one-dimensionality toward two-dimensionality and from two-dimensionality toward three-dimensionality, as we have noted of other types of phenomena, is of the essence of fractals.

Calvino's "The Invasion of the Starlings," in *Mr. Palomar* (1985, 61–66), has our protagonist observing the unruly "cloud of flying wings." This should be a reassuring sight, this "passage of migratory birds" associated in "our ancestral memory with the harmonious succession of the seasons." But Palomar "feels something akin to apprehension. Can it be because this crowding of the sky reminds us that the balance of nature has been lost? Or is it because our sense of insecurity finds threats of catastrophe everywhere?" (62). The customary image of migratory birds is that of "a very orderly and compact flight formation, which furrows the sky in a long host or a right-angled phalanx, like a bird-shape made up of countless birds" (62). In contrast, the disheveled aggregate of starlings makes up "an airy crowd that seems always about to scatter and disperse, like grains of a powder suspended in a liquid; instead, it thickens constantly as if from an invisible conduit the jet of whirling particles continued, never managing, however, to saturate the solution" (62).[5] As Palomar continues obsessively to observe the birds, he

> feels caught in a weft whose continuity extends, uniform and without rents, as if he, too, were part of this moving body composed of hundreds and hundreds of bodies, detached, but together forming a single object, like a cloud or a column of smoke or a jet of water—something, in other words, that even in the fluidity of its substance achieves a formal solidity of its own. But he has only to start following a single bird with his gaze and the disassociation of the elements returns; and the current that he felt transporting him, the network that he felt sustaining him, dissolve; the effect is that of a vertigo that grips him at the pit of the stomach. (Calvino 1985, 63–64)

In the first place, it bears mentioning that Palomar's name is not coincidental—*Palomar:* a dwelling place of doves. In the second place, Palomar feels, kinesthetically, as if he were part of a vertiginous mini-maelstrom. In fact, he *is* part of it. He is within it, it is within him. While he is focusing on one solitary bird, a sense of vertigo overcomes him, as if he were the one being swirled about, without will, like a leaf in a hurricane. The complex swirl of Palomar and the collection of fowl is dynamic: the flutter of each pair of wings, as a bird twists and turns during its flight, describes its own double, irregular corkscrew within converging and diverging lines making up the fuzzy not-quite-toroidal form. Each bird swimmingly describes its own giddy path within the whole. The whole itself is an intertwined mass of knotted, interweaving trajectories. While attempting to read this whole, Palomar is in the best of all possible worlds neither subject nor object, neither inside nor outside, neither knower nor known. There is neither determinable pattern nor knower of the pattern in its act of becoming—in the becoming of its being and at the same time the being of its becoming. There is simply an act of patterning. Or perhaps we should better say: "It patterns." This is much like the impersonal "It rains," comparable to the subjectless "It writes," "It pains," "It dances," which integrate the personal pronoun, *he* or *she,* with the act of becoming: there is neither author of the act nor completed act; there are only becoming acts, acts in their act of becoming, becoming as the emergence and passing away of acts.

A switch from Palomar's ornithologizing to the painter's act of ornamenting his or her canvas may provide us with a more comprehensive model of fractals, chaos, and becoming processes.

Art from Chaos

Paul Klee, one of the most profound avant garde artists of the twentieth century, in both his works and his meditations on the nature of the painted canvas alludes to *chaos, disorder, order,* and *cosmos*.[6]

Chaos as the antithesis of order is not utter chaos but, rather, a concept related to the notion of cosmos, while order should be viewed as the opposite of disorder. Utter chaos is much like the "emptiness" alluded to previously and of which Buddhist philosopher Nāgārjuna speaks in reference to *Sunyata*. It cannot be put on a scale but remains forever unassailable and unmeasurable: it is both nothing and everything, both the ultimate of chaos and the ultimate of order inasmuch as chaos and order can be related (Huntington 1989; Kalupahana 1986). In other words, utter chaos can be no more than a nonconcept, like a point that is really not a point but a nowhere-existent some-*thing* or a somewhere-existent no-*thing*, completely free of the idea of opposition. In Klee's words:

> If we express it [chaos] in terms of the perceptible (as though drawing up a balance sheet of chaos), we arrive at the concept grey, at the fateful point between coming-into-being and passing-away: the grey point. The point is grey because it is neither white nor black or because it is white and black at the same time.
>
> It is grey because it is neither up nor down or because it is both up and down. It is grey because it is neither hot nor cold; it is grey because it is a non-dimensional point, a point between the dimensions. (Spiller 1961, 4)

Klee illustrates his notion in three dimensions with the Cartesian coordinates on a two-dimensional plane. First, the distinction between cosmos and chaos is presented in terms of figure 7, where chaos and cosmos are viewed as opposites, the union of which becomes "antithetical chaos," that is, not utter chaos but, rather, a locally conceived state of chaos. Utter chaos can never be named, conceptualized, and made intelligible, for it is totally unweighable, unmeasurable, and outside any and all categories and lines of demarcation. Antithetical chaos, in contrast, as a locally conceived phenomenon, is set apart from its opposite, order, and named as such.

The transformation of cosmos to chaos is presented as figure 8, which is quite comparable to the hierarchy of attractors as described

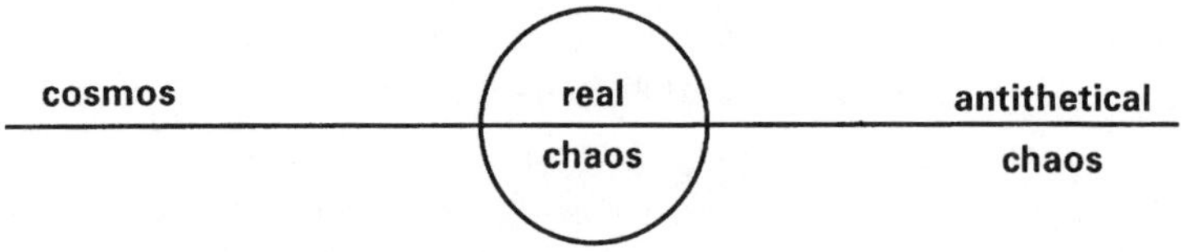

Fig. 7. The beginning of distinctions. (Adapted from Spiller 1961, 3.)

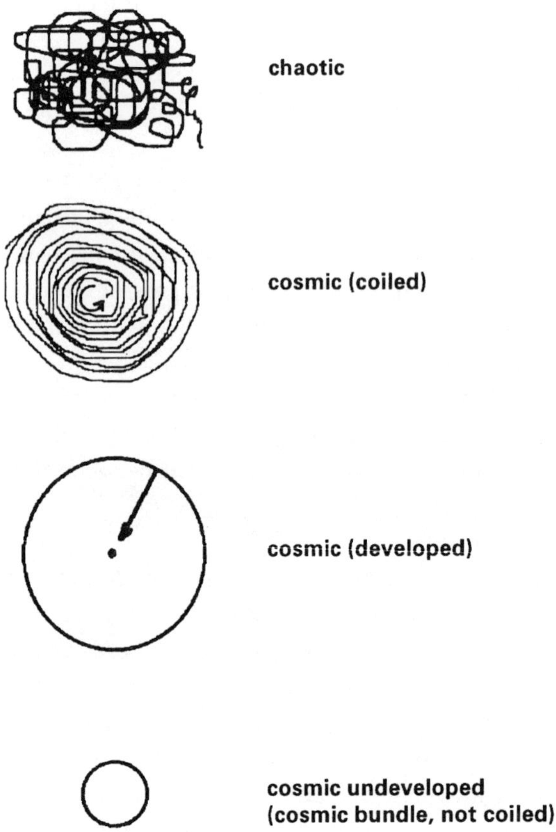

Fig. 8. From chaos to cosmos. (Adapted from Spiller 1961, 2.)

previously: from fixed point (cosmic, developed) to strange (chaotic). Klee then goes on to pattern gray as a vague image of antithetical chaos disseminated outward in three-dimensional space (fig. 9). In terms of the Cartesian coordinates depicted in figure 9, gray spreads, cloudlike, throughout space, and, as the product of a struggle between black and white, colors are "cosmogenetically" produced, a graphic depiction of which Klee offers in an image much like that depicted in figure 10. Klee tells us that he begins "logically with chaos," which is quite natural, and he is "at ease" with this mythical beginning, "because at the start I may myself be chaos." Chaos evolves into ordered cosmos by a "natural movement" from white to black, which "is not unregulated, but unarticulated." It is regulated "by comparison with chaos, in which light and darkness are still undivided. It has the natural order of an unbroken flow from one pole to the other." The swirl of activity as a result of the tension or struggle of complementarities is "infinitely subtle. The minute particles can scarcely be distinguished. A definite orientation is not possible. Position cannot be sharply fixed, because the flux, the subtle flow, takes certainty, gently but surely, away with it" (Spiller 1961, 9). The dynamic two-dimensional hustle and bustle of atoms making up diffusive gray centrifuges into a three-dimensional rainbow of tints and hues.[7]

And figures 5*a* and 5*b* once again make their presence felt! The Harlequin lifts itself up by its own bootstraps, distinguishing itself in terms of intermittent and asymmetrical patterns of black and white, by gray's being separated now into white, now into black, through the action of the $\sqrt{-1}$ Escher operator. Ultimately, we can imagine colors, as evidenced by a close inspection of Tobey's work at the lower left-hand portion of figure 5*a*'s rectangle, emerging into the light of day. Through the apparently timeless oscillations between the one and the other, linear time, because of myriad series of differences that make a difference, becomes essential to the entire process. The becoming of the Harlequin constitutes time's fluctuating, dancing, ephemeral beingness; its beingness gives the becomingness of its fluctuating, dancing activity form, however ephemeral.

What we have here can also be patterned quite effectively by the image of flow. Take a motionless body of water behind a dam for example. It is akin to Klee's antithetical chaos; it is emptiness that contains within itself the possibility of everything. Now suppose we poke a small hole in the dam so as to allow a steady stream of water into a trough

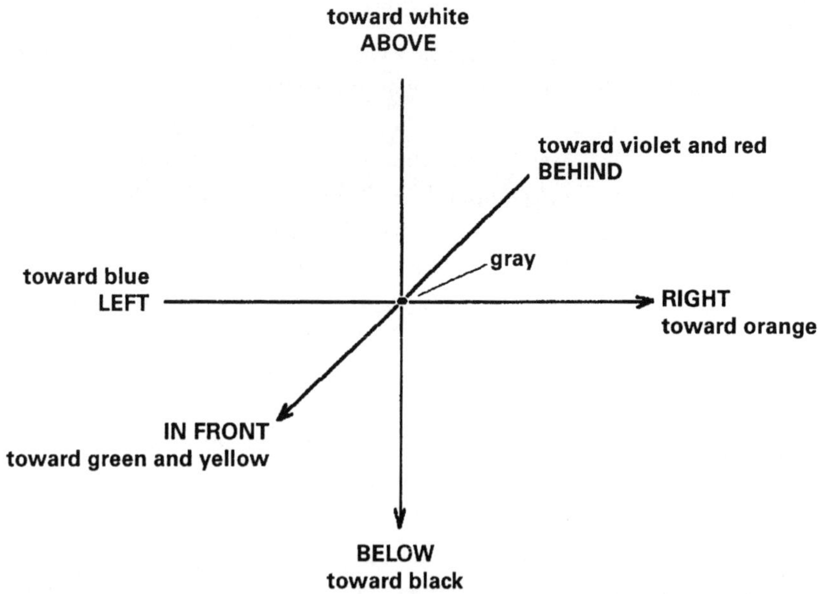

Fig. 9. Colors out of gray. (Adapted from Spiller 1961, 4.)

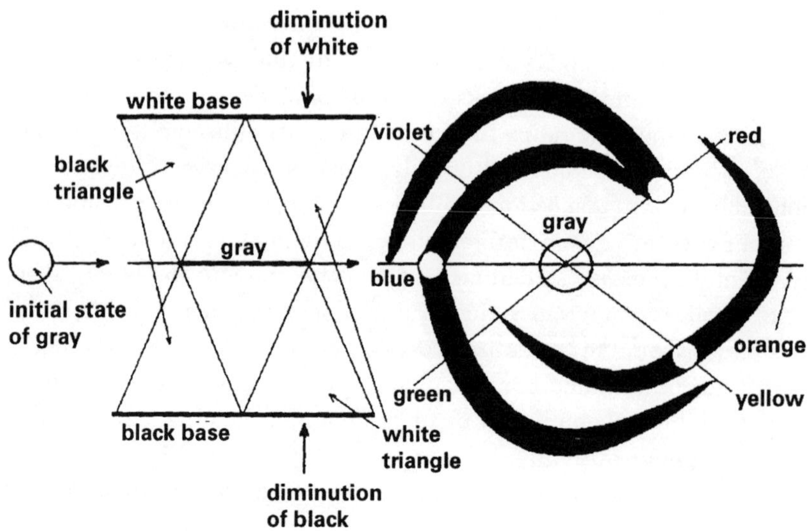

Fig. 10. Dynamics of colors out of gray. (Adapted from Spiller 1961, 9.)

below, and we place a rod at the midpoint of the flow. The velocity of the stream remains constant with time, with the rod acting as a fixed point (see fig. 11, after Kadanoff 1983; also Kellert 1993, 6). If we enlarge the hole in the dam, with the increased flow some turbulence ensues. But there is no cause for alarm, at least not yet, since the small eddies indicate that the rod now acts as a limit cycle attractor. Making the hole larger still increases the turbulence, and the rod's role as a quasi-periodic attractor becomes evident, with the eddies becoming more complex—eddies within eddies, in the order of a fractal. Then, if we continue to ream out the hole, eventually the flow will be overtaken by chaos. That is to say, apparent chaos, "antithetical chaos," since we see it as chaos against the conception of order. Then the dam breaks and wreaks its devastation below. Actually, the destruction wrought by the fractured dam on the towns downstream would not ordinarily from a human perspective be termed order, yet, viewed from a larger perspective, it can be said that the tension has been released and the water flows with a minimum of perturbation, dissipation, fluctuation. It has realized its potential for pulling itself into some new form of order from the erstwhile chaos. And assume it does so, after it enters the ocean and settles down into a massive eddy some miles off the coast: order from apparent chaos.

Relating figure 11 to the notion of fractals, it can be said that the lines of flow of the steady state brought about by the fixed point attractor can be mapped out quite effectively in terms of one dimension. Lines in the periodic limit cycle attractor cannot but be slightly more than one dimensional, yet they are less than two dimensional, since only at certain points are the lines required to jump slightly above or tunnel below other lines in order to avoid a collision. As the quasi-periodic motion on the two-dimensional surface of the torus becomes more and more unstable, the lines will slip into slightly more than two dimensions. Then, when a chaotic attractor erupts, lines will be shoved into the beginnings of a third dimension and beyond. The onset of greater degrees of turbulence leads to the piling up of incommensurable lines giving rise to ever greater degrees of freedom, of eddies within eddies, and of greater turbulence. Then finally, with a cathartic heave, the repressed mass behind the dam loosens and breaks its bonds: at long last, it is free, it has the liberty to find its own form of order.

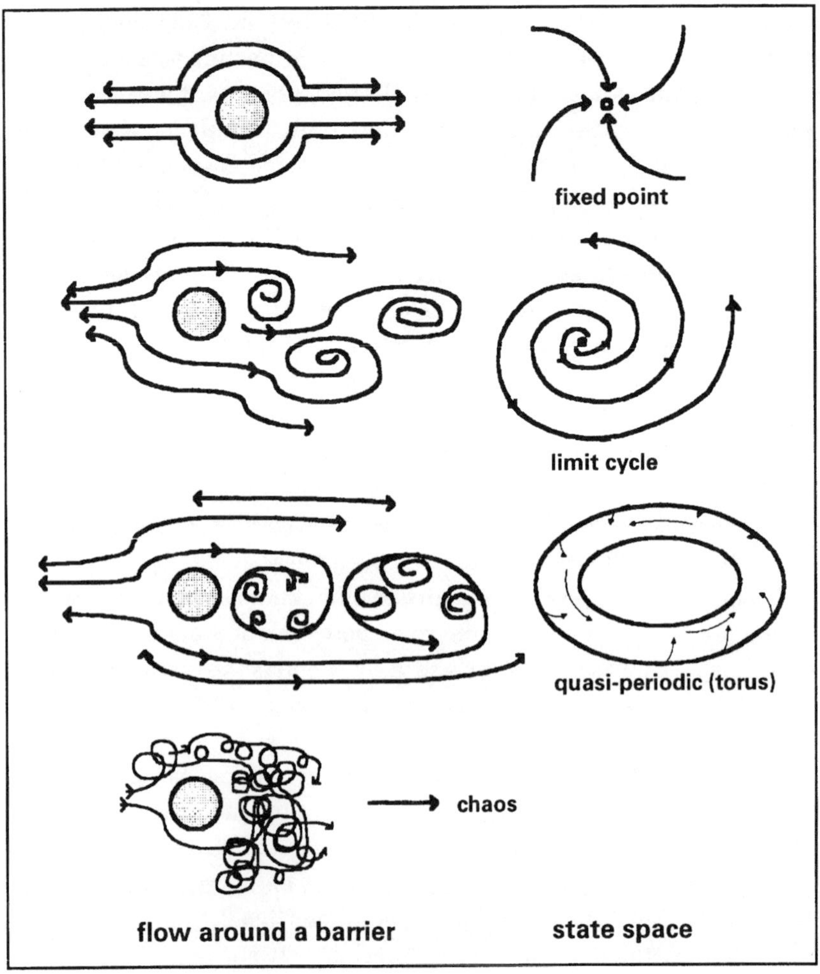

Fig. 11. Attractors (Data from Kadanoff 1983; Kellert 1993)

Simplicity from Chaos?

Now let us relate this situation to its "quantum level" for a moment. The notion of the "probability amplitude" of a quantum state before it has "collapsed" into a particle is not a matter of probability in the classical sense, that is, in the sense of dice throws, gas molecules in a flask, or the second law of thermodynamics. It is more akin to the complex square root of a probability.

This involves what are known as complex numbers, which integrate natural numbers and real numbers with imaginary numbers. A geometrical depiction of complex numbers could be something in the order of the Cartesian plane mapped onto figure 4, where we stretch the oval out to a circle and bisect it with a horizontal line, the intersection of the two lines on the plane representing 0. Now suppose we stretch the 0 point of the graph in both directions, forward and backward orthogonal to the plane, and in the process construct another plane that intersects our first plane. We would have something like figure 12. The axis of imaginary numbers has a negative side and a positive side, the same as the axes of whole and real numbers. This plane, a combination of whole integers and imaginary numbers, is called an "Argand plane," named after the self-taught Swiss mathematician Jean-Robert Argand. Argand noted that an operation through the "imaginary" axis could turn +1 into –1. First, +1 is multiplied by $+\sqrt{-1}$, which rotates it to an orthogonal position—by a 90° shift outside the domain of the whole integers. Repeating the operation takes the imaginary number down and back to the horizontal axis, and the product is –1. Carrying out the procedure again, but this time by multiplying –1 by $-\sqrt{-1}$, takes the number down under and to the opposite orthogonal position. A final reiteration returns the number to its original value, +1. This set of operations describes a circle, the Argand circle. According to the dancing, scintillating, vibrating play of opposites represented by the Argand circle depicted in figure 12, intermittent patches of white and black, and, when the patches are small and diffused, patches of gray, emerge from the to-and-fro "*now*-White" . . . "*now*-Black" . . . "*now*-White" . . . "*now*-Black" . . . *n* oscillation.

In terms of Goethe's theory of colors, the grays, like Tobey's grays that contain, within themselves, the colors of the entire spectrum, are the product of a perpetual Heraclitean struggle between the opposites, black and white. The grays are there, as if lying in wait for the opportunity to spill out and then display myriad colors according to the allowances and the limitations of the moment. But this cannot occur without our participation, without our constructing some world, *our world,* from the virtual infinity of possible worlds. The *(our) world's* spectacle that spreads before us, according to Goethe, is not that of balance, harmony, and symmetry but of some semblance of order that we as participants in the world's unfolding help bring about. But lest I be accused of waxing inordinately romantic, I must return to the more comfortable and customary sphere of technicalities.

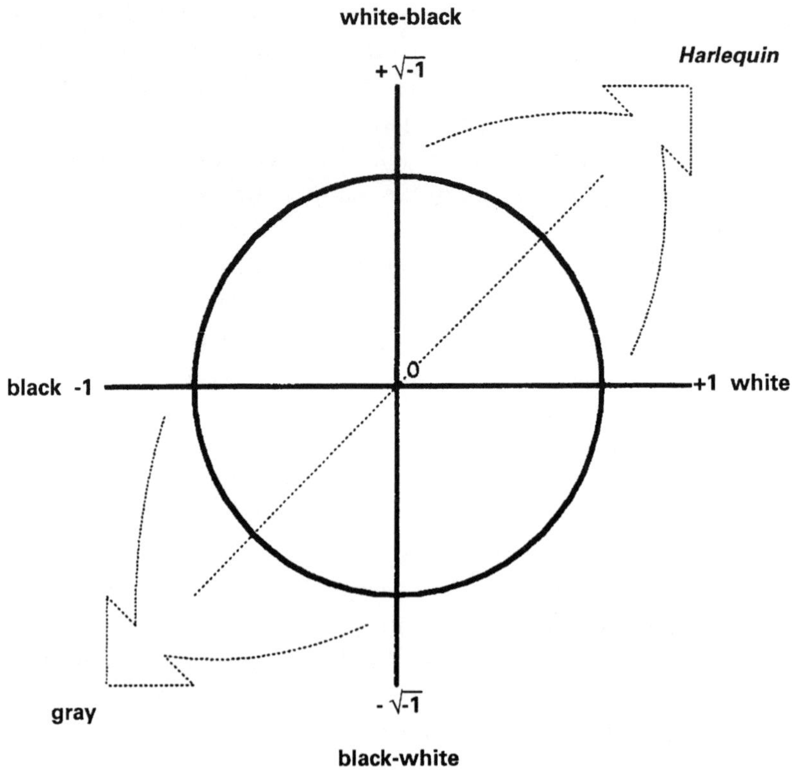

Fig. 12. The plane argandized

More specifically regarding complex numbers on the Argand plane, let us designate the imaginary number, $\sqrt{-1}$, as *i*, in order to dissolve the self-contradictory symbol in a more general medium—recall the function of *i* in figure 4 following Sagredo's suggestion. Now let us plot two values on the plane, one a real number and the other the imaginary number multiplied by a real number, and we have a complex number of the form $a + ib$.[8] In other words, if a is 2.0 and b is 1.5, we have the complex number: $2.0 + \sqrt{-1} \times 1.5$. Plotting this number on the Argand plane, one moves to the right on the horizontal axis to +2, then up the vertical, imaginary axis to 1.5, and records a point. With additional points on the plane, a geometrical figure can be constructed. Addition, subtraction, and geometrical operations on the plane are quite conventional, except that the numbers are "ambiguous," "undecidable."

Complex numbers enter into calculations in both relativity theory and quantum theory. In quantum physics, the possible positions of a particle, once it has "collapsed" from its complex probability amplitude (its wavelike character), can be mapped onto the complex plane. If we take the Argand plane, consisting of the *Y-Z* axes, as the depiction of the possibilities of occurrence of a quantum event and the *X* axis as the linear movement of the plane from past to future, then we have a series of events as illustrated in figure 13—following Penrose (1989, 244). The fine line represents the "worm line" or "curve" tracing out the path of the actualizations of a particular subnuclear particle. This is the "history" of the particle, or, if you will, it can be a minute glob of paint from an artist's brush on the canvas plane—or ourselves for that matter, for like particles and globs, we are signs. As a result of the oscillation of values represented by $\sqrt{-1}$, the plane can take on now a speck of *this,* now a speck of *that,* and a minuscule patch of something else. Gray can become differentiated into blacks and whites. And between the two, from "nothingness" to every-*thing* or from every-*thing* to no-*thing,* colors can emerge, interact, fold back upon themselves, and merge into one another. And eventually, a world is constructed.[9]

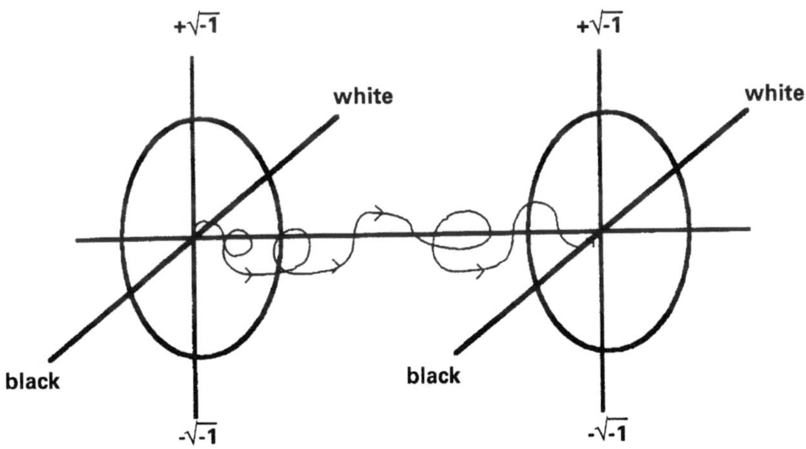

Fig. 13. The argandized plane extended

Significantly enough, a block depiction of the whole of this process is provided by Klee in an image akin to figure 14. If on this figure we construct another metaphorical plane, with its respective axes, and if we stretch gray in one direction and the Harlequin in the other direction, we can include figures 12 and 13 within the figure 14 scheme. The important point of all this is that there is no real incompatibility between the block representation of colors, on the one hand, and the dynamics of temporal processes and irreversible, nonlinear, fractal, order-out-of-chaos events involved in creating a world on canvas, on the other.

A wave pattern of oscillating blackness and whiteness, from gray upward and toward the right side of figure 5*a* can result in Vasarely's apparently static depiction. By a comparable token, in real life situations Ian Stewart and Martin Golubitsky (1992) demonstrate how strictly mathematical fluctuations, as in standing wave patterns, can produce the equivalent of a zebra's stripes, a giraffe's spots, or a Holstein cow's irregular blacks and whites. This phenomenon also brings to mind William Blake's "Tyger, tyger" and its "fearful symmetries." However,

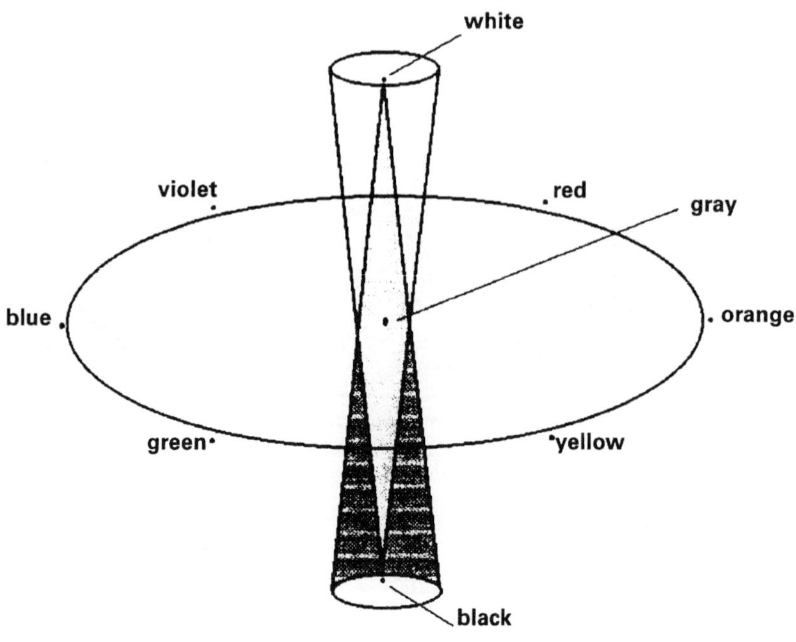

Fig. 14. Colors from another view

there should be no misunderstanding. The standing wave pattern is actually not static at all, but, rather, it is as if the zebra's stripes, or the blacks and whites in Vasarely's Harlequin, were an incessant Siva's dance giving rise to the emergence of what only appears static. In reality there is movement of the most violent sort, minus irreversible time and hence without appreciable change from one moment to another.

A variation of the standing wave theme, which includes the entire universe as well in holographic fashion, is depicted in Borges's "The God's Script" (which appears in *Labyrinth, Selected Stories and Other Writings*) with which we are now somewhat familiar. Tzinacán is within his half of the hemispherical prison, that ovoid, its two halves separated by a line bisecting them, and on the other side is the jaguar upon whose hide there is a jumble of standing waves giving rise to unintelligible ciphers. This affords us yet another image of the observer who is inside and hence nonobjective, the subject who is a nonobserver in the sense that what is observed is none other than the observer observing the observed, which is part of that selfsame observer. As we noted in chapter 2, Tzinacán, with lingering hopes of discovering the cipher to end all ciphers, the cipher encompassing the entire universe, finally concludes that it can be found in the patterns on the jaguar's hide. After prolonged meditation of the living hieroglyphics carried by the animal, in a moment of insight Tzinacán believes he experiences a moment of epiphany. He sees "the mountain that rose out of the water, I saw the first men of the wood, the cisterns that turned against the men, the dogs that ravaged their faces. I saw the faceless god concealed behind the other gods. I saw infinite processes that formed one single falsity and, understanding all, I was able also to understand the script of the tiger" (Borges 1962, 173). It hardly needs saying that in this holographic image, this set of standing waves, we have, once again, a pearl in the Buddhist string of pearls that mirrors the whole, and so on.

However, just as a small part of the holograph is fuzzy, yet it contains the essence of the whole, albeit vaguely, so also fractals, which, once crisp, become frizzled as the algorithm continues to repeat itself. The pattern developed by a Mandelbrot fractal hot patched on a T-shirt is a sight to behold. But as the reiterated pattern becomes smaller and smaller, the line becomes less and less precise. In another manner of putting it, the algorithm begins by repeating itself—as Derrida says of the initiation of *différance*—and differentiation arises; then it reiterates its iteration, and then again, and again . . . until finally, the process moves toward finer and finer differentiations bringing about the tenuous suggestion of the undifferentiation, of the equivalent of grays from blacks and whites; and at the infinite stretch it reaches utter vagueness or, better, Nāgārjuna's "emptiness" or perhaps Klee's undifferentiated "cosmic bundle." It is that which precedes all differentiation (though we cannot really say "precedes" for there is no time, and we should not really say "differentiation," for it implies space and there is no space as such). With such an image of pure confusion, or pure complexity as it were, let us reconsider Klee.

But, When a Line Is Not Simply a Line?

Paul Klee tells us that a concept is not thinkable, and much less articulable, without its opposite: there is no cosmos without chaos, no order without disorder, no outside without inside, no discontinuity without continuity—no simplicity without complexity. In other words, there is no such thing as a concept in itself, that is, a concept that can in and of its own be made intelligible *for us.* A concept needs its counterconcept.

At the outset this would remind us of Borges's citizens of the planet Tlön for whom there is neither science nor reason and on which planet any and all acts of classification imply falsification. On the other hand, sciences and modes of reason and systems of classification do exist, "in almost uncountable" numbers. In order to include the existence and nonexistence of such sciences, modes of reason, and systems of classification within the whole package that goes by the name of Tlön, there is an ironclad necessity that throws our own need for stable bearings into vertiginous loops: every theory must include its countertheory, every proof its refutation, every metaphysical doctrine its blasphemous opposite, every text its own countertext (intertextuality). Indeed, for the Tlönians, "metaphysics is a branch of fantastic literature," for they know "that a system is nothing more than the subordination of all aspects of the universe to any one such aspect" (Borges 1962, 10).

Now this is Popperian falsification with a vengeance![10] Yet, when we come to think about it, contemplate it, and sense it, truly sense it, we must somehow acknowledge that in good Tlönian fashion dualism should not be treated as such in the ordinary sense but as unity, that is, as complementary pairs ultimately forming unity. Opposites, differences that make a radical difference, exist solely in the constructive eye of their beholder, whereas the process of the becoming of the beingness of all us Harlequins of order and progress is no more than the process of the beingness of our becoming. We, all things that enjoy some fleeting form of existence, in the manner of that most fundamental and at the same time the most metaphysical of sciences, quantum theory—according to John Archibald Wheeler following Niels Bohr—organize ourselves in a process that is codependent with the self-organization of everything else: the universe, ourselves included, lifts itself up by its own bootstraps.

Back to the issue at hand, we must ask the question Klee asks: How can this sort of self-organizing come about through the emergence of colors by means of a simple line that separates something from something else? In Klee's conception, from the initial flux a point arises, and then it becomes a *cut,* a mark of distinction.[11] From that mark, everything stands at least an outside chance of emerging into the flow of things. In this manner, Klee the nonmathematician outdoes many of those philosophers whose tirades against logocentrism reveal their own linguicentric bias. Klee makes a gallant effort to avoid language altogether (of course in order to communicate, like all of us, he is forced to commit the sin of linguicentrism at least to a minimal degree). A point on a plane is the instantiation of an actual entity, not a Platonic ideal, Klee reveals. It is not an ideal in the abstract sense of infinites and infinitesimals that are imaginary and sayable but nonactualizable. The point, as a concrete event—the combination of an act and what that act is producing—is not dimensionless, but, rather, it contains, within itself, an infinitely small elemental plane, and the plane contains an infinity of planes. Moreover, though apparently at rest, the point contains the possibility for motion in an infinity of directions. Carrying out a particular motion gives birth to a line. Or better, the line is the yield of a "tension between one point and another" (Spiller 1961, 19). And another point, which could well create a tension with another point outside the line, yields yet another line, and hence the beginning of a plane.

This essential origin of things—the concrete painted image, of course, since Klee was a painter—is thus the result of a "reciprocal tension, a striving for two dimensions" (Spiller 1961, 19). However, for the painter, the spatial character of the point, the line, and the plane, although concrete and seeable before they can be sayable—however incompletely sayable—are never completely free of the imaginary. In order for the point to become a line and the line a plane, the dimensionalities within which they were engendered exist in the mind before they exist on canvas. This serves to create a conflict also in the painter, especially regarding the third dimension, which lies beyond the two dimensions upon which paint is deposited, for

> [h]e does not wish to treat the third dimension illusionistically. Today the flat effect is sought in painting. But if different parts of the plane are given different values, it is hard to avoid a certain effect of depth. If everything remains perfectly flat, we might under certain circumstances have a good carpet. If it does not remain flat, we come to the formal problem of the third dimension. One of the artist's basic problems is how to enlarge space. (Spiller 1961, 49)

The artist can, Klee argues, introduce the third dimension by means of "overlapping planes" to "create the illusion of larger and smaller planes in depth" (Spiller 1961, 49). This, of course, we have seen in Pollock, in fractals, and in strange attractors. In this manner, just as fractal dimensionalities between point and line and plane are imaginary, yet they are imaged by the paint on canvas, so also extension from the plane toward the next dimension is implied, imagined, yet suggested by the lines, swirls, and spots and smudges in the plane. The painter, it would seem, is living proof of the emergence of order from chaos, complexity from simplicity.

"But this is impossible," one might wish to say. "Planes stacked on planes in two dimensions? This can't be, for there is no 'outside' space from which to bring in the planes that are to be placed on the already existing planes." However, the apparent impossibility stems from the limitations of concrete instantiations of dimensionality in terms of paint on a plane, although in terms of mathematical symbols—depicting fractals and strange attractors—the impossible is made possible as a matter of course. Nevertheless, Klee brings about the apparently impossible by a sheer act of the imagination that precedes the concrete act: planes side by side, overlapping and interpenetrating to produce the (imaginary plus concrete) effect of difference, borders, and depth—we need only recall the action of $\sqrt{-1}$ in figure 12, that of the imaginary number.[12] This effect, a convergence of the imaginary and concrete acts, is illustrated through what Klee calls "exotopic" and "endotopic" treatment to create fuzziness, vagueness, a flooding and tapering of colors, hues, and shades. Endotopic treatment is brought about by shading inside the curve to highlight the inside against the outside; the inverse is the product of exotopic treatment. A closed curve treated exotopically brings it into relief; it appears to stand out. In contrast, a closed curve treated endotopically causes it to appear recessive. It is all a struggle for space, of negative treatment and positive treatment of relief applied to curves containing intersections.[13]

Consequently, the third dimension asserts itself, though it continues to operate with what are for all intents and purposes flat values that simply appear either to stand back or come forward. The struggle, then, exists at the border separating one thing from another thing. Depending upon what is subject to exotopic treatment and what is subject to endotopic treatment, the inside or the outside takes on a certain value. During the operations the border itself fuzzes from a line toward a plane, and the plane fuzzes toward the third dimension. This process can be seen by a sort of two-dimensional depiction of a Möbius strip (fig. 15). The endotopic accent (a) becomes the exotopic accent (b) by means of a transform, along the same line as it folds over and upon itself (c), thus inverting the nature of the border. We have, then, Klee's counterpart to partial dimensions as a necessary component of fractal geometry. The important point is that the yin/yang struggle between light and dark, black and white, some-*thing* and some-*thing* else, like the hairline of "emptiness" in figure 4, potentially gives rise to everything, to every-*thing,* every-*thingness.* Mathematical and abstract and concrete and intuitive borders become vague, dimensions lose their wholeness and become fractured. "Imaginary"—abstract and mathematical—values come into play. Ultimately, what is real and what is imaginary, or vice

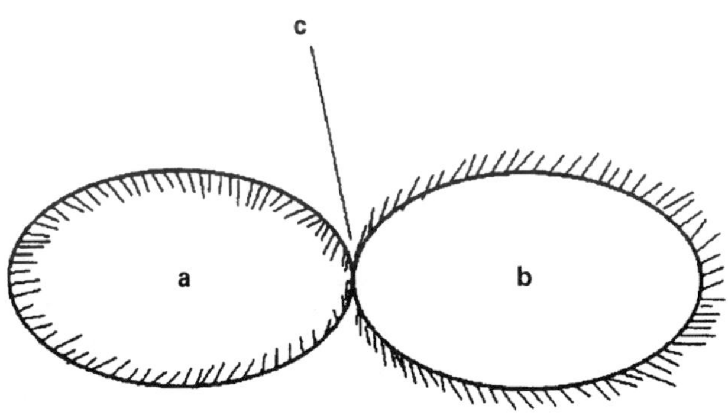

Fig. 15. Endotopic and exotopic spaces

versa, is itself placed in question, and by and large the very nature of our perceived and conceived world is apparently up for grabs. Ultimately, the world, *our world,* is more a construct than a gift from the gods.

Thus, we have, through Klee, an image of chaos becoming order, and all this happening in what is in essence phase space. This entire process necessarily occurs in time; hence it is irreversible and it entails the prior emergence of asymmetry, or so it would seem.[14]

A brief turn to this issue, then.

CHAPTER 6 • Rending and Mending without Ending

Deep in the human unconscious is a pervasive need for a logical universe that makes sense. But the real universe is always one step beyond logic.

—Frank Herbert

I behaved stubbornly, pursuing a semblance of order, I should have known well that there is no order in the universe.

—Umberto Eco

To annihilate the law of contradiction is perhaps the highest task of the higher logic.

—Novalis

To become truly immortal a work of art must escape all human limits: logic and common sense will only interfere.

—Giorgio de Chirico

Dimensions and Fractures

As has been mentioned in the preceding, classical science made an effort to depict time as if it were another dimension of space, plotting position on one axis and time on the other one. But things are more complex than classical science would like to have them. Indeed, we might say they are infinitely more complex than the classical model had them.

Let us return to our billiard ball example. When the ball is in motion, it moves along in its six-dimensional space, depending on position and momentum, and it traces out an elaborate and very slightly curved path. This goes against the grain of our ordinary perception and conception of the ball simply moving along a Euclidean straight line from point *A* to point *B* where it collides with another ball and ricochets to the left or to the right, all on a two-dimensional plane and occupying a second or so—which corresponds to a natural sort of gut feeling that has actually been inculcated into us by classical mechanics. This more practical way of seeing the ball's behavior accounts for only two of its variables: the classical Cartesian coordinates on a plane. The ball's path in phase space, in contrast, includes the six dimensions of space as discussed previously. In addition, if we wish to be properly technical, it includes two spin variables that must be taken into consideration regarding billiard balls and, indeed, all objects, since they never fixedly face forward and only forward when flying through space. All eight variables are virtually impossible to visualize, and for that reason billiards is quite difficult to master—think how challenging, in their own way, are games such as handball, Ping-Pong, and tennis, where the ball moves in three-dimensional space rather than on a two-dimensional plane. For a possible illustration of this added complexity, imagine that you are piloting an airplane and due to a technical failure you go into a tailspin. The plane travels through three-dimensional space with a certain velocity that varies from one moment to another; it also spins in a particular direction and with so many revolutions per minute, which also varies. With this many variables to keep tabs on it is no small wonder that pilots have difficulty coming out of a spin.

Calvino's "The Chase," which appears in *T Zero* (1969, 112–27), affords another image of nonmechanical—asymmetrical, nonlinear, virtually chaotic—systems. At the outset the narrator tells us he is being pursued along a freeway by his hopeful assassin. There is, as would be expected, rapid-fire action, with a precarious balance perpetually established between the zooming vehicles. If any form or fashion of organization exists, it is surely some well-nigh unintelligible self-organization as a result of reckless abandon on the part of the worn-out yet hyperattentive drivers on their way to and from somewhere. There is a dynamic equilibrium that verges during each moment on total chaos. To be sure, there is symmetry in the sense of Newtonian mechanics where time can be given a plus sign or a minus sign, it makes no difference. And all the cars are interchangeable, as if they were billiard ball atoms indistinguishable from one another, as if the entire group were commutative, along with the principles of reflexivity and nontransitivity as an accompaniment to symmetry.

Yet there is a certain tension, a certain imbalance, a certain fluctuation, that constantly threatens to plunge the dynamic system into chaos. Significantly, our hero decides to become himself a pursuer and render the buzzing confusion asymmetrical. But he then speculates that since everybody on the freeway is actually both pursued *by* somebody and a pursuer *of* somebody, the system as a whole must be symmetrical, though each finite individual is incapable of surveying this whole. So it all appears to him a hopeless muddle, for he is caught within the whole hubbub apparently without the capacity to know his way about with any degree of certainty. In the end, as pursued now turned pursuer, the protagonist manages to kill his intended victim. At the same time, he escapes the bullet of his own pursuer—it would be quite useless, we are told, if the narrator were eliminated. Irreversibility has triumphed. And fortunately, our hero will live to pursue another day. Yet, he reflects, "I am still a prisoner of the general system of moving cars, where neither pursuers nor pursued can be distinguished" (Calvino 1969, 127). So in the final analysis it is back to symmetry, with a Brownian-like movement of all cars in apparently disorderly fashion. Yet from the view of the whole somehow there is order, *the Order.* The infinitely chaotic is tantamount to infinite degrees of order, paradoxically.

Phase space, it is now quite obvious, takes us outside the Euclidean notion of space and the intuitive notion of time with which we have become familiar. In the classical concept of things, the idea of dimensions was clean and crystal pure. A point supposedly has zero dimensions; it is pure location without extension. Recalling Borges's example in "The Book of Sand," a line, of infinitesimal thickness, has one dimension, a plane has two, a cube has three, and so on. When phase space comes into the picture, however, this unity takes on infinite diversity, that is, complexity, much like the distinction between our experience of walking from *A* to *B,* on the one hand, and the impossibility of doing so according to Zeno's logic, on the other. Mr. Palomar's shore at a given instant appears smooth and easily definable insofar as it is set apart from the water conceived as a static state of affairs. However, in the case of coastlines, mountains, clouds, ferns, sunflowers, and, in fact, all natural phenomena, we have observed that lines are not simple at all, nor do they simply trace out a path in one-dimensional space: they are infinitely complex. Indeed, Mr. Palomar's consternation is deeper than we would initially suspect, and he observes that when he closes his eyes for a moment and then opens them to the square he has bracketed from the entire panorama of waves, his perspective has invariably changed:

> Now, in the overlapping of crests moving in various directions, the general pattern seems broken down into sections that rise and vanish. In addition, the reflux of every wave also has a power of its own that hinders the oncoming waves. And if you concentrate your

> attention on these backward thrusts, it seems that the true movement is the one that begins from the shore and goes out to sea. (Calvino 1985, 7)

If the shore constituting the background of Mr. Palomar's perceptual grasp of the phenomenon under study is of infinite complexity, how much more so regarding the incessantly transient waves of the sea!—one might recall that painting *The Great Wave* (1823–29), by Japanese artist Katsushika Hokusai, which marvelously captures an ebullient form of Heraclitean flux within the confines of a static frame. Quite understandably, Mr. Palomar left the beach "even more unsure about everything" than before.

Palomar's infinitely complex and subtle meandering through perceptions of lines, contours, and shapes calls for a new concept of the very idea of dimensions that abruptly departs from the dictum with which Borges opens "The Book of Sand." To reiterate the preceding assertions in this regard, it is not a matter of a simple 0, 1, 2, 3 . . . enumeration but of dimensions of magnitude somewhere between the whole numbers. What is the dimension of a tetherball suspended by a thin rope? From a distance we cannot see the rope, and the ball appears to be a mere point. As we approach the point, it grows into what seems to be a two-dimensional shape connected to a one-dimensional line. Closer up we notice that the ball is actually three dimensional. Upon even closer inspection the rope takes on increasing thickness to occupy three-dimensional space as well, but it consists of numerous interwoven threads, each of which is nearly one dimensional but not quite—recall Klee's remarks on the "thickness" of lines and planes. Dimensionality, in the final analysis, depends largely upon the perspective.

Regarding the dimensionality of fractals from a closer look, take 100,000 grains of sand from Mr. Palomar's beach and spread them out on our billiard table. If we draw a straight line in any direction from one side of the table to another, it will likely pass through the approximate center of quite a few grains. Now if we weave the line in and out in order that it may pass through as many grains as possible, it will become increasingly complex. Will it still be one dimensional or not? Such complex lines have no well-defined slope in the classical sense, and their length is enormous. In fact, if we could place 6×10^{23} grains of sand on the table, a line passing through the center of a given set of grains in forming a curved trajectory from one corner to the other of the table would be of incredible length. This is comparable to the case of Palomar's beach or the coastline of Great Britain or of Maine. From a distance it might appear quite smooth; then upon a closer look we see that it is actually jagged. As we zoom in further, the obliquely connected increments affording the jagged appearance become smaller and smaller, until theoretically there will be an infinity of them, each of infinitesimal length (Mr. Palomar's coastline, then, is literally infinite!). By the same token, a line connecting an infinity of grains on our billiard table would fill the entire plane. The line would now be more than one dimensional! In other words, as it became more and more complex, its dimensionality would have increased, very gradually, from 1 toward 2. Contemplate, in comparison, the complexity of merely one of Palomar's waves in isolation from the rest. Upon considering this wave to be coterminous with and tantamount to all the waves within his perceptual grasp and, in fact, to the whole ocean, the numbing complexity of the world, *our world,* begins to take effect.

If we heed the conclusions of Benoit Mandelbrot, coiner of the term *fractals,* we must concede that all conventional descriptions of the world are illusory and have little to do with the "real world" of clouds, mountains, rivers, hurricanes, dripping faucets, and even clocks—that is, clouds are not simply round, mountains are not triangular, rivers are not smooth squiggles, hurricanes are not exactly self-repeating cycles of air masses, and dripping faucets are not entirely predictable. We begin to sense that the richness of the world lies far beyond the Cartesian-Newtonian fiction, that grand dream of reason. But our awakening from the dream does not leave us merely with unreason. On the contrary. We open our eyes to a completely new world of beautiful, albeit irregular, images. This is the world of chaos, which, we are told, is far closer to reality than the convenient fictions of classical science. The world we experience is not that of elegantly predictable mechanical collisions and reversible, symmetrical reactions. It is a world of process, of time-bound self-destructive and self-organizing tendencies. It is a world of "almost chaos"—a counterpart to Klee's "antithetical chaos." In other words, disorder rather than order is the norm, if norm there be.

And yet, our intuitive side cries out that there is something terribly wrong here. Surely there must be some sort of overriding order to everything, however chaotic it might appear on the surface. And surely this order must have evolved in a quite linear and orderly fashion, however many fits and jerks and setbacks there might have been along the way.

As a matter of fact, Borges's "The Garden of Forking Paths," which appears in *Labyrinths, Selected Stories and Other Writings* (1962, 19–29) seems to afford the image of a nonlinear, infinitely bifurcative, unwinding of multiple possible worlds. However, this is not the case, not necessarily that is. Granted, the "forking paths" universe is nonlinear and bifurcative, and it is probably determinate, but it is unpredictable, at least by us finite, fallible beings suspended within it. In a certain manner of speaking Borges discovered bifurcation theory before science did (Weissert 1991, 223).[1] There are actually in the "garden" four different levels, three of mathematical dynamics and one of apparent subjectivity. The first level is a simulation model, the author's representation of a fictive world as an alternative to what we ordinarily consider to be the real. The second is a pure abstraction of mathematical logic, the story of Tsui Pên and his construction of the infinite, self-reflexive, self-contained book with the same title as that of Borges's story. The third is the topological depiction of Tsui Pên's book, a story that contains Borges's story that in turn contains the story of the book. And the fourth entails the subjective, time-bound grasp of the story or stories at whichever level. The whole is static, consistent, and deterministic, much in the order of the set of all sets. Anything short of the whole, such as an attempt conceptually to grasp it, runs the risk of becoming tantamount to a member of the whole taken as the whole, or the whole as a member of itself, which was categorically barred from set theory and logic by Bertrand Russell (1910). Yet, Russell's prohibition is of hardly any consequence from within a finite, dynamic, ongoing processual perspective, especially when considering the whole to be not static but perpetually in a process of self-organization—a logically impossible but somehow an existentially or a phenomenologically and pragmatically feasible operation.

From another angle, the idea of sets and sets within sets, like Chinese boxes or Russian dolls, never ceases to seduce the mind. Moreover, the idea affords the comfortable feeling that, after all, there is a substratum of order that gives meaning to the world's chaos. As an illustration of this idea, within fictive worlds, we have the fabled book in Eco's *The Name of the Rose* (1980) that holds the key to deciphering the mysteries surrounding the main characters in the tale. By a comparable token, Borges's massive compendium contained within his Library of Babel in the short story by that title, which appears in *Labyrinths, Selected Stories and Other Writings* (1962), was the "formula" (algorithm?) for all the books. But this compendium is not really necessary, for according to one argument, the vast library is actually useless. A single volume would suffice, "a volume of ordinary format, printed in nine or ten point type, containing an infinite number of infinitely thin leaves. . . . The handling of this silky *vade mecum* would not be convenient: each apparent page would unfold into other analogous ones; the inconceivable middle page would have no reverse" (58). This allusion is far from insignificant. Borges often referred to the Schoperhauer-Buddhist dictum that just as one nightingale is all nightingales and one tiger is all tigers, so also one book can be all books.[2]

The middle page of Borges's single book would have no reverse, for, like 0 separating one infinity of integers from its mirror image, the inverse infinity of integers, the middle page would be nothing, *zero.* Moreover, if the book is infinite, then it will play another trick on us. For, let us turn the book over to see what there is on the back side of the last page. To our surprise, nothing will be there to meet our eyes; there will be nothing to see. There is no "last page," since the book is of an infinity of pages. So also the inconceivable middle page would not be seen either—in another way of putting it, the middle page would be everywhere, and the first and last pages would be nowhere: like Borges's definition of the library as *"a sphere whose exact center is any one of its hexagons and whose circumference is inaccessible"* (Borges 1962, 52).[3]

Moreover, the chimerical middle page of Borges's book would be like the hairline in figure 4, which contains nothing. It is pure emptiness. Yet it reflects every-*thing.* The black portion of figure 4 contains every-*thing* and reflects no-*thing:* it projects emptiness. Like the empty set of set theory, and unlike the emptiness of Buddhism according to Nāgārjuna, the black portion is the "noticed absence" of some-*thing.* In contrast, the white portion contains no-*thing* and reflects every-*thing;* it projects the fullness of things, yet, since it is not really any-*thing* actual (actualized) it is not necessarily any more or any less than the black, which is at least noticed absence. So we have emptiness, which is, well, just that, emptiness. And we have noticed absence, which, like the empty set, is something that once contained some-*thing* but is now empty, and if it never contained any-*thing* at all, it would still be some sort of box that is there and waiting for a deposit of some-*thing:* it would not be pure emptiness. In other words, however we wish to cut the cake, there is either emptiness, which is neither order nor chaos but simply emptiness, or there is the absence of order since there is no-*thing* at all.

There is no more than that infinitesimal line between Klee's cosmos and chaos. There is no substratum of order; there is no-*thing* at all. The world, *our world,* to repeat, is what we make of it, what we have constructed. Apparently we cannot have our piece of cake, nor can we eat it, according to our customary notion of "our piece of cake." If we once had it and ate it, nothing but the empty set would remain; if we held onto it, we would really have no-*thing* except that some-*thing* that could not be any-*thing* without its relation to the pure emptiness from which it emerged. Either way, to repeat, there would be "nothing" at all, except what we make of it.

"But," you say, "all this is exceedingly vague. We must have some handle to grab onto." Well, OK. Let me once more try your patience with a reconsideration of $\sqrt{-1}$, but this time in light of Calvino's short story, "T Zero," which appears in the book by that title (1969, 95–111), where we read that "the universe does nothing but pulsate between two extreme moments, forced to repeat itself forever—just as it has already repeated itself infinite times and just as at this second where I now find myself is repeated" (100). This pulsation—that is, the universal standing wave pattern consisting of Siva's vibrant dance alluded to previously—can be taken as a matter virtually of nanoseconds, according to Buddhist doctrine and certain strains of quantum theory as well (Capra 1975; Hayward 1987). The universe is a scintillating, undulating, pulsating, dancing glitter of activity (i.e., physicist David Bohm's [1980] *holomovement* consisting of the implicate [enfolded] order perpetually in the act of explication into the actual [unfolded] universe as we know it). Or it can be taken in terms of billions upon billions of years from the big bang to the great crunch and back again as one monolithic pulsation (Wheeler 1980). Either way, Nietzsche's eternal return is suggested.

Significantly, in this regard, Calvino's narrator of "T Zero" finds himself at

> a random space-time intermediate point of a phase of the universe; after hundreds of millions of billions of seconds here the arrow and the lion and I and the bush have found ourselves as we now find ourselves, and this second will be promptly swallowed up and buried in the series of the hundreds of millions of billions of seconds that continues, independently of the outcome, a second from now, of the convergent or divergent flight of the lion and of the arrow; then at a certain point the course will reverse its direction,

> the universe will repeat its vicissitude backwards, from the effects the causes will punctually arise, so also from these effects I am waiting for and I don't know. (Calvino 1969, 100)

The narrator is the hunter-hunted, with bow and arrow poised and ready, and the lion as the hunted-hunter is in readiness for the initial act of pouncing on its victim. All is in that instant as if it were suspended between the +1 and the –1. Time apparently stands still—recall Hladik of Borges's "The Secret Miracle," which appears in *Labyrinths* (1962)—yet there is a violent sort of scintillating, vibrating, pulsating activity. No time, but violent kinetics; nothing but space, yet incessant change as if all were a matter of successive temporal bundles of energy. Emptiness, it would seem—and in an anomalous way of putting it—looms large.

As the black marks on the printed page of Calvino's story roll by, we occasionally find the narrator in his timeless present alluding to a comparison between the series of static moments within which he is caught and the frames of a movie film. The frames emerge spontaneously, one by one and at a rapid pace, to give the illusion of continuity generated from discontinuity. Each moment is closed in on itself and incommensurable with all others, as if it existed in a static, autonomous film frame. But there is no promise of any jump to the next frame, for the narrator has decided to inhabit forever the moment within which he finds himself, which is, precisely, t_0. From this static "now," he has "ample leisure to look around and to contemplate" his moment "to the full extent" (Calvino 1969, 107). He is, significantly, at the point where the grand splits of the universe occur, at $\sqrt{-1}$. It encompasses "on my right a river blackish with hippopotamuses, on my left the savannah blackish-white with zebras, and scattered at various points along the horizon some baubab trees blackish-yellow with toucans" (Calvino 1969, 107).

The narrator is aware that each second, each individual moment, is a universe and that "the second I live is the second I live in, la seconde que je vis c'est la second où je demeure, I must get used to conceiving my speech simultaneously in all possible languages if I want to live my universe-instant extensively. Through the combination of all contemporaneous data I could achieve an objective knowledge of the universe-instant t_0 in all its spatial extension, me included" (Calvino 1969, 108). In other words, each second of his past is interrelated with the past seconds of the hippopotamuses, zebras, trees, toucans, and so on, and the same is to be said for all future moments in the interconnectedness of things. In order to map out all these moments and lay claim to an objective grasp of a single universe-instant, the narrator must establish with utter precision the coordinates of all points, of all interrelated entities, and must calculate certain constants that obtain for all of them.

But two problems—at least two—become imminent. First, the sensitivity of chaotic systems with respect to initial conditions will quickly throw the narrator's predictions into disarray, that is, unless his computation of the initial conditions at t_0 has been infinitely precise, which is an impossibility in the finite kingdom of this world. Second, in order for the narrator to establish an objective configuration of his present moment, t_0, he must already have somehow moved to t_1, and to establish this configuration he must have moved to t_2, and so on. This is impossible from within Zeno's overpowering framework, unless, like Laplace's Superobserver, the narrator adopts, by an act of subjectivity from an omniscient view of the whole of the universe, past, present, and future, a conscious awareness of all movement.[4] So in the final analysis, if the narrator as knower in the full-blown sense is to remain within one increment of time, he must paradoxically move with time, and to become objective, he must contradict himself by remaining subjective. He realizes that even if he abandons his idea that "time knows no repetition and consists of an irreversible series of seconds each different from the other," each second, nonetheless, "happens once and for all, and living in it for its exact length of one second means living in it forever" (Calvino 1969, 110). For each second contains as a singularity, wrapped tightly within itself, all its past moments and the past moments of everything with which it is interrelated, which is everything, and the seeds of all future moments—at least, if that second could be known with absolute precision, which it cannot for our finite and helplessly and hopelessly fallible narrator. We have here, it will immediately be recognized, the problem that has plagued the West at least since the Greeks and will undoubtedly continue to do so into the foreseeable future, and beyond.

Thus we are reintroduced to chaos and the part that contains the whole that is itself self-contained, somehow (the Aleph), like that individual cipher containing implications of the entire universe (Tzinacán) or the solitary book that is all books and hence the universe (Library of Babel, Garden of Forking Paths). The whole consists of parts the existence of each of which is dependent upon its interrelation with all other parts, and when emptied of all its parts, that whole must be the premier empty set, the noticed absence of all the parts. But where did this empty set come from in the first place? The answer from Nāgārjuna might be, Before there was an empty set, there was-is-will be simply emptiness.

This is a far cry from the age-old Western idea of a labyrinth, that essential, objective, lifeless whole with a predefined center that can be known by the persistent seeker: the labyrinth as mirror of the universe. Yes, the labyrinth and its strange relation to chaos and order.

Is Ariadne's Thread of Infinite Length?

It can be said that a labyrinth is designed chaos, static rather than dynamic chaos; under certain conditions it is even necessary chaos. And it is ultimately chaos with a purpose. Its purpose includes intentionally induced confusion and complexity, the paradoxical collusion of unity and multiplicity, clarity and confusion, simplicity and complexity, order and disorder (Doob 1990, 1–3, 8–9). This notion of a labyrinth entails a focus on process rather than product, on paths—infinitely diverging, converging, involuting and convoluting—rather than any totalizing pattern. Yet, once one sees the labyrinth in its entirety, "elaborate chaos is transformed into pattern. This potential conversion of the labyrinth from confusion to order . . . is a common theme in later writings, particularly those dealing with metaphorical mazes of epistemological or of literary texts" (24).[5]

From our consideration of Lönnrot's proposed Zenoesque line-labyrinth—or the linear time labyrinth of the prisoners—we progress to Lönnrot's planar labyrinth—or ourselves or the rat in its maze for that matter—and from there to Scharlach's three-dimensional labyrinth at Triste-le-Roy. In essence, what we have in the cases of Lönnrot and Scharlach, we and the rat, and the lawyer and her incarcerated client-prisoners, are three types of labyrinths corresponding to a typology set up by Eco (Rosso 1983). First, there are linear labyrinths of the Greek sort, typically that of Theseus, which do not allow the weary traveler to lose his or her way, though the journey might appear interminable. This labyrinth consists of one entry point and a winding narrow path, without bifurcations or devious detours, that eventually reaches the center. Then, from that central point, the pathway leads outward and ultimately to the exit. The labyrinth's uncertainty consists in the legendary Minotaur that inhabits the center: one does not know when one will reach the center and therefore at what point one will be confronted by the terrifying, unpredictable beast; otherwise the journey would hardly be more than a comfortable stroll along the corridor. The second type of labyrinth is "mannerist." It is a labyrinth of bifurcating linear paths, evincing a treelike structure, with many alternate routes but almost all of them offering no more than false promises. The uncertainty regarding this sort of labyrinth consists in the choice available at each of the numerous bifurcations in their sum offering the possibility of many roads, with only one of them leading to the desired goal. Since a traveler can soon become lost in a mannerist labyrinth, Ariadne's thread is helpful and, in cases of more complex labyrinths, a necessity. Otherwise, the traveler is confronted by hardly more than a trial-and-error affair. The third type of labyrinth is nonlinear and rhizomic, following Deleuze and Guattari's (1987) concept of the *rhizome.* Each path of the rhizome labyrinth is connected to every other path, with nothing that can be called a true center, no clearly distinguishable periphery, and no possible way out, for the helpless and hopeless traveler is immanent: he or she is within, and a part of, the labyrinth itself. Rhizomic labyrinths are potentially infinite in extension. They are open mental spaces whose physical manifestation cannot but be an exceedingly impoverished model. Although they are not exactly structureless, they are incessantly in the process of structuration without the possibility of any fixity of structure.

We find a poetic combination of the first and second classes of labyrinth and a premonition of the third class in Borges's "Two Kings and Two Labyrinths," which appears in *Obras Completas, 1923–1972* (197, 607–8). The king of Babylon instructed his wisest architects to construct a labyrinth of the mannerist sort so complex as to confound even the most brilliant minds. And they did so. One day an Arab king visited his court, and intent on demonstrating the superiority of his own culture and the naïveté of that of his counterpart, the king of Babylon enticed his guest into entering the fabled construction. After some time, lost and confused, the king of Arabia implored of his host that he might be offered aid in finding his way out. Help was forthcoming, and the desert dweller, now safe and sound, thanked the Babylonian for the interesting adventure and invited him for a visit in order that he might experience the Arabian form of labyrinth. The visit was made, and the Babylonian king, at the proper time, was led to the middle of the desert in a caravan. Then he was abandoned, without food or water and with the information that this was an alternative to his Babylonian labyrinth. Soon thereafter, we would expect, he perished on the desert sands.

The Arabian labyrinth is a strange combination of a one-dimensional and a two-dimensional labyrinth: the pathway is unilinear, but at each point there exists an infinity of directions that can be taken. As with Lönnrot's Zenoesque proposal, if the Babylonian king had perchance happened to travel half the distance to his destination from the spot where he was abandoned, he would then be required to traverse half of the remaining distance, then half of that, and so on. But, of course, the forlorn, helplessly lost traveler would not travel along a rectilinear, properly Euclidean path. Like any human being when lost, he would be destined to describe an intricately interwoven set of one-dimensional loops on the two-dimensional floor of the desert, with no end in sight (as if his trajectory described a strange attractor). So it might appear that Zeno is of no consequence after all. In the desert labyrinth there are choices, an infinity of choices, without solution or salvation. Moreover, it is precisely as Zeno constructed his world: passage through all possible paths after having exercised all possible choices would be a matter of traveling along a path that eventually passes through the center of each and every grain of sand on the desert floor, so the sum of all paths, the Path, is of infinite length; hence there is no way out. Half the Path, and the weary wanderer has covered the first leg of the trajectory; half that half, and an additional fourth of the trajectory has been traversed, and so on. No end. If the Greek labyrinth's uncertainty rests in the surprise upon having reached the center and the mannerist labyrinth's in the traverler's ignorance regarding which of the bifurcating possibilities is the right one, the Arabian labyrinth adds to them the impossibility of a determinate beginning, end, or center. It is like the rhizomic variety of labyrinth, with an infinity of possible tangents that can be taken. So the one-dimensional trajectory entailed by the desert labyrinth actually affords a glimpse beyond, to labyrinths of the third class. In reiteration of Sagredo's words, what appears to be a simple line becomes, upon one's contemplating its complexity, a labyrinth of the sort that, in Lönnrot's words, "so many philosophers have lost themselves that a mere detective might well do so, too."

In short, we have labyrinths the trajectories of which correspond to our familiar one dimension, two dimensions, and three dimensions. But as we have noted in the case of fractals, in the world of our living and breathing existence, dimensions rarely come in whole numbers. For example, no matter in which direction the king of Babylon strikes out, he will never arrive at his destination, just as if the cat and mouse game proposed by Lönnrot were to continue forever. Ultimately, his path will describe a closed curve on a plane, which divides the plane into two domains, inside and outside. This condition stipulates that all points of the plane fall into two classes, those outside the curve and those inside the curve, such that any pair of points of the same class can be joined by a curve that does not cross the initial closed curve, while any curve joining a pair of points belonging to different classes will cross the initial curve. Simple enough. But not so simple upon considering increasingly complex forms. It is simple with regard to circles, ellipses, triangles, rectangles, and relatively noncomplex polygons. But it is complex regarding the lines and curves of an intricate labyrinth of the rhizome sort and even of a complex bifurcating or Greek labyrinth. If we contemplate the twisting, contorting, centerless polygon-labyrinth in figure 16, we begin to realize that, just as in chaos theory, the line of demarcation becomes vague and fuzzy. Which points are exactly inside and which outside? At the liminal zones a precise answer becomes problematic (compare fig. 16 to the previous example of grains of sand on the billiard table and a line passing through the center of each of them).

Fig. 16. A "fractal" maze

"At any rate," one might wish to conclude, "according to traditional wisdom a labyrinth is a mirror, model, or metaphor, if you will, of the universe, and since one can usually get a reasonably successful grasp of labyrinths, perhaps they can help one to understand the universe, at least at some intuitive level." Wishful thinking again, however. Models are not quite the reliable key we would like them to be, as Palomar painfully learned. According to classical dictates, Palomar speculates, an ideal model is "that in which nothing has to be changed, that which works perfectly." The problem is that reality "does not work and constantly falls to pieces, so we must force it, more or less roughly, to assume the form of the model" (Calvino 1985, 109). This sounds reasonable enough. Models cannot be more than extremely limited (i.e., "minimalist") abstractions, and as they are abstractions, something must be left out of the *modelandum;* that is, it must be whittled down somewhat in order to fit the *modelans.* Palomar strives to construct a model with classical impassiveness, detachment, and objectivity, but no sooner than the serene harmony of his model appears to be at hand "irrelevant accidents" pop up in his modeled world. A "delicate path of adjustment" is constantly required, "making gradual corrections in the model so it would approach a possible reality, and in reality to make it approach the model" (110). Soon he needs a great variety of models. With the continuing proliferation of models it becomes apparent that a model of his models is imperative. So he constructs one, and then many, and then he must construct a model of his models of models. After things become of mind-numbing complexity, Palomar finally concludes that "what really counts is what happens *despite* them" (111). So he erases all models, and models of models, from his mind. His mind has now become, so to speak, an empty set.

Now he is genuinely free to face reality, which consists of no more than fleeting fragments of experience, much like, we would suppose, the world of that inveterate supernominalist, Borges's Funes the Memorious—as we shall observe shortly. In spite of his previous ups and downs, Palomar now feels he is the owner of some pretty fine thoughts. That is to say, they are potentially fine thoughts, for to be thought and said, they must be put into systematic linguistic form, and thus they will threaten to become models of one sort or another. So Palomar decides to "keep his convictions in the fluid state, check them instance by instance, and make them the implicit rule of their own everyday behavior, in doing or not doing, in choosing or rejecting, in speaking or remaining silent" (Calvino 1985, 112).

However, all is to no avail, for in this state Palomar's world is flux, fluctuation, disequilibrium, instability, uncertainty. Yet there is not merely no-*thing* at all. The problem is that Palomar's mind as an empty set cannot help but be a mind noticing the absence of some-*thing*. Of what? Where? When? These questions imply the existence, whether potential or actual, of some-*thing*. So his mind is not really empty. It expects some-*thing*, and that some-*thing* must surely be of some nature or other, which corresponds to Palomar's past observations of the nature of things. The moral to the story is that just as Palomar cannot think or see every-*thing* at once, so also he cannot not think and not see any-*thing*. In other words, pure emptiness is entirely beyond him, immersed as he is in traditional Western modes of reasoning and seeing.

That Supreme Labyrinth, The Universe

But, then, after all is said and done, in our everyday living it would seem that Palomar's way is the only feasible way. Life is quite uncertain for most of us, almost always uncertain for some of us, and absolutely certain for none of us. In other words, if the universe is more disorderly than orderly, so also with life.

The gods appear to be rolling the dice, and the dice may even be loaded, but if so, their loading is definitely out of our favor. Or so it would seem. In ancient cultures, the cosmos was considered to be chiefly a roll of innocent dice—Sagredo revealed as much. Now, in the age of the "new physics" and especially the "science of complexity," it is once again a matter of chance. But not quite pure chance. For even if things are unpredictable, at the same time they are according to chaos theory at least in principle determinate, although determinacy at increasingly larger domains will in all likelihood continue to elude our finite, fallible minds into the indefinite future. Perhaps the best we can do, then, is find what bits and pieces of order and harmony we can and get on with life.

This was apparently the only alternative for poor Funes the Memorious. Funes, we will recall, could not forget and could not generalize. His form of numbering and his form of arithmetic were without form and without order as we know them. For Funes, to know the number series was to know each individual numeral as an individual without any necessary relation to any other numbers of the series (in other words, there was for him no ordinality). In fact, Funes once invented an alternative system consisting of arbitrarily selected terms that, given his unlimited mnemonic capacity, was equally effective. He could in the blink of an eye multiply, say, "Plata" (= 1,275) and "Quebracho" (= 836) and respond with "Rosario" (= 1,065,900). In other words, in contrast to our ordered number series, Funes's numbers consisted of an unordered system, yet, like an "idiot savant," Funes was able somehow unconsciously to hold all of them in his mental surveillance in a timeless instant.

We could speculate that a comparable contrast between unordered parts and their collection into a whole exists between the Zahir, which entails an unordered series of gazes and glances, and the eternal Aleph, contemplation of which entails a monolithic orthogonal perspective—via an operation enabled by $\sqrt{-1}$—affording a grasp of the whole in one fell swoop. That is, Funes's world is that of myriad lines of unordered points, in fact, potentially an infinity of points, that are, nonetheless, always already there, in his mind's eye, though, knowing them intuitively, he is quite unconscious of them. In like fashion, the Zahir evinces the image of a world of irrational numbers in all directions along a two-dimensional plane, while the Aleph provides a vision of all this from an orthogonal view from within a third dimension, giving rise to a fourth dimension that consists of abstract time as another dimension of space, which is all there en bloc and open to the gaze of an impossible outside observer, sub species aeternitatis. This view from "nowhere," like that view of the supercondensed Aleph, is, precisely, a view of everything as if it were a hyperblock (which is comparable to that hoary sphere from within which Scharlach traced out the path of the helpless and hapless Lönnrot).[6]

For Funes, language, like his conception of numbers, also consists of an unordered concoction of signs very loosely related, if at all, according to a rather haphazard sort of *ars combinatoria*. But once again, a perspective limited to the Zahir is hardly more advantageous: it consists of a series of perceptual grasps, which, even though presumably ordered, can conceal any order, hence no one and only order. The Zahir is an arbitrarily chosen sign, as are the labels for Funes's numbers. Consequently, both the Zahir and Funes's number signs could have been one of an indefinite, and potentially infinite, number of other signs, and the series could always be different for any one of an entire collection of different individuals. In a world like that of the supersentient though painfully limited Funes, or like that of the anguish-ridden observer of the Zahir, how could one hope to find some sort of order? If one cannot hope to know the whole, is a chaotic form of complexity destined to prevail?

Our guts most likely tell us we must respond with an emphatic "No!" Our concepts, we would like to believe, are tied up with the stability of the furniture of our world to which they refer. We see a lemon and properly classify it as a "lemon" because we are quite familiar with "lemonness," "yellowness," and so on. Or at least our ability to so use language is one of our last great hopes that words can correspond to the world; and we would like to fear not the possibility of our signs and our world becoming hopelessly chaotic. We would therefore be quite homeless in Funes's world, in the world to which the Zahir refers, or even in the Tlönians' nebulous mind-dependent reality for that matter. Perhaps such a world would be as uncertain as that of the vacillating, wandering hero of Flann O'Brian's *The Third Policeman* evinced by his description of some criminals he presumably witnessed:

> In colour they were not white nor black and certainly bore no intermediate colour; they were far from dark and anything but bright. But strange to say it was not their unprecedented hue that took my attention. They had another quality that made me watch them wild-eyed, dry-throated and with no breathing. I can make no attempt to describe this quality. It took me hours of thought long afterwards to realize why these articles were astonishing. *They lacked an essential property of all known objects.* I cannot call it shape or configuration since shapelessness is not what I refer to at all. I can only say these objects, not one of which resembled the other, were no known dimensions. They were not square or rectangular or circular or simply irregularly shaped nor could it be said that their endless variety was due to dimensional dissimilarities. Simply their appearance, if even that word is not inadmissible, was not understood by the eye and was in any event indescribable. That is enough to say. (O'Brian 1974, 117)

Given such a chaos of perceptual grasps, of drab grays with hardly any distinctions, would it be possible to talk of perception *of* any-*thing* as a general at all? Is O'Brian's helpless character not tantamount to Funes, with his teeming world of disconnected particulars? To see some-*thing* as such and such instead of any one of a range of alternatives requires one's distinguishing it from every-*thing* else and interacting and participating with it. Then to talk about it requires interaction-participation with other members of a community of individuals. Dialogue in a community of like-minded—hence "right-minded"—speakers is essential. Otherwise, what is "blue" for you might be "grue" for somebody else, and what is for him or her "grue" today might be tomorrow's "green" for you. What is a goose for one would be a gander for another. Each individual "I" would be the measure of all things, much like the "I" of each Tlönian, whose thoughts and percepts construct a world, his or her own world. Such rampant solipsism is enough to elicit knee jerks in even the most liberal minded postmodern traveler in that hopefully intelligible labyrinth, the universe. What is more disconcerting, if what lies behind the "I"'s seeing and saying is nothing but an empty set, and beyond that, pure emptiness, one's gut inevitably suffers from another violent contraction.

Must we eventually concede that there is no absolute guarantee—that is, *absolute guarantee*—that we have much certainty at all? As a matter of fact, given the Funes-Zahir-Tlön-Palomar quadruple, it is possible to imagine and invent new terms of all sorts at the drop of a hat, and, like Beckett's Unnamable, we will never cease to be surprised, awed, dumbstruck, and even shocked and numbed by the plethora of new signs we receive and we find ourselves spewing forth. This is, after all, one of the lessons Mr. Palomar brings to our attention: many of the things that will happen will happen, whether we want them to happen or not, and in however uncertain and unpredictable a fashion. Thus, the $\sqrt{-1}$ operator of figures 4, 5*a*, 6, 12, and 13, put into use by the artist or as a result of everyday happenings according to the apparent whims of nature, brings about the spectacular emergence of colors as a result of the incessant struggle between grays and blacks and whites.

The happenings of the world apparently in spite of our wishes to the contrary also bear on Nelson Goodman's (1979) "new riddle of induction," which stands tenuously on the relationship between our concepts, our expectations of the world of sensory data, and what actually happens (or at least appears to us) to pop up. Goodman invents a new color, the color "grue" cited previously, meaning "green" before time t_1—the present moment being t_0—and "blue" after t_1 (compare to the narrator's situation in "T Zero"). Take emeralds, for instance. We dub them "green," and we take it for granted that they have always been "green" and will remain "green" into the unforeseen future. But an imaginary group of apparently misguided souls has adopted the color term "grue." According to these "Grueworlders," emeralds, before t_1, are like our own emeralds, "green," but emeralds, after t_1 has transpired, will be the equivalent of our "blue," the color we would ordinarily attach to sapphires or any other "blue" object. Our world is as far as we are concerned nontransient, while that of those strangers from a strange land, the Grueworlders, is irrationally alterable. But, actually, for our strangers the shoe is on the other foot. Their world is as far as they are concerned as stable as can be, while they consider our world to be inordinately transient: the color predicate we attach to emeralds, from their perspective, suffers an irreversible transformation from t_1 onward.

This would be like two characters suffering from "Funes's syndrome," each of whom constructs his or her own number system: they would most likely be unable to agree on so simple an equation as 3 + 2 = 6. It would be like two Tlönians gazing at an animal they vaguely see before them: one mind constructs a gazelle while the mind of the other constructs a gnu. If we could imagine a view of things from "nowhere" that would include our world as well as that of the Grueworlders, of the victims of Funes's syndrome, of the Tlönians, and of all other possible worlds to boot, we might have a situation not unlike that of Calvino's archer in "T Zero" and Funes wrapped into one. Each instant would be an atom contained within the whole, yet it would be tantamount to that whole. That instant would give way to a succeeding instant ushering in an entirely different atom, that is, an alternate universe in all its complexity, which would not necessarily be related to anything that went before. There would not necessarily be any permanence at all, and with the emergence of each moment, a surprise might be in store, for the unexpected could always pop up. Strange.

But perhaps not so strange. Perhaps our world is not so uncertain as all that. Are emeralds not for us "green," quite clearly and distinctly? But Goodman asks, How is it that we could come to be so comfortable with our "Emeralds are green" convention? He tells us that it is because all the emeralds we have seen, and presumably all anyone has seen, have been "green" in the past, they are "green" now, and there is no reason to suspect that they will be anything other than "green" in the future. In fact, if we wake up tomorrow morning and discover that an emerald in our midst is not "green," we will most likely suspect we have been conned by the local jeweler who sold us the gem rather than question the ways of nature or of our perception. However, Goodman goes on to suggest that our expectation that emeralds will be "green" tomorrow depends more upon how we describe emeralds than on our perceptual grasps of emeralds in the past—take, in this regard, "Mount Everest is the highest point on Earth," which we would hardly question, though I assume none of us has verified this proposition empirically. If, following Goodman, the utterance "Emeralds are green" depends upon our descriptive more than on our perceptual modes, then there is no reason why we cannot invent any number of new terms at will. We might even be able to out-Funes Funes himself by creating the term "gree" for "that emerald on the table," "geu" for "that emerald attached to a band on your finger," "goo" for "an uncut emerald in the wilds of Brazil," "gro" for "that emerald on my mother-in-law's necklace," and so on. What can stop us?

But actually, our situation would be very much unlike that of Funes, who is surprised that a dog early in the afternoon is the same dog in the late afternoon. Upon creating our words at will, we would in a strange manner of speaking surprise, dumbfound, and amaze ourselves, for a new word would pop into our heads and slide out our mouths as a label for whatever might happen to exist before us. This situation, then, would be more like that of Borges's Tlönians, for whom the physical world does not exist and for whom what happens to appear before them does so because they see it and articulate it in such-and-such a way. For them, continuous existence is impossible, but unlike with Funes, the discontinuity of their world depends upon them and them alone, while Funes the inveterate hypernominalist merely sees what is there at each moment, and that is that.

"Quite trivial, all this," one might be prone to say with a glazed look in one's eyes. Yet, perhaps not so trivial. For while we actually have none of the preceding set of bizarre words in our language, it is not *necessarily*—I must emphasize *necessarily*—because of the fact that our past experience of colored things rules them out of our language; rather, it is because we never happen to have changed our color terminology in such radically distinct ways. But quite conceivably we could have done so, just as some peoples have relatively few color terms with which to cut up their world and others have many, and just as for some peoples what is "blue-green" might be distinctly "blue" for others and "green" on no uncertain terms for still others—thus "turquoise" could be either "blue" or "green," or we might say it could be both "blue" and "green," or neither "blue" nor "green."[7]

Goodman's answer to the problem is this. Our world appears to us as stable as can be because of our "entrenched" linguistic practices. The reason our practices are entrenched is because, given our particular conventions, we have learned to project the predicate "green" on emeralds instead of "grue," or some other term. "Green," simply put, has become a more effectively entrenched predicate: it has worked for us in the past, and we believe it will do so in the future. Emeralds, then, are "green" because we are prone to follow the lead of our past projections. We see an emerald and run a search for the record of our past projections, and invariably "green" seems to come up. Quite obviously, then, "'green,' as a veteran of earlier and many more projections than 'grue,' has the more impressive biography. The predicate 'green,' we may say, is much better *entrenched* than the predicate 'grue.'" The fact of the matter is that for "every time that 'green' either was actually projected or—so to speak—could have been projected, 'grue' also might have been projected" (Goodman 1979, 94). That is, whenever there is assent to the general statement "All emeralds are green," there could equally have been assent to its apparently contradictory statement "All emeralds are grue."

Moreover, if we consider each and every possible occasion in which assent could have been forthcoming regarding either one of the two statements, then we must concede that in principle at least they have equal status. The difference between them emerges only when considering the actual occasions in which "green" was granted assent rather than "grue." "Green" enjoys the advantage of a history of occasions when it was given preference, while "grue" does not. However, in another world inhabited by other beings, "grue" could have been granted preferential status. Preference for "green" over "grue" has been the result of our habitual projection, rather than our habitual projection as a result of our preference. The reason "green" has been so fortunate as to have become entrenched is because well-entrenched predicates have become the "right ones" for the members of the English speaking community. Goodman concludes that our "agreement with regularities in what has been observed is a function of our linguistic practices. Thus the line between valid and invalid predicates (or inductions or projections) is drawn upon the basis of how the world is [perceived, conceived] and has been described and anticipated in words" (Goodman 1979, 121).

Over the long haul, we read from Goodman that our linguistic practices and our everyday living are so interdependent that they determine the way we see and think of the world. This world, *our world,* is from this view hardly much better than that of Borges's enigmatic Tlönians for whom what is is what they see and what they make of it *(esse est percipi)*—to be discussed in the following. What these inveterate idealists see exists insofar as they see it and/or think about it, and when not, it dissolves. Consequently, for them, continuous existence is an impossibility. Their existence is a sort of Buddhist flash-dance: now it is *this,* now it is *that,* and on and on. For them also, every instant can always already be different, like the world Funes sees—the very important difference being, of course, that Funes's supernominalism includes existents "out there" while supernominalism of the Tlönian sort is purely mental. So as far as the Tlönians are concerned, what will be isn't yet, and since it isn't yet, then once it is pulled into the realm of what is, it's already something else. So possibly any-*thing* and every-*thing* can actually be. But the picture need not be as dire as it appears at the outset. At a given moment the entire community of Tlönians could quite conceivably actualize, each in his or her own mind, the same sign related to some event, and at future moments a comparable sign could be evoked, so in the best of all possible worlds they would make up a rather coherent community. Yet there is no absolute guarantee that at the drop of a hat this well-wrought package might not disintegrate and leave each and every Tlönian in his or her own solipsistic world. Once again solipsism comes to mind, as well as the idea that *our world* is what we make of it—the next step, perhaps even an inevitable step, according to some onlookers, is nihilism.[8]

But things are perhaps not necessarily so calamitous. Let us inject the notion of *becoming* into the equation, thus apparently evoking a quite non-Tlönian universe. Whatever some-*thing* actually happens to be in the here-now, it is in the process of becoming what it will be, a becoming that marks the dissipation of what it was in the last moment and the anticipation of what it will be becoming in the next moment. So at each moment it is what it is, but actually it is not what it is, because it is always already passing on and transforming itself into something other than what it is, that is, what it is becoming, what it will have become. And so on. All this is becoming inordinately confusing and complex, granted, but it is necessary fusion, confusion and complexity. For if the universe and every-*thing* in it is self-organizing, as the signposts currently seem to be indicating, then every-*thing* is incessantly in a process of becoming without the possibility of that becoming becoming being for all time. In other words, what there is is emptiness and becoming; other than that, there is construct, *our* construct, that is also always becoming something other than what it was.

If one wishes to forgo this conclusion, life's uncertainties will most likely continue to exercise their force from within the erroneously conceived labyrinth as mirror of the universe.

But Then, What Is in a Book Anyway?

Unfortunately, from yet another vantage, the world traveler's helpless, hopeless plight might well be more serious than he or she thought, perhaps even more serious than the traveler is capable of thinking. It is apparently more than that of which the traveler is capable of thinking because it is in essence comparable to Borges's Library of Babel.

The narrator tells us the monolithic aetheneum might seem infinite, hence unfathomable, at least in concrete terms. However, from another view, since the library contains all the possible permutations of the twenty-five orthographic symbols, and even though the total number of books must be monstrous—as we have noted—nonetheless, it is necessarily finite. Those who judge the library to be limited

> postulate that in remote places the corridors and stairways and hexagons can conceivably come to an end—which is absurd. Those who imagine it to be without limit forget that the possible number of books does have such a limit . . . . If an eternal traveler were to cross it in any direction, after centuries he would see that the same volumes were repeated in the same disorder (which, thus repeated, would be an order: the Order). (Borges 1962, 58)

The solution to this quandary, the narrator suggests, stems from the fact that a finite system can apparently become infinite if it is infinitely perpetuated. That is, after all the permutations are exhausted, then they repeat themselves limitlessly and hence become, rather incomprehensibly, a finite but unbounded system. This premise of the eternal return, we are told at the end of Borges's story, stems from the effort to order the disordered, the compulsion to find meaning in the universe in its entirety (i.e., the grand unified theory); that is, it is tantamount to Mr. Palomar's desire to grasp everything at once.

Interestingly, an impious group exists that maintains that "nonsense is normal in the Library and that the reasonable (and even humble and pure coherence) is an almost miraculous exception" (Borges 1962, 57). On the other hand, others hold a superstition about a "Man of the Book." There is supposedly a book that is "the formula and perfect compendium of all the rest: some librarian has gone through it and he is analogous to a god" (56). But how could one hope to locate the unique hexagon housing this book? A Zenoesque regressive method was proposed: "To locate book A consult first a book B which locates A's position; to locate book B, consult first a book C, and so on to infinity" (56). The narrator confesses that he has "squandered and wasted" his years in such adventures, and now it seems unlikely to him that "there is a total book on some shelf of the universe" (56–57). Yet, he persists in his search for harmony amid the apparent chaos. He prays to the unknown gods "that a man—just one, even though it were thousands of years ago!—may have examined and read it [the Book of Books]" (57). But the narrator is never allowed this epiphanous moment of insight. The most he can do is "venture to suggest this solution to the ancient problem: The Library is unlimited and cyclical. . . . My solitude is gladdened by this elegant hope" (57).

And so all the questions remain. Is the library finite or infinite? Ordered or disordered? Of infinite and unfathomable complexity or governed by a simple and decidable algorithm? Even though the orthographic symbols making up the books of the library might be graphically identical to themselves in all their uses, a potential infinity of possibilities is open to them, in view of the potentially unlimited number of textual contexts and contexts of readings. A given collection of symbols combines into words (whether meaningful or nonsensical), words into sentences, sentences into paragraphs, paragraphs into texts. Comparably, according to some optimistic accounts, the universe—which is another name for the library—consists of the combination of a few simples, which make up larger and more complex particles. These are the symbols of the language of nature, which are strung together with their own grammar rules—quantum theory and ultimately, one might suppose, DNA. Eventually they make up sentences (molecules). According to this hopeful dream:

> Soon we have books, entire libraries, made out of molecular "sentences." The universe is like a library in which the words are atoms. Just look at what has been written with these hundred words! Our own bodies are books in that library, specified by the organization of molecules. The universe as literature is, of course, a metaphor—but the universe and literature are organizations of identical, interchangeable objects; they are information systems. (Pagels 1982, 255)

The ancient topos of the universe as a book—or better, as a text—became an obsession with the Kabbalists and the alchemist Paracelsus,[9] whose "book of nature" could presumably be deciphered if read in the proper manner. The concept was revived in science, especially with Galileo (1957, 237–38), who declared that

> [T]his grand book, the universe . . . stands continually open to our gaze. But the book cannot be understood unless one first learns to comprehend the language and read the letters in which it is composed. It is written in the language of mathematics and its characters are triangles, circles, and other geometrical figures without which it is humanly impossible to understand a single word of it, without these, one wanders about in a dark labyrinth.[10]

Notably, Galileo envisages the book of nature to be written in mathematical language. However, whether in a formal language that inexorably loses its precision when interjected into the sphere of human intuitive faculties or in a figurative or natural language irreducible to formal conciseness, the book of nature, as Heisenberg remarks concerning atomic physics, has taken on the trappings of an archetype (Miller 1986, 259).[11]

However, if the book metaphor is embraced, problems arise, problems of the sort regarding the labyrinth as mirror of the universe. Is not the closed book a closed case insofar as its black marks on white are fixedly there? If the book is a model of the world, is the world, like the book itself considered in its entirety, not equally fixed, static? And if so, then where is process to be found? The age-old contrasts between change and stasis, flow and fixity resurface. They must be addressed, somehow, if any lesson is to be learned from literature, science, and painting.

When the Gaps Are Finally Filled

A preliminary stage in tackling this issue entails a better image of changelessness. It might go something like this. Suppose we put a series of two types of letters into a block sort of marching order. The two letters are used to build up successive layers, each letter with its corresponding subscript. What I have constructed in figure 17 is a sample of the whole, which is of mind-boggling complexity—that is, considering that all the *ns* trail off to infinity. The entire block—or better, hyperblock—is of binary construction, whether considered either in terms of 0s and 1s or *as* and *bs*—recall Simplicio's remark with respect to Sagredo's initial presentation of figure 1.

Let the terms in figure 17 be called "puncts," and let the straight lines between the terms, whether actual and at 90° angles or diagonal, imaginary lines between them, be called "rects." The combinatory rule for connecting puncts bars any angle other than 90° or 45°. We can have a_1-a_2-b_2-b_3, for example, which makes up a possible "word" and consists

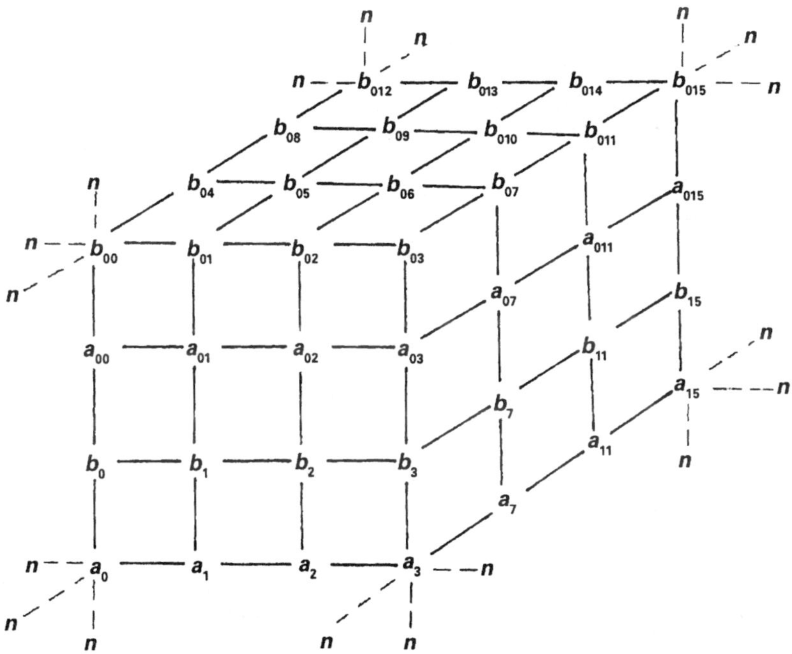

Fig. 17. The signifying block

of two horizontal and one vertical rect, and a_0-b_1-a_{02}-b_{03} is another "word" composed of diagonal rects. But a_1-b_2-b_{03} is an illegitimate "word," since it violates the rule of combination. And so on. Now, what if we squeeze our cube together more tightly so that the diagonals projecting upward are of considerably less than 45° with respect to the horizontal lines? And what if we continue squeezing the "terms" closer and closer together until there is no longer any perceptible distance between them?—that is, the cube is on its way toward becoming a vertical plane. Do we now have a continuum? A plenum? Not really. That is, not really if we are regarding the perceivable and conceivable nuts and bolts of our physical world. But yes, really, if we place figure 17 within the sphere of nonalgebraic transcendental numbers, given what some mathematicians considered shaky footing by Georg Cantor.

Cantor revealed the self-asymptotic nature of series of numbers that converge toward a rational limit. Such self-asymptotic sequences generate a new kind of irrational numbers or real numbers, examples of which are found in π, $\sqrt{2}$, and the like, and whose total population is infinitely greater than that of the whole numbers. The study of real numbers following Cantor was characterized by groping, guided by "a sort of implicit faith in the *absolute nature of the unlimited*. This faith found its extreme expression in the finale: the new theories of the *continuum*" (Dantzig 1930, 162). In other words, at the end of the tired trail, when all the gaps had been filled in and the real numbers had realized their plenitude, continuity would have at long last been created. The puncts and the rects in figure 17 would have become one, a static plenum: the ultimate book, the book of the universe.

How can we reconcile the appearance of numbers on a sheet of paper or the monitor as "secondary qualities" with their presumable nature as dynamic, ceaselessly changing "primary qualities"? How can we account for physicist Arthur Eddington's (1958) writing table as the solid, granular, brown, wood scented object of our appearances and a senseless rush of microscopic particles in chiefly empty space according to prevalent physical theory? How can Achilles subvert Zeno's logic by overtaking the tortoise and in the process actually passing through an infinity of segments? The library must be finite, for in our nitty-gritty world it could hardly be otherwise; but as a transcendental being it is surely infinite, like a fully grown irrational number.

What Eddington's table seems to us is from the theoretical view virtually nothing at all; what appears to be in motion is in reality nothing but a series of infinitesimal Zeno jerks. In comparable fashion, we would like to believe there is a distinction (gap) between one book and another in the library, but in reality it is filled in with an infinity of books the differences between which become infinitesimal, thus making up a continuum of books—that is, like figure 17 if the puncts were to be multiplied infinitely. Between any two books there is another book (and the excluded-middle principle is subjected to irreparable damage). A book and what it is not, some other book—its counterbook—become an infinite stream of other books such that there is for practical purposes and for us hopelessly finite individuals no appreciable difference between one book and some other possible book (and the noncontradiction principle is virtually trashed). Moreover, since both the narrator and the readers of "The Library of Babel"—and the author as well, we must presume—are in the library, they must be, themselves, read. They are books among books, part of the continuum made discontinuous solely by cuts and divisions that are fashioned rather than found, invented rather than inherent to the system.[12] It is once again a matter of what Sagredo called "sludge" (a meltdown of fig. 17), arising from the conflict between our intransigent ideals and our real capacities, what we would like the world to be and the best we can make of it as we live and breathe it.

Regarding the library's raw physical structure, Borges's choice of the hexagon as the form of the galleries (rather than the cubicle structure of fig. 17) is, I would suggest, far from a mere whim. It is not coincidental that the hexagonal pattern is perhaps the most frequently occurring ornamental design in human cultures throughout the world.[13] A prime example of this pattern in nature is realized in the honeycomb of the beehive. A colony of worker bees, which are of nearly identical size, build their cells around themselves by gyrating about a point while secreting semifluid wax. The form of the cell, one might assume, would be circular. But the capillary action of the wax adhering to itself causes it to cluster at six equidistant points along the circumference to form a hexagon, which is naturally the densest and most economical packing of the parallel cylinders. If a collection of marbles is placed in an inclined tray, they will arrange themselves in a triangular pattern. Three marbles (or circles on the plane) form a triangle and seven a hexagon, with one marble occupying the center. If we stack successive layers of marbles upon the first, it is not difficult to visualize the accumulation of these individual cells becoming spherelike and to comprehend why the narrator tells us that the library is a "sphere whose exact center is any one of its hexagons and whose circumference is inaccessible" (Borges 1962, 52). If there is a central hexagon, its center must be the center of a central sphere, which in turn will mirror and duplicate the entire sphere. The hexagon, then, is the imperfect facsimile of divine perfection: a sphere. This imperfection, in the library, profanes the sacred, taints the pure.

Interestingly enough, the hexagonal structure of each gallery in Borges's monolithic library, which is reduplicated indefinitely such that the library appears to become more and more spherelike as the sides of each hexagon become smaller relative to the whole, creates an image of fractal geometry. Traditional geometry treats regular forms: straight lines, smooth curves, shapes with perfect symmetry. Yet nature, according to chaos theory following fractal geometry, rarely displays such harmony. Rather, edges are ragged, surfaces at a closer look become tangled networks, and, regarding bilateral symmetry, the left side is never a faithful mirror image of the right side. A common and relatively simple example of the generation of fractals is the "Koch curve" consisting of the infinite reiteration of a recursive operation. Starting with an equilateral triangle, the first step yields three identical triangles on each side, making a six-pointed star, the second gives twelve more triangles, and the third an additional thirty-six to produce a "snowflake" image. At the theoretical upper limit we would have a continuous "curve" with an infinite number of infinitesimal barbs or excursions (fig. 18). Like Borges's library, the curve would be monstrous. It would be impossible to visualize in terms of a precise image. Possessing no predetermined tangent, it might change direction at each and every point, thus rendering it infinitely irregular, or infinitely regular, take your pick. As a recursively generated pattern, the Koch curve could be used to create a pattern comparable to Borges's library seen from afar. However, the library's inhabitants are not as fortunate as we are, with our ability to take in the two-dimensional pattern in one perceptual gulp. Inextricably caught within the library's confines, they are capable of perceiving no order, though the desperate hope remains that somehow the library must eventually evince some modicum of order.

The library-labyrinth and the universe: Borges's model is intriguing, enticing, and enigmatic. Ultimately, however, the model affords no more than a static image. It is a hyperblock somewhat comparable to the

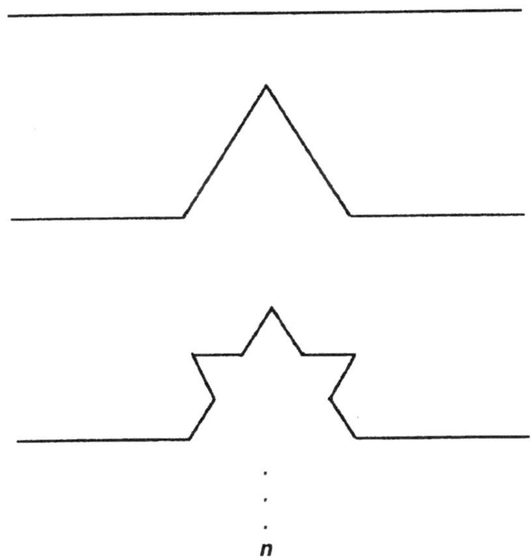

Fig. 18. Koch's curve

space-time continuum of relativity theory as viewed by some superobserver from an outside vantage point. We have seen this Parmenidean permanence often: the binary block world of puncts and rects illustrated in figure 17, the infinitely regressive fractal images (fig. 18), the Aleph (after all, it is a fixed point in space-time), the timeless orb of Borges's "The Secret Miracle" and Calvino's "T Zero," and even the rhizome, which, as a vast intertextual set of interrelations, is always already there, in one fell swoop. In contrast, we have the library, labyrinth, rhizome temporalized and dynamized by the hands and minds of their frantic seekers of knowledge, each of which traces a lonely, minuscule world line within the whole. The trajectories of these hopeful knowers at least serve to squeeze time, subjective time, out of the block. This we have also seen in the Heraclitean flux of consciousness illustrated through the supernominalist, Funes the Memorious, the Tlönian mental nominalist-solipsists, the linear sequence of images experienced by the owner of the Zahir, the schizophrenic and inductively impotent Grueworlders. Actually, the universe, it is becoming increasingly apparent and as suggested in the preceding, is a dynamic, self-organizing entity, as are all the islands of order tentatively arising from the vast ocean of disorder. This notion favors transience over stasis, change over immutability, novelty over mechanical reiteration, complexity over simplicity. It is order out of chaos rather than order for all time, but above all, it favors becoming over being, from the emptiness to the interdependent arising of every*thing* that we have made, that has made us: *our universe.*

We must look further into this notion of a self-organizing universe, and its relation to the hyperblock image, it would appear. But not before exercising a further incursion into the problems inherent in the notion of timelessness.

PART 3

In chapter 7 the universe (or labyrinth or library) as hyperblock, with the inclusion of time, complex time, is related to Calvino's "The Count of Monte Cristo." This unavoidably takes us back to Zeno's trap and to the disconcerting paradoxes of infinity such as that found at the heart of Borges's "The Book of Sand" and "The Disk." Chapter 8 offers a sort of preamble on Prigogine's far-from-equilibrium dynamics and life processes. A brief interlude follows as an exemplification in human life and human communities of Prigogine's process physics. This interlude consists of the living chaos (complexity) of life's processes in the Caribbean as viewed in Antonio Benítez-Rojo's *The Repeating Island,* coupled with a brief sojourn through Beckett's consciousness as the asymptotic collapse of chaos to an infinitesimal point. The works of Benítez-Rojo and Beckett are, each in its own way, living illustrations of rhizomicity. In this vein, Deleuze and Guattari's rhizomic, schizophrenic, deterritorialized nomad is placed in the spotlight in regard to figures 1 through 5*a* and 5*b* from the introduction of this inquiry and in support of the Benítez-Rojo–Beckett thesis. Chapter 9 gives further consideration to entropy and its relationship to simplicity-complexity, order-disorder, and statistical determination and indeterminacy. Human limitations regarding knowledge and the limitations of life are foregrounded, all of which bear on tragic flaws deriving from asymmetry, which is the general norm of life and of the universe, in spite of the crystalline purity of which dreams have traditionally been made.

CHAPTER 7 • Infinite in All Directions?

Physics is mathematical not because we know so much about the physical world, but because we know so little: it is only its mathematical properties that we can discover.

—Bertrand Russell

The little bug, which runs in all directions under a glass, believes to see before it an unrestrained freedom.

—Wassily Kandinsky

The complexity of art is due to the fact that different degrees of its evolution are present at one and the same time.

—Piet Mondrian

The Image of Eternity, but Where Is Time?

In an effort to bring about a meltdown of the block conception of space-time patterned in Borges's library and figure 17, one must first somehow contend with that formidable opponent, timelessness, which was dear to the heart of the classical view of the universe. Let us continue our efforts in this regard with further consideration of Calvino's "The Count of Monte Cristo" (in *T Zero* [1969, 137–52]), which creates a hyperblock image and, in addition, offers another vivid example of what goes by the name of complex time.

The protagonist is confined to a cell with a tiny barred window "at the end of a shaft that pierces the thickness of the wall: it frames no view; from the greater or lesser luminosity of the sky I can recognize approximately the hours and the seasons; but I do not know if, beneath that window, there is the open sea or the ramparts or one of the inner courtyards of the fortress" (Calvino 1969, 137). This affords him no more than a tunnel perspective or, better, a funnel perspective. Since the funnel is so long, it is possible that the walls could contain another cell between his cell and the outside, and between this inner cell and the outside and the inner cell and his own cell, there could be two more cells, and so on. This situation is not really that of the rects in figure 17 collapsed to form a virtual continuum of puncts in such a way that the cubes become more and more slender until they approach the infinitesimally thin. Rather, it is reminiscent of what is known in fractal dynamics as a "Sierpensky sponge," a cube that is perforated with smaller and smaller square holes until it is apparently nothing but an infinity of holes; yet it has an infinity of walls the sum of which make up a wall of infinite length (see Stewart 1989; the Sierpensky sponge will be discussed briefly in chap. 10).[1]

This being the case, the number of cells in Calvino's prison is unforeseeable, as is the total length of the walls of all the cells contained within it. Moreover, if the Sierpensky sponge is complete and its virtually innumerable walls to all appearances finally dissolve into space and the sponge remains completely empty, our prisoner has no way of knowing how much of the prison consists of walled cells and how much of unconfined space. That is to say, if the Sierpensky sponge prison is extrapolated to its ultimate, it will consist of space and nothing but space; hence it will be no prison at all. The walls will be infinitely thin, no more than a mathematically imaginary line. Yet, the sum of the area of all the walls of the cells will be infinite. So at the infinite stretch of things the prison, this cosmic prison, like the Library of Babel, or the book of the universe, will in a sense be absolutely confining, paradoxically.

As should be deemed most appropriate, the narrator's perception, from within his dank, dark dungeon, is more auditory than visual, with intractable elements of tactile, olfactory, and gustatory sensations. He pricks up his ears at the sounds indicative of jagged spaces and forms around him. He tries to infer "the network of the corridors, the turns, the openings, the straight lines broken by the dragging of the kettle to the threshold of each cell and by the creak of the locks," but he succeeds "only in fixing a succession of points in time, without any correspondence in space" (Calvino 1969, 138). He becomes obsessed with escape. In order to realize his goal he must first know the whole plan, the pattern, of the edifice. But perhaps he can never know the whole, for, like the human library rats in Borges's Library of Babel, he is caught within a minuscule portion of it.

Abbé Faria, another prisoner, has perforated the walls of the edifice in every direction. His itineraries wind around themselves like a ball of yarn—or like a strange attractor. As a consequence, he lost his sense of direction long ago: he no longer recognizes the cardinal points, and zeniths and nadirs have no meaning for him. His trajectory is comparable to a world line in the four-dimensional space-time continuum. At times the narrator can hear him, perhaps scratching at the ceiling; plaster falls; a hole opens up, and

> Faria's head appears, upside down. Upside down for me, not for him; he crawls out of his tunnel, he walks head down, while nothing about his person is ruffled, not his white hair, nor his beard green with mold, nor the tatters of sackcloth that cover his emaciated loins. He walks across the ceiling and the walls like a fly, he sinks his pick into a certain spot, a hold opens; he disappears. (Calvino 1969, 141)

Each cell is for Faria much like Escher's *Relativity*—or the central point, the Penrose triangle, in figure 5*a*—which evinces stairways going up or coming down, or to the right or to the left or forward or backward, according to the vantage point. Faria even has trouble distinguishing one cell from another. As such, the fortress has no favored point, no privileged center. Or better, as with Borges's library (or fig. 17), every point can be conceived as the center, and its circumference can therefore be construed as nowhere. In other words, it appears to follow diverging, converging, convoluted, involuted, nonlinear lines to nowhere and everywhere. Hence every hypothesis regarding the whole is immediately falsified; there is no difference that makes a significant difference between the real and the imaginary. The narrator no longer knows if the person he hears chipping away with his pick is the real Faria or merely some chimerical being.

The procedures of the narrator and Faria in their attempts to break out of their confines differ. Faria confronts a problem, studies a possible solution, tests it, encounters a new problem, and the process begins anew. As he proceeds, chipping away the fruitless solutions—that is, falsifying them—ultimately he hopes to know the whole so that he then can plan his escape. The narrator, in contrast, operates on the initial assumption that the fortress is a perfect, self-contained, self-sufficient whole and, as such, escape is impossible. Yet, escape may somehow, and perhaps miraculously, be realized if a solitary sign (model) can be found within it that will provide insight into the whole. This is most likely a futile dream, but a dream that must be pursued, for otherwise existence loses all its meaning.

Consequently, the models of the fortress constructed by the narrator and Faria become increasingly differentiated. Faria begins with a simple figure that takes on more and more complexity as it is placed alongside other figures, the conjunction of all of which will eventually compose the self-contained whole—or at least that is the hope. The narrator, in contrast, seeks in the garbage dump of phenomena surrounding him isolated clues giving him a picture of the prison as a self-contained entity, the whole. Thus:

> I, setting out from the jumble of these data, see in each isolated obstacle the clue to a system of obstacles, I develop each segment into a regular figure, I fit these figures together as the sides of a solid, polyhedron or hyperpolyhedron, I inscribe these polyhedrons in spheres of hyperspheres, and so the more I enclose the form of the fortress the more I simplify it, defining it in a numerical relation or in an algebraic formula. (Calvino 1969, 144)

In a sense, then, Faria's method is chiefly inductive; the narrator's is by and large deductive. The one is based on simplicity made to account for complexity; the other is predicated on the idea of complexity made simple. The problem is that just as Faria's task is interminable, since he cannot exhaust the universe in all its complexity, so also the narrator's project is destined to fail, for his model of the fortress—a hyperblock—is, though presumably finite, unbounded. In this manner, there is no end of the line, no knowing the prison in terms of all its intricacies. For:

> each cell seems separated from the outside only by the thickness of a wall, but Faria as he excavates discovers that in between there is always another cell, and between this cell and the outside, still another. The image I derive is this: a fortress that grows around us, and the longer we remain shut up in it the more it removes us from the outside. (Calvino 1969, 145)

In other words, the cosmic prison is a self-organizing, expanding universe! If it grows with the speed of time—that is, of light—the narrator would have to travel at a speed greater than that of light in order to reach the extremity, a physical impossibility according to Einsteinian physics. Yet, conceiving that this might be possible, upon penetrating the shell of the fortress to go outside, the narrator would complete the circle and enter the moment when he began his trajectory within the confines of his universe. He has no recourse but to conceive of the prison

> either as a place that is only inside itself without an outside—that is, giving up the idea of leaving it—or . . . as a place with no relation to me inside or outside; that is, I must study an itinerary from inside to outside that precludes the import that "inside" and "outside" have acquired in my emotions; valid, that is, even if instead of "outside" I say "inside" and vice versa. (Calvino 1969, 146)

Discovery of a center is apparently impossible, since the center is everywhere, and a break into the outside is equally impossible, since the circumference of the sphere is unlimited, unbounded, in essence nowhere. In a strange way the whole concoction appears to be rhizomic. And time within this cosmic prison, this hyperblock universe, is not that McTaggart (1927) A-series time as a present flowing from past into future along the knife-edge of the now, but rather, it is B-series complex time, mathematical or imaginary time based on the function of $\sqrt{-1}$ (that is, on complex numbers as discussed in chap. 5—which are actually timelessness from the perspective of our own subjective temporal grasp).

Finally, Faria opens a breach in the wall—or perhaps the floor or the ceiling—and bursts into Alexandre Dumas's study, where he is in the process of writing the novel about the narrator in the fortress! As it turns out, in this concentric fortress, Dumas's desk contains the narrator himself and all the other prisoners as well, along with the treasure and the "supernovel" *Monte Cristo,* "with its variants and combinations of variants in the nature of billions of billions but still in a finite number" (Calvino 1969, 150).[2]

Textually speaking, Faria had set his heart on the accumulative grasp of the pages giving him the final answer. The narrator desired to take all the sheets into account in one massive conceptual gulp. Faria's arrangement of the pages, one after another, constructed a spiral extending ever outward, ultimately as a positivity, whereas the narrator's successive concentration of the pages took the spiral inward, in the sense of negativity. Either way, at the most extreme stretch, the two trajectories would meet. That is, Faria's project entailed inductive gatherings and rejections of myriad items of experience until, hopefully, their sum might pattern the real. In contrast, the narrator wished to conceive of an imaginary fortress that would be either the same as the real one or different. If his mental construct turned out to be the same, then surely there could be no escape, but at least he could "achieve the serenity of one who knows he is here because he could be nowhere else." If it proved to be different, then the mental prison would resist escape even more vehemently than the real one, which would indicate that there might be, after all, some possible escape from the latter. In this case, "we have only to identify the point where the imagined fortress does not coincide with the real one and then find it" (Calvino 1969, 152).

In other words, if a figment is by chance a perfect model of what the facts should be but the facts do not coincide with the figment, then the facts must be somewhat less than perfect. They are flawed; hence in the flaw, the *aporia,* the *mise en abîme,* the gap between the otherwise continuous stream of signs, there may be room for escape into the diaphanous light of freedom, somehow, somewhere, somewhen.

Advancing or at a Standstill, Wherever We Are?

But we really should alter our frame of reference, once again, in the direction of the notion of timelessness if we are to bring about a meltdown of the block idea, which is apparently the emerging image within science, literature, and painting insofar as they are discussed within the covers of this book.

As we have noted, Lönnrot's (i.e., Zeno's) alternate labyrinth within which he would have Scharlach entrap him the next time around apparently could not have been simpler at the outset. Yet the fact that "so many philosophers have lost themselves in it" testifies to its numbing complexity. Following Sagredo's teachings, we need no more than a simple line to illustrate the most complex of complexities, as mathematicians Richard Dedekind and Georg Cantor have also demonstrated in their respective theories of the continuum (Huntington 1917; Dantzig 1930). Zeno's (i.e., Lönnrot's) fallacy consists in mixing continuity with discontinuity, of postulating the fracturation of a dense continuum into an infinity of infinitesimal increments. But, to repeat, an infinity of dimensionless points is merely a point; that is, it is dimensionlessness. On the other hand, an infinity of segments, no matter how minute, is infinite in length.

The same can be said for time, as we also observed in the introduction and more recently in Palomar's meditations on an individual moment. Achilles traverses half the distance in one-half of a minute, then half of that in one-quarter of a minute, and so on. And just as he never overtakes the tortoise, so also he never sees the end of the particular minute he inhabits. So is time actually continuous or discontinuous?[3] If it is broken up into moments, and if each moment has the fleetest duration, then there is no beginning or end of time. On the other hand, if time is a series of durationless points, then the sum of those points is durationless. It seems to follow, taking a renewed cue from Zeno, that (1) no finite space interval can be traversed in a finite duration of time, (2) a physical object cannot endure from one finite time interval to another, and to top it all off, (3) Achilles could never have been able to lurch out of the starting blocks in the first place (Grünbaum 1967). In this ideal sense, the slug from Scharlach's pistol could not have passed from one spatial segment to another between assassin and victim, nor could its existence have continued from one segment to another—that is, Funes's dilemma. It couldn't even have left the pistol in the first place if motion and time are judged to be discontinuous. For any given instant of a dense temporal or spatial order, there is no next instant or increment either before it or after it, since any two instants or increments are separated by others that are between them, and so on. This dense sort of betweenness exists in contrast to the betweenness that can be assigned to a discrete series according to which the elements are ordered consecutively, such as 1, 2, 3 . . . *n*. Zeno's paradoxes illustrate by means of a reductio ad absurdum that the denseness or continuum postulate grafts onto time and space, and hence motion, a property quite alien to them.

We also get an uncanny feeling: for an instant we sense the *mise en abîme* regarding the impossible endpoint of the infinite stretch of things during a reading of Borges's "The Book of Sand," in the book by that title (1978, 117–22). The narrator is introduced to the chimerical book by an unnamed stranger who told him the book was so called "because neither the book nor the sand has any beginning or end" (119). The narrator is then asked to find the first page:

> I laid my left hand on the cover and, trying to put my thumb on the flyleaf, I opened the book. It was useless. Every time I tried, a number of pages came between the cover and my thumb. It was as if they kept growing from the book.
>
> "Now find the last page."
>
> Again I failed. In a voice that was not mine, I barely managed to stammer, "This can't be."

> Still speaking in a low voice, the stranger said, "It can't be, but it is. The number of pages in this book is no more or less than infinite. None is the first page, none the last. I don't know why they're numbered in this arbitrary way. Perhaps to suggest that the terms of an infinite series admit any number." (Borges 1978, 119)

As if the stranger were thinking aloud, he then remarks: "If space is infinite, we may be at any point in space. If time is infinite, we may be at any point in time" (Borges 1978, 119). The infinite, of course, knows of no beginning, middle, or ending.[4] It simply is as it is, and if it is "cut," somewhere, at an arbitrarily selected point, the cut is *our* cut: in fact, we are ourselves cuts that never cease to exercise cuts in the continuum of possibilities before us. The infinite is tropologically recapitulated in Renaissance philosopher Nicholas of Cusa, the circumference of whose sphere—which is God—is nowhere and whose center is everywhere. Or, tropologically speaking, it is like Borges's Library of Babel, or the dungeon into which the narrator of Calvino's "Count of Monte Cristo" has been thrown. Up, down, forward or backward, here or there, this "center" or that "center," it's all the same. It is "infinity in all directions," if I may avail myself of Freeman Dyson's (1988) provocative phrase.

The narrator's predicament in Borges's "Book of Sand" also reminds us, once again, of Cantor's work with infinity. Cantor teaches that we can add 1, 10, 100, or 10^{10} integers to our initial sign depicting infinity and it will be that much larger, though the number contained within that infinite set is still the same, infinity.[5] If the number of pages in the book were less than infinity, and yet if that number were virtually unfathomable—as was the number of pages in the book for the narrator—then there could be no way one could know without a shadow of a doubt whether the book is of infinite or finite length. What is more, there would be no knowing where beginning and end are, or where one finds oneself when opening the book to a particular page. This is the same dilemma that confounded the inhabitants of Borges's Library of Babel. For ages it has also been the dilemma the West has confronted in the labyrinth it calls the universe. Like the preceding question regarding the continuity (the infinite divisibility) or discontinuity (the finite divisibility) of time, one might ask the age-old question anew: Is the universe infinite or finite? Orderly or disorderly? If it is infinite and either orderly or disorderly, one cannot know that it is so until and unless one reaches the infiniteth point, which one cannot do, and so one cannot know. If it is finite and orderly, one can in principle know it is so, but according to the latest reports it is more disorderly (chaotic) than orderly. If it is infinite and orderly one must trace the disorder that seems to be its chief character out to the end and therein finally discover some semblance of order, but theoretically one cannot do so according to the limitative theorems, so the universe will continue to appear of overriding disorder. Apparently it is a no win situation, paradoxically.

Significantly, in this light, astrophysicist David Layzer (1975) has put forth the radical hypothesis that not only are we unable, from our microcosmic vantage, to know where we are in the universe, but that such information would be impossible even at the macrocosmic level, for

> [t]he order is unknowable even in principle. Imagine an unbounded stack of playing cards, topless and bottomless, deck piled on deck without limit. Information about the order of the cards in one section of the stack is of no help, because any given sequence is repeated an infinite number of times elsewhere, in the same way that patterns of stars and galaxies are repeated throughout the universe. It is meaningless to say that you are at such and such a place in the stack, even when you have full information about the order of the cards and that place. You still don't know whereabouts you are in the stack, any more than the typical observer knows whereabouts he is in the cosmos. The Cosmological Principle says he doesn't know and he can't know. (Layzer 1975)

In fact, an infinite sequence of spatial increments or temporal intervals has no last term; neither does it have a first term, for the finite human agent at least.

Samuel Beckett's opus taken as a whole bears witness to this sort of phenomenon. Each work, each chapter, paragraph, sentence, word, is sublimated from Beckett's gray matter with increasing excruciation, until hardly any words are forthcoming at all. In Beckett's trilogy *Molloy, Malone Dies, the Unnamable* (1955), Molloy, caught up in a journey toward the whereabouts of his mother, ends up in a ditch, face down. Malone's pencil scratches fewer and fewer words on paper as the seemingly interminable moments drag by. The Unnamable longs for, but never encounters, "the story . . . that I should never have left, that I may never find again, that I may find again, then . . . it will be I, it will be the place, the silence, the end, the beginning, the beginning again, how can I say it, that's all words, they're all I have, and not many of them, the words fail, the voice fails, so be it" (413). The protagonist of *How It Is* (1964) painfully inches forth in the ooze within which he finds himself, having departed from an irretrievable beginning and with no possible end in sight. If that last word could have come gurgling from his mouth, from the depths of his body and mind, it would have been virtually no word, perhaps a mere syllable, perhaps even less, surely even less. It would have been well-nigh nothing, no-*thing.* At the end of the Zenoesque infinitely stretched asymptote, the concluding term is infinitesimal. How could it be any-*thing* other? How could it have any *other?* It would be itself and no-*thing* but itself. It would be one pearl of the entire string of pearls each of which mirrors the whole; so it would be tantamount to the whole, paradoxically and rather unthinkably. It would be a cosmic black hole or a "naked singularity"—or the Aleph—containing every-*thing* and every event in the entire history of the universe.

But, There Is No End of the Road, for the Finite Knower

The notion of the impossible last term in an infinite series is also beautifully illustrated in Borges's "The Disk," which appears in *The Book of Sand* (113–16). The narrator comes upon a Saxon coin, "Odin's disk," whose obverse side is strangely nonexistent, for it cannot be seen. Overcome with greed, he kills the traveler who allowed him to see it, but not before the coin is thrown to the floor in the struggle. After removing the body from the hut where the struggle took place and disposing of it, the narrator returns to retrieve the coin, but try as he may, he cannot locate it. This coin is surely a metaphor for infinity, and as such it can have no reverse side, for infinity according to its interpretation as a potential is incomplete and incompletable. The reverse side of the coin must therefore be invisible. In this manner the front side of the coin represents the first term in the series. It is visible; at least ideally it is visible. Since there is no final term at the infinite horizon, for a finite observer it can be none other than nonexistent and invisible. The coin, while flip-flopping through the air, oscillated between *something*-ness and *nothing*-ness until finally, to the chagrin of the narrator, it ended up face down, its *derriere* revealing no-*thing* but the nonontological *nothing*-ness of a black hole.[6]

In a footnote at the conclusion of Borges's "Library of Babel," in *Labyrinths, Selected Stories and Other Writings* (1962), we read of a comparable phenomenon. It consists of the hypothesis, as we have noted, that "rigorously speaking, a single volume would be sufficient" (58). The middle page of this solitary, all-encompassing volume, like Odin's disk, would have no reverse because it would be the equivalent of 0—or in terms of previous allusions, emptiness—that marvelous state of absolute nothingness, whose origin lies in Buddhist philosophy and the Sanskrit language, separating the positive integers from the negative integers. There is yet another aspect that Borges does not reveal in "The Library of Babel" but that he does explore in "The Disk." The disk, like the inconceivable middle page of Borges's "single volume," and unlike ordinary coins, has no reverse side precisely because there is actually nothing (no-*thing*) to see, since an infinite series will not tolerate an existent final term.

In the final analysis, what can 0 be but a state of absolute equilibrium between the push of the plus sign and the pull of the minus sign, of one thing and its opposite, of blind antagonistic forces? It is that never-never land between the + and the – schizophrenically dancing within the spaceless space of $\sqrt{-1}$, the white-black-lessness of the hairline in figure 4, or the center of figure 5*a*. It is the space of emptiness, the prisoners' timeless slice between before and after as conceived by their lawyer, that instant in Hladik's life miraculously stretched out to a year. It is patterned in Beckett's characters' tense effort to reach a state of passive acquiescence, paradoxically.[7] It is also imaged in Oedipa Maas's fixation, discussed in chapter 1, on that strange symbol consisting of a trumpetlike figure made up of a line, a loop, a triangle, and a trapezoid. Oedipa is caught up in (1) semantic overload, overdetermination, many things that are uncontrollably coming out as if by chance and (2) differences between signs and their possible meanings that become smaller and smaller until they are virtually indifferent with respect to one another. The first condition, product of a conjunction of opposing and even contradictory tendencies, oscillates, vibrates, scintillates in an interminable Siva dance, with each and every possibility threatening to actualize and fly into space in any direction. The second leads toward acquiescence, indifference, quietude, another tendency evinced by Oedipa as she becomes increasingly worn out in face of the vertiginous onslaught of signs before her.

That same condition is magnified in the case of Jacob Horner of John

Barth's *The End of the Road* (1967), whose schizophrenic oscillation between the *either* and the *or* occasionally dumps him into catatonia, into a completely static state of equilibrium, when all action, reaction, and interaction become impossible. He often reaches this point on "my weatherless days [when] my body sat in a rocking chair and rocked and rocked and rocked, and my mind was as nearly empty as interstellar space" (36). Horner's rocking is a limit cycle or periodic attractor characterized by almost identical reiterations around a central point, which, like pendulum swings, may eventually come to rest at a sink, where stasis prevails and catatonia takes over. Horner reaches this final resting point on March 17, 1951, at the Pennsylvania Railroad Station in Baltimore: "I was sitting on one of the benches in the station with my suitcase beside me. I was in an unusual condition: I couldn't move" (73). He has attained the state equivalent to that infinitesimal dividing line—like the white-black-lessness—the moment at the peak of the syncope, when time ceases (Clément 1994). This "zero point" is tantamount to John Cage's silence, Mallarmé's blank page, an empty white canvas before Malevich fills it in with textured white paint. It is also evocative, at a deeper, ineffable level, of Buddhist or Nāgārjuna emptiness *(Sunyata),* mindful mindlessness, all-encompassing selflessness, groundless unknowing knowing: that teeming spaceless space, that energizing timeless time potentially giving rise to whatever wherever and whenever.

It also brings to mind that infernal nesting of words, images, narrators, imaginary readers, and the world in Italo Calvino's *If on a Winter's Night a Traveler* (1981). In this novel, fictional readers and writers are posed in the text itself and thus draw the actual reader in as a potential writer him- or herself. Stories are embedded within stories—like the nested Chinese boxes or Russian dolls of which Simplicio spoke—and characters suffer a mirrorlike entrapment within texts of their own making. Ideally this novel must be read by quietistically letting the world, all worlds actual and possible, fade from the eyes, the heart, the mind, and the gut, and without any expectations whatsoever—or with the Beckettian expectation that nothing will happen, and if it happens, the unsurprising surprise that it did so. This is ultimately the resting state between Lönnrot and Scharlach: the indifferent difference between victim and victimizer, pursued and pursuer—like the dreamer dreaming a son while being dreamed by another in "The Circular Ruins," which appears in *Labyrinths* (Borges 1962, 45–50)—becomes increasingly vague and finally vanishes altogether.

But, of course, one cannot read a book with such ideal quiescence. Reading is inexorably an act, an activity, an acting on and interacting with (Roland Barthes's [1975] *readerly* and *writerly* texts). We cannot have the text before us as if we were hovering somewhere out in free space and detachedly viewing the panorama before us with some God's-eye all-embracing, neutral perspective (i.e., Palomar's supreme goal). Neither can we split the world into a convenient set of *thises* and *thats,* for the hairline divisions inevitably become smudgy: pursuer becomes pursued in Borges's "Death and the Compass" (in *Labyrinths* [1962]), and so on. We cannot have things either two ways or no way, we cannot have either every-*thing* or no-*thing,* nor can we be squarely in the middle of things: we have no sextant with which to navigate through space and time and no stars to tell us the way. Significantly, in this regard, at the end of "Death," only Zenoesque distinctions in time and space separate Lönnrot and Scharlach. If those distinctions are conceived as mere points, then nothing separates them; if the distinctions are conceived to be incremental, no matter how small, then a gaping chasm of infinite proportions separates them. In this world of ideal forms, it is a matter of either *all* or no-*thing* at all. It is the same story: once again, the jaded knower becomes oppressed by the hyperblock view, a world of binary puncts and rects (fig. 17), of infinitesimal fits and jerks (fig. 18), of the asymptotic move toward the Sierpensky sponge, of the plethora of cubicles in the architectonic library or the mechanical generation of the books therein, of the infinite "Books of Sand" or the "disk," all of which bear the image of timelessness, suffocating timelessness. But, after all, that is the ideal world, the world of good logic and good reason, the comfortable, benign world the mind usually wishes soporifically to embrace, especially the mind in tune with Simplicio's conceptual framework. We have heard from the mouth of Sagredo, on the other hand, that our actual world is something else entirely. Between our intractable ideals and our paltry real capacities, there is sludge, and plenty of it.

For this alternative to the hyperblock view, I very briefly cast a glance at a particular cultural manifestation of this theme of failed dichotomies and fuzzy boundaries: the Caribbean. But not before we have heard a word from the author of "dissipative structures" and of the living flame of process physics, Ilya Prigogine.

CHAPTER 8 • Patterns That Connect

The true, strong and sound mind is the mind that can embrace equally great things and small.

—Samuel Johnson

The silences of the infinite spaces terrify me.

—Blaise Pascal

Nothing is more real than nothing.

—Samuel Beckett

The value of a work resides precisely in what it is not.

—Pablo Picasso

Life Is Change, Flow

The things of living substance are actually not things qua things but the result of something that interacts with something else; they are events, not things; they are eddies that appear, whirl around for a while, then dissipate, not rocks strewn along the bed of a placid river; they are happenings, not things that bring about happenings as a consequence of linear cause-and-effect sequences.

Ilya Prigogine's task of wedding quantum theory to classical thermodynamic principles and at the same time accounting for life's processes appears by its very nature paradoxical, and, if the task were to be carried out in all honesty, perhaps it could not but culminate in the Heraclitean dictum that, simply put, everything flows.[1] And so it did, or at least that is the implication. Of course, the very idea of life was shunned by thermodynamic equations. Thermodynamics is the science of closed systems. Life calls for openness. Closed systems must have things their own way, the way of thermodynamics toward tension-free—and hence dead—equilibrium, and if things go awry, that is because there must have been some sort of infiltration by corrupting forces from outside. Life, in contrast, thrives on struggle, on far-from-equilibrium conditions, and if the context changes, more often than not life can find a way to adjust. A living organism takes in food and information, grows, replaces some of its own parts, reproduces, and all this without the assistance of any monkey-wrench–wielding repair person. In other words, life systems can by and large take care of themselves. Thermodynamic systems and machines, in contrast, need some (real or imaginary) outside agent to construct them and keep them going in order provisionally to postpone heat death. To make matters worse, the microscopic world of quantum events is so foreign to the macroscopic world of living organisms that the two, like thermodynamics and life processes, have also been considered simply incompatible, two distinct ways of accounting for the physical world. But that is another story. The focus here must remain on thermodynamics and Prigogine's order out of chaos.

Prigogine noticed that closed mechanical thermodynamic systems and open and lifelike, yet equally thermodynamic, systems are distinguished primarily in terms of a movement toward equilibrium in the first case and perpetual far-from-equilibrium conditions in the second case. Closed systems are usually destined to end in a homogeneous soup of chaos, maximum entropy; open living organisms and societies of organisms may verge on chaos, to be sure, but when they are able effectively to sustain life, they fluctuate on the knife-edge between chaos and order, managing to take advantage of the situation by grabbing a little energy there and organizing it for their self-perpetuation here. (The battling gladiators on the floor of the basketball arena are an incessant effort of two quintuples: one quintuple attempts to order its defense and confuse and complexify its opponent's offense; the other quintuple strives to order its offense and throw its opponent into disarray. The point guard of one of the teams looks for an opening, finds it, threads the needle with a bullet pass to the streaking small forward, who catches the ball on his or her way up for a slam dunk in the face of the helpless defender. They are in rhythm; they move as one; they live. Now they must go to work on the opposing quintuple, throwing its offense into confusion in order hopefully to steal the ball or grab a rebound and begin their own ball-court ballet anew. They must do so in order to stay alive, for they are up against a formidable five. Life: a flow of agonistics.)

But the flow is not limited to the phenomena we ordinarily conceive as life processes. The magic trick of give-and-take with the environment so as to bring about increasing order and even growth is found in many inorganic processes as well. From water gushing out of the nearest tap to cloud formations to fluctuating air masses to the birth and growth of stars and galaxies, the watchword is *flow,* out of fluctuating, far-from-equilibrium conditions. In the best of all worlds the flow moves toward that undulating, palpitating, heaving sort of behavior that is able to sustain itself, at least until final dissipation and "death" ensue. Apparently complex disorder becomes relatively simple order that in turn becomes organized complexity, which finally falls into a relatively inert, that is, dead, state of virtually chaotic simplicity. Simplicio at the outset was right, and so was Sagredo, depending upon the perspective; and Simplicio's breakthrough at the end of their chat—itself the product of order arising out of what was verging on cerebral chaos, I would suspect—placed him in a new framework where he was then also right, as was our now perplexed priest of complexity, Sagredo.

Examples of Prigogine complex dissipative structures with the capacity for self-organization abound. There are "Bénard cells," for example. Under the proper conditions, and at certain critical temperatures, a bit of water in a shallow pan heated from below will give rise to convection currents. These currents will at some point organize themselves in hexagonal cells comparable to a beehive structure in order to facilitate movement upward of the molecules of greater kinetic energy and downward of the molecules with lesser kinetic energy. There is also the "Belousov–Zhabotinsky reaction," named after two Russian researchers who in 1951 happened to stumble onto it. At specific temperatures, a mixture of malonic acid, bromate, and cerium ions in solution with sulfuric acid brings about the creation of spiral cells that pulsate. They move outward in the form of concentric circles, then collapse, then move outward again. The process is self-perpetuating, and it perpetuates its fluctuating behavior in quite stable fashion under carefully controlled conditions. In both cases, that of the Bénard cells and the Belousov–Zhabotinsky reaction, countless molecules in virtually random motion at some juncture begin orchestrating their movements in such a way that they create order from chaos. They seem to have a will of their own; from indeterminate origins they engage in what appear as purposeful acts in order to survive in a dog-eat-dog universe. They evince the characteristics of basic life forms, though they are strictly inorganic chemicals and devoid of DNA.

To paraphrase Alvin Toffler, who wrote the foreword for Prigogine and Stengers's *Order out of Chaos* (1984), it is as if we had a few million white Ping-Pong balls chaotically mixed with a few million black ones, and we set them to bouncing about randomly in a huge tank. From a distance the entire concoction would appear as so much gray matter. Then, all of a sudden, a collection of black balls forms a circle and moves outward, incorporating other black balls as it does so in order to sustain its growth. Inside the expanding circle of black balls there is an expanding circle of white balls, and inside that circle a smaller circle of black balls, and so on (compare this process to the emergence of difference from blacks and whites to bring about the Harlequin in fig. 5*b*). These living concentric circles expand and contract with pulses of the same periodicity, and they continue to do so until we grow tired of the display and take our leave in search of a more interesting pastime. Indeed, the process is quite monotonous and repetitive. But it is spectacularly lifelike. It sports a characteristic that simply should not exist, from the classical view of the universe, for it breaks all the rules. Yet it exists (recall Boltzmann's conclusions).

Regarding life processes at the most basic level, a colony of termites building a nest offers up for us a remarkable display of order out of chaos. At first the termites deposit bits of moistened earth in an apparently random disarray, as if each bit were a molecule in classical fashion unrelated to the adjacent molecules. There is apparently no coordination of movements, no harmony. However, since the earth is moistened with a chemical the termites secrete to attract other termites, they begin clustering minute globs of soil and chemicals. Then, at some critical point, apparently orderless activity becomes coordinated movement. A pair of columns appears, the columns grow, and at a certain height, as if the termites were endowed with extrasensory perception (ESP) capacities, the towers begin curving in toward each other and finally connect in the middle. These and other inorganic and biological systems are proof to dissipative structure scientists that self-organizing domains of activity emerge everywhere: in geometric cells, vortices, colonies of insects, the animal population, the growth of cities, social and political movements, and the life and death of stars and galaxies.

The term *dissipative structures* evokes the apparently contradictory image of disintegration and chaos along with its opposites, structure, balance, harmony, equilibrium. And contradiction it is: the process entails far-from-equilibrium conditions containing forces and elements in a constant struggle with one another; then at some unforeseeable point they erupt into irreversibly evolving, self-organizing wholes, that are capable of maintaining their identity by remaining open to the flux and flow of their environment. Migration of inorganic chemicals and functions of living organisms collaborate to bring about order from far-from-equilibrium, fluctuating, virtually chaotic situations. The second law of thermodynamics stipulates that in a closed system entropy will increase and islands of organization will dissolve; the system will eventually settle down to a tensionless sameness. Open dissipative structures, in contrast, incessantly expel entropy into their environment so as to create order and organization within and at the same time stave off the onslaught of entropy buildup. There is a constant input of matter and energy that is organized within and an expulsion into the environment of chaotic matter of lesser organization—excrement, urine, air lacking oxygen, sweat, and so on. A warm-blooded animal maintains a constant body temperature in spite of wide variations of temperature outside. It does so by creating flows within that ingest matter and energy, using them to produce flows of heat, and emitting flows into the environment. All is process, nonlinear movement, constant exchange, interaction—very much unlike the images evinced by Borges's library, Calvino's prison, figures 17 and 18.

The watchword is now *nonlinear flow.* Linear laws of cause and effect break down in far-from-equilibrium, nonlinear situations. In nonlinear systems, rather than a tunnel-minded singular focus, there is a multiplicity of possibilities at each juncture. Rather than one response to virtually every problem situation that presents itself, there are many possible answers to each problem situation. Rather than a race of the knife-edged "now" from past to future, the weight of the past is always present, and both past and present collaborate and contrive to pattern the future. What we have in essence is, once again, Borges's "Garden of Forking Paths," where the protagonist for a few split seconds became aware of the results of parallel worlds, of past bifurcations that were not of his world but took on a history of their own. Many alternate worlds could have been actualized but were not, for the protagonist's world is one and one only. But his world could always have been another world from among a plethora of possible worlds. These worlds are interrelated in a twisted, knotted ball, all going their own way yet somehow continuing to mutually influence one another, if not physically, at least in terms of which paths among many paths might be chosen at any and all future moments.

All this is still inordinately obscure and obtuse, I fear. A healthy injection of chemical reactions and mathematical equations could clear up the matter considerably. That clarity would entail an overtly conceptual-intellectual level, however. What I am shooting for, in contrast, is a general feel for the relationship between simplicity and complexity, and order and chaos, with respect to Prigogine physics. The conceptual-intellectual grasp strives toward the crispness of the Canadian Rockies, while a tenuous, tentative, and vague feel for that which has been sharply specified through formulas and equations is a South American jungle. The task at hand might appear formidable, but perhaps it is not impossible.

A brief digression through a radically complex example of human life worlds might be of help.

Cultural Simplicities and Complexities

Throughout history, order has been conventionally construed as unicity, disorder as plurality. Simplicity or order entails balance, harmony, symmetry, equilibrium; complexity is the opposite of these qualities. In Deleuze and Guattari's terms (1983), fascism—or the paranoid mindset—is monolithic monism, with all signs pointing toward ultimate power at the apex of the pyramid; nomadism—the schizophrenic mindset—is hyperconnectivity spread out along lines and planes and hyperplanes making up cubes and hypercubes.

Antonio Benítez-Rojo's *The Repeating Island: The Caribbean and the Postmodern Perspective* (1992) draws rather disparately from Deleuze and Guattari and chaos theory to arrive at the picture of the Caribbean as a bridge, an other *other,* the other as difference that makes a different difference, between North and South America. The geographical area consists of a discontinuous conjunction, a disjunctive continuum, a complementarity of agonistics, a fluctuating hodgepodge of subalterns—all of these in a contradictory mix—evinced by perpetually opposing forces. It is a mess of "unstable condensations, turbulences, whirlpools, clumps of bubbles, frayed seaweed, sunken galleons, crashing breakers, flying fish, seagull squawks, downpours, nighttime phosphorescences, eddies and pools" (Benítez-Rojo 1992, 2). It is of radically uncertain signification and significances. The entire area repeats itself again and again, in an ongoing, obstinate series of recursive loops. But unlike with the simple Zenoesque algorithm suggested by Sagredo regarding Simplicio's agenda, with each reiteration there is necessarily a difference that makes a different difference, however slight. And the combination of all reiterations makes up a monstrous strange attractor, where self-repeating hurricanes have for eons gyrated about the

Bermuda Triangle, causing the unexpected huffs and pitches that capsized galleons in olden times and that have devastated ramshackle sprawls of poverty in both rural and urban areas in modern times. If, as Sagredo illustrated, Zeno's ideal world consisting of simple reiterations along a line, a one-dimensional labyrinth, lands us in sludge with respect to our actual capacities and practices, then I leave your imagination to wrestle with the infinitely more complex labyrinth Benítez-Rojo's Antilles throws at us. Each return of a line, any line wandering through the Caribbean, potentially ends up every-*where* and no-*where,* pointing in an infinity of directions toward every-*thing* or perhaps no-*thing* at all—that is, like a complete meltdown of figure 17. Yet it consequently turns back on itself to continue along the tired irreversible push toward some undefined and undefinable end. The Gulf of Mexico, sporting its strategically placed emerald gems of tropical lushness, has weathered the devastation of time, of geological, meteorological, and ecological catastrophes, and of repeated Western world interventions. Yet, in spite of the downturns and disappointments, the unfilled promises and shunted dreams, the area has remained vibrant, a superposition of sand, soil, rock, and delicate life and of languages and cultures. It evinces a titillating, scintillating, shimmering, effervescent flutter of activity, like the glimpse of a hummingbird poised in midair while inspecting one of the gala array of flowers offering themselves up to it. It is a flux of virtuality in expectation of some-*thing* that never quite comes to pass, that always lurks in the uncertain future that remains beyond the receding horizon.

Benítez-Rojos's Antilles knew of no forbidden contradiction in the sense of classical logical principles. What is *A* here and now is sure to be not-*A* there and then or perhaps vice versa. All is a cascade, a continuous outflow, a vertiginous rush, of semiosis. Nor does the excluded-middle principle manage to wield its terrible swift sword: choices of all or nothing, this or that, we or them, I or other are never exactly an either-or affair, but they are a matter of either neither-nor or both-and and at the same time of neither the neither nor the nor and both the neither and the nor. Catholicism or Afro-Cuban Santeria (a vibrant syncretic mix of African religions and Catholicism)? Mysticism or magic? Synthesis or syncretism? Juxtaposition or fusion? French or Creole? Bourgeois capitalism or socialism or paternalism? Anglo-American or Puerto Rican? Or Hispanic or French or English or Dutch or Afro-Caribbean? The marvelously and uncannily irreal or hard-nosed reality? It is not a matter of the one or the other but of a continuum of shades without borders or clear divisions. Whatever ephemeral borders and divisions there may be, they are the product of human minds, hearts, and hands. There is tension, for sure, which intermittently erupts in violence. Yet there is a certain free-flow harmony perhaps unknown elsewhere in the world. There is dissymmetry with a vengeance, yet fearless symmetries manage to prevail as well. A move toward equilibration is always present, yet dynamic far-from-equilibrium processes perpetually maintain the entire region open to well-nigh every-*thing* and any-*thing*. Smooth lines and surfaces reveal, upon closer inspection, the fissures and cracks, nooks and crannies, broken edges and punctuated contours of myriad fractal cultures. The Antilles could well be intuitively viewed as living proof of the new physics of complexity.

The situation is compelling regarding the remainder of Latin America as well.[2] There are important differences between the Caribbean and the continent, to be sure. The Caribbean was the first to suffer violation by the cross and the sword of European outsiders, most often motivated more by dogmas and grand designs than by their experience of the immediate surroundings, more by lust and power than by compassion and understanding. And it was first to taste the bittersweet glory of a successful independence movement, in Haiti, 1804, a country that as of this writing ironically and unfortunately has yet to see the completion of a presidential term created by democratic means. At the same time, some areas of the Caribbean were last to gain independence from their colonial exploiters: Cuba, Puerto Rico, Trinidad, Jamaica. The similarities between the Antilles and continental Latin America outweigh the differences, however.

The peoples and cultures—I put the terms in plural, for there can be no singularity here—do not and cannot enjoy any stable form of identity, for they are one of the most pluralistic, if not *the* most pluralistic, of cultural conglomerations in the history of the human species. It is not simply a matter of the customary dualistic labels: Amerindian/Spaniard, African/Spaniard-Portuguese, America/Europe, country/city, civilization/barbarism, nature/culture, and so on. Between Aztec and Mayan and Quechua and Guaraní and a host of other sedentary and nomadic cultures there are commonalities. But there are also differences greater than those between Greek and Roman and Moorish and Spanish and Germanic and Saxon. What is the Latin American? Native America in its multiple forms? But native America is itself of mind-numbing diversity. Is it part African? What part? Bantú or Sudanese or something else? Is

Latin America a product of the Renaissance or of a Spain and Portugal still having medieval roots and blended with Arabic cultures? Or is it more a product of the postcapitalistic, postindustrial era? What is the future of the area? Back to the pre-Hispanic roots? To European civilization? Or is its future one of forging ahead toward the grave new world of postmodernism, whatever that ultimately may be? The Latin Americans thrive on pluralism, they breathe it, they consume it, they expel it. They are living pluralism, living complexity.[3]

Now, my motive for bringing up the "repeating island" issue is surely being questioned at this point. "Why should you, how can you, plunge headfirst into this exceedingly vague and confusing topic when you ought to be discussing the trials and travails of simplicity and complexity? After all, is establishment of the relationship between the pair of terms not what you proposed in the beginning and have yet to deliver? How can you, then, exercise a quantum leap to such an unruly global domain in one fell swoop. In other words, why don't you just say what it is you have on my mind?"

Well, I must concede that this criticism certainly appears on track. I should be more direct and forthright. On the other hand, my own rather nomadic, multitangential digression has been, I would suggest, more necessary than whimsical. For I believe it implies, and may access a feel for, a most compelling point: the very nature of chaos, of complexity, is bound within language and culture. That is to say, the meaning of the message is in the feel not the form, in kinesthetic meanderings not precise mental constructs, in a language of vagueness not of that pristine logic making up a security blanket to which under most circumstances we need desperately to cling. Unfortunately, limited time and space do not allow me the luxury of entering into the niceties of those Latin American narrative texts that outline the nature of the situation—of Latin American pluralism—about which those narratives owe their very existence. But, then, I have no intention of dishing out a hash of stories, anecdotes, and cute examples in the hopes of explaining what the relationship between simplicity and complexity—or simplicity and chaos theory for that matter—is all about. Popularizations and vulgarizations stand or fall by and large according to the rhetorical aplomb of their creators, and I really have no special talent in that regard. What I have attempted throughout this inquiry, and what I will continue attempting, is a mapping of lines of interrelation between radically distinct disciplines and, in the broadest sense, between distinct cultural and subcultural and academic and nonacademic practices, all of which bear on the notions of simplicity and complexity, order and disorder, the harmony of the spheres and chaos.

I fear I continue to digress inordinately. And I suppose I digress because, quite frankly, regarding the topic that has been thrown into the spotlight, I cannot simply say what is on my mind, for it is not a matter of simplicity at all but complexity of the most complex sort. Well, I will try narrowing my focus by another radical switch, from a collectivity expressed through a plurality of individuals of the repeating island sort to individuality as a microcosm of some undefined and undefinable collectivity. I will do so by lending an ear to that supreme deconstructor of consciousness, Samuel Beckett. Perhaps he holds a necessary key to the puzzle before us.

Differences Are By and Large What We Make of Them

It has been said that Beckett's fiction is somewhat of a jazz performance: self-organizing (Federman 1965, 23). Self-organizing, yes, but solely with respect to the parts, for the very idea of self-organization is strained when considering the whole of Beckett's work.

As the implications of this whole approach infinity, meaning approaches zero. Nevertheless, whether focusing on infinity (chaos) or zero (chaos), basically the same inheres: either in the absolute of absolutes or the absolute absence of absolutes, assuming either one of them could be achieved, there would be no meaning at all. Here, in a nutshell, we have the strangely inverted link between the collective, Caribbean image of order out of chaos and the solipsistic, self-conscious image evinced by Beckett's individuals. What is individual, "in here," is a victim of the entropic push toward stasis and meaninglessness, yet Beckett's characters cannot remain silent nor can their actions cease, so they continue to talk and to walk, even though there is no rhyme or reason for their so continuing. What is collective, "out there," is a dynamo of restless activity that is at every moment threatened by entropic disorder and meaninglessness, yet there persists, strangely enough, a certain permanence, a resistance to change. But let's get on with our story about Beckett's supreme antihero: the Unnamable.

The Unnamable's desire is to "speak and yet say nothing, really nothing." At the same time, "it seems impossible to speak and yet say nothing" (Beckett 1955, 303). The Unnamable speaks, with brave trepidation, with grave and vacillating uncertainty, of things that "are no longer, or are not yet, or never were, or never will be, or if they were, if they are, if they will be, were not here, will not be here, but elsewhere" (301). He hopes it will all soon end, but at the same time he is afraid to go on, fearing the unknown, the unordered, chaos. Most of Beckett's characters, like the Unnamable, simply can't get the necessary words out of their system. They feel something in their bones, though they know not rightly what it is and can hardly even begin to say it. They claim they have somewhere to go, something to say, but all roads lead to chaos. Still, they "have to go on, as if there was something to be done, something begun, somewhere to go" (335). What they can do is occasionally speak or write with a few fits and jerks, much repetition, many lapses of silence, and an inordinate amount of hesitation. There is perpetual oscillation between the eithers and the ors, and as a consequence, there is frequent stuttering. The venerable principles of classical logic have obviously lost their power. This incessant oscillation patterns, I would suggest, the incessant action at the heart of figures 4, 5*a* and 5*b*, 6, 9, 11, 12, and 13, which also, by the way, on a massively complex scale patterns Benítez-Rojo's repeating island. In the sense of Deleuze and Guattari (1983), this oscillation is also a paradoxical fusion of inclusive disjunction, of conjunctive discontinuity, of the principles of noncontradiction and excluded-middle dispersed. In another way of putting the matter, Beckett's particularistic one-*thing* (individual consciousness) centrifugally spreads out vaguely to include every-*thing* (all consciousness): one becomes many. In contrast, the Caribbean Sturm und Drang as a whole, as every-*thing* in terms of a particular geographical entity, centripetulates, collapsing into one-*thing* of the most general sort: many become one.

Throughout Beckett's narrative, things and their opposites are in near simultaneity embraced and paradoxically affirmed in their move toward generality. At the same time there is a certain longing for the "nothing" between extremes: "Nothing is more real than nothing" (1955, 165). Like the Argand plane of figure 12 or $\sqrt{-1}$ of figure 5*a*, it is as if there were a dynamic struggle between the pluses and the minuses, the yeses and the nos, and blacks and whites (in the sense of Goethe). It entails a struggle evoking the image of Jean-Jacques Lecercle's (1990) "violence of language" in its most vicious incarnation: "Words and images run riot in my head, pursuing, flying, clashing, merging, endlessly. But beyond this tumult there is a great calm, and a great indifference, never really to be troubled by anything again" (Beckett 1955, 198). From absolutes at the extremes, toward the absolute absence of absolutes—or 0—between the infinite stretch above and the infinite stretch below: that is the site of the quivering, scintillating, undulating, vibrating rhythm giving rise to life's processes. In Benítez-Rojo's Caribbean, the inverse inheres. Rather than a move toward no-*thing* in order to embrace every-*thing,* there is a natural gravitation toward every-*thing,* which cancels itself out, leaving not no-*thing* but virtually every-*thing* regarded as a complex mix of differences (grays) that gives rise to a set of differences, affording substance to life's very processes.

Beckett's early essay on Proust (1965) sets up the contrast between Dante (virtually the absolute) and Joyce (virtually the nonabsolute). It might seem that we have here a virtual reality containing a static plenum, of every-*thing* all at once—the block, figure 17—that can be none other than lifeless, dead, and the absolute absence of any-*thing* at all without there being any possibility of being. But that is not the case. Not at all. For the nothing that is more real than the real itself is the locus of $\sqrt{-1}$, somewhere, somewhen, somehow, without which there would not be any-*thing* at all. But $\sqrt{-1}$'s role is not dialectical, for God's sake, nor is it synthetic in the customary sense, or anything of the sort. It is the foreplay that gives rise to any-*thing* and every-*thing.* In a certain manner of speaking, it is Parmenides against Heraclitus, but certainly not classical logic against Hegel (even if we discard the absolute). As an outgrowth of the tension between every-*thing* and no-*thing,* it is vaguely patterned in the words of Søren Kierkegaard (1944, 78) that "the genuine subjective existing thinker is always as negative as he is positive, and vice versa. . . . He is conscious of the negativity of the infinite in existence, and he constantly keeps the wound of the negative open, which in the bodily realm is sometimes the condition for a cure. The others let the wound heal over and become positive; that is to say, they are deceived." The positive is the body and the wound is its own negativity, the absence of part of what the body would have been were it whole. Something is always missing in the equation; there is never any completeness of all things, there is always the wound, testifying to some vague some-*thing* which is apparently missing. And to believe in the realization of a cure is to be deceived; it is falsely to believe that wholeness, the plenitude of things, is up for auction and available to the highest bidder: "[I]s that the possible thing at last, the extinction of this black nothing and its impossible shades, the end of the force of making and the silencing of silence, it wanders, that voice which is silence, it and me, it and him" (Beckett 1967, 139). In the ultimate absence of the black no-*thing* and the white every-*thing*—or is it vice versa?—and in the absence of any and all gray shades in between, there is nothing but simply nothingness, emptiness: Beckett's impossible goal at the end of the infinite, Zenoesque, asymptotic stretch that can hopefully be approximated but never reached.

What we also have is a strange concoction, a mishmash, of the comic and the tragic, the ironic and the pathetic. Or better, there is neither exactly the one nor the other but always something else that was always already residing there, in the included middle. The yes and the no, optimism and pessimism, hope and despair, Beckett's "I can't go on, I'll go on" of affirmation and negation bring with them the vague suggestion that there is some-*thing* in between the two horns of every dichotomy. Indeed, Beckett's axiom, in the final analysis, is *Let everything cancel itself out.* But there is never really the grand absolute nor absolute absence, neither monolithic being nor nonbeing. There is only the quite modest $\sqrt{-1}$. Oscillation between the +1 and the –1; between what is and what is not: that is the question to which there is no answer, yet through some, many, countless affirmations and negations, there might, hopefully, be some vague approximation toward some nebulous response, whenever and wherever that might be. I repeat: in Beckett, as the number of words approaches infinite and silence becomes virtually nonexistent, meaning approaches zero. This leaves nothing but the tension. It is the tension between differences that make an insurmountable difference and ultimately a move, as the distinctions between the +1 and the –1 diminish, toward *indifference.* The approach toward indifference is an approach toward that point where any and all distinctions threaten to collapse into oneness (and in the process they fake a pass and run around right end for a touchdown in the face of the noncontradiction principle). On the other hand, it is a matter of differences, differences that become so slight that they are rendered indiscernible (while in the process blitzing the excluded-middle principle's offense for a touchback).

Beckett's ultimate goal of the virtual sameness of things, the apparent lack of differences that make a difference, like a pile of sand, dust, ashes, or millet seeds, is the ineffability of what is, for here, there are no clear-cut differences about which logically cogent predications can be forthcoming. This image bears on the so-called sorites paradox: plucking a hair from the hirsute man doesn't make him bald, ergo plucking the $n + 1$ hair—whichever it may be and wherever it is—should not render him bald either. The sorites conundrum operates in Beckett's "little heaps." It is, "though more unutterable, like the crumbling away of two little heaps . . . of unequal size, but diminishing together as it were in ration, if that means anything, and leaving behind them, each in its own stead, the blessedness of absence" (Beckett 1955, 222). The dustlike particles keep trickling away, and from one second to another things apparently remain the same, yet everything changes, somehow. Between the eithers and the ors there is an infinity of infinitesimals, yet the eithers and the ors remain separated by a yawning chasm, and words teeter about somewhere in between. As millet seeds are removed, one by one, the heap tends toward nothingness and meaninglessness, but it never gets there, no matter how many seeds lose themselves from the mass. This is the inverse of Benítez-Rojo's Caribbeans, whose massive effervescence tends toward absolute flux, pure disorder, noise, yet there is always some undescribable, overriding meaning to it all.

The to-and-fro shilly-shallying of Beckett's characters between distinctions coupled with tenuous, tentative wades through the muck of virtual indiscernibles is not only evident in their language; it is manifested in their actions as well. From Murphy to Molloy their comings and goings are robotized, yet they are asymmetrical; they appear to be mechanical, yet they entail radical uncertainty. As a consequence, these hapless souls cannot help but continue on, ambulating with a jerky, asymmetrical gait. In *Watt* (1953), Knott's method for putting on his shoes, moving about the room, or relating to the objects he finds in his environment affords a picture of continuously affirmative yet disjunctive, and of disparate yet conjunctive, action. Now the yeses now have it over the nos, now they don't; between any two entities there is now a difference, now there isn't. All is in the fluxes and flows of transition. In spite of the contrived appearance of a universe governed by mechanics, life's processes go on. Watt himself is even more caricaturesque. His elaborate formula for walking consists of a "funambulistic stagger" that reminds one of early computer simulations of human movement:

> Watt's way of advancing due east . . . was to turn his bust as far as possible towards the north and at the same time to fling out his right leg as far as possible towards the south, and then to turn his bust as far as possible towards the south and at the same time to fling out

> his left leg as far as possible towards the north, and then again to turn his bust as far as possible towards the north and to fling out his right leg as far as possible towards the south, and then again to turn his bust as far as possible towards the south and to fling out his left leg as far as possible towards the north, and so on, over and over again, many many times, until he reached his destination, and could sit down. (Beckett 1953, 30)

Watt's twitching, lugging, and snapping, in conjunction with his rolling and tossing movements, combine physical form and fashion and their linguistic expression and content into a strange whole that tends somehow to balance itself out. Yet, when considered within time, his movements are constantly out of kilter, off-balance, asymmetrical.

In Molloy's case, because he has one infirm and inflexible leg, the very act of walking, or of riding a bicycle—his preferred mode of travel—is a perpetually dissonant act. And significantly, it is a purely physical, tacitly engineered maneuver: "It is difficult to think riding, for me. When I try and think riding I lose my balance and fall" (Beckett 1955, 26). In other words, the body has its mode of action of which the mind is unaware. Consequently it remains aloof and always a step behind. But that is not quite what I wanted to write, for actually, out of the pencil of Malone we read that "my body does not yet make up its mind" (198). As if the body had a mind of its own; and it does, it is the automatic pilot of the mind, mindless activity a portion of which can be made mindful at the drop of a hat, but the majority of which—for example, lung heaves, heartbeats, gut peristalses, and such—is destined to remain mindless to all minds save those that have been adequately prepared—mystics, Deleuze and Guattari's schizophrenics, and such. The same mindlessness applies to speaking: "[H]ow can you think and speak at the same time, how can you think about what you have said, may say, are saying, and at the same time go on with the last-mentioned, you think about any old thing, you say any old thing, more or less, more or less, in a daze of baseless unanswerable self-reproach" (374). And: "Where do these words come from that pour out of my mouth, and what do they mean, . . . and one listens, that stet, without reason, as one has always listened because one day listening began, because it cannot stop" (370).[4]

So what, actually, does all Beckett's characters' incessant palaver mean? No, I apologize. That is the wrong way of putting it. What do

Beckett's signs do? Yes, that's better, I think, or at least it is probably as good as I can do. Well, Beckett's characters' words very tentatively approach the infinitely receding point of doing every-*thing,* which is also to say that they approach the infinitely receding point of doing no-*thing.*[5] This is comparable to the distinction between abstract expressionism (of, say, Pollock) and abstract reductionism and even "minimalism" of the most simple sort (of, say, Kandinsky). The first threatens to fill the canvas with a virtually undifferentiated every-*thing,* which becomes ultimately the equivalent of no-*thing;* the second threatens to represent some-*thing* with hardly any-*thing* at all, say, a solitary image in an absolutely abstract code. It is a sign of every-*thing,* but since it thwarts all freedom and all possibilities, it is a sign signifying virtually no-*thing* at all. At the one extreme we have white and black oscillated in figure 6 by the $\sqrt{-1}$ to yield Pollock in a delirious dance toward the lower left-hand corner, and Kandinsky ever so carefully walking on eggs toward the upper right-hand corner. But neither artist goes very far: each in his own way remains close to his origins, at $\sqrt{-1}$, the origins of life, wherever that happened, whenever it was. Both are at a certain reach from the center, which is everywhere and nowhere toward the receding outer boundary, which is nowhere and everywhere. They extend, becoming a well-nigh random mix of squiggles and colors on the one hand and a lonely line only slightly and very tenderly and tentatively tinged with color on the other hand. Then they retract. That's all. They cannot go any further, or they would risk all; all could possibly be lost; or nothing could be lost, for it was all really a matter of no-*thing* in the first place, that is, of every-*thing,* every-*thing* and no-*thing.*

In complementary fashion, on the one hand, Beckett's characters also ever so tentatively stretch their limbs, minds, and signs out a little, test the environment, and, like the pod of an amoeba, retract them, but perhaps not all the way. On the other hand, gradually limbs, minds, and signs are extended in all directions simultaneously, becoming intricately tangled in the process, forging *n*-dimensional lines of interconnectivity. In short, they describe a syncopated, asymmetrical, vacillating gait of ambulation, cogitation, and articulation into unknown territory: their universe is, properly speaking, rhizomic, not arborescent. But their universe is rhizomic in a manner complementary with Benítez-Rojo's Antillian rhizomicity. Beckett's characters feel *it*—whatever *that* is—but can't get *it* out of their system. So they stutter and stammer, caught in a bind of discontinuities and incessantly in the process of starting. Starting? Starting what? Who knows what, or where for that matter? The Antillians ebullient counterpoints to Beckett's indecisive characters also feel *it* in their soul—however one wishes to define *that*—and they liberally dance around it with a perfusive, profusive outpouring of words, gestures, art, and literature, manifested through continuous, rhythmic, fluid activity as if by their so doing *it* might be converted into perpetual motion. In either case, as a consequence, to reiterate the well-worn theme, contradictions are engulfed in the tide, and excluded middles, now capsized, will hardly be able to evince even so much as a vestige of their past glory.

For Beckett's characters no-*thing* is more real than no-*thing;* for the Caribbean mind every-*thing* stands a chance of becoming every bit as real as virtually any-*thing* else. For the former, "little heaps," of sorites vintage, highlight the sameness of things, until finally, indifference brings peace to the curious seeker who finally has become incurious.[6] For the latter, differences are proliferously highlighted; they tend toward mergence into one another, thus yielding a continuous, hybrid whole. For the former, disequilibrium arises from out-of-balance existence; for the latter, disequilibrium is the product of differences accumulating to the extent that every-*thing* becomes a bloated whole whose tension intermittently erupts in sharp bifurcations, violent dissipations, new disorderly orders. For the former, the either-or fails to exercise its force, since between any pair of differences another less distinguishable difference can emerge, often at the least expected and perhaps most undesirable of occasions: the latter, in contrast, pays diminished respect to oppositions, for every-*thing* possible is always already there, awaiting its being drawn into the light of day at any moment.[7]

Upon speaking of such complexities, we find the rhizome asking for another moment in the spotlight.

Orderly Chaos, Chaotic Order

A rhizome, with which we are now quite familiar, consists of a centerless, wandering, chaotic (or perhaps even virtually random) network. It is the antithesis of arborescent structures the likes of Linnaean taxonomies, Porphyrian trees, and Chomskyan $S = NP + VP$ diagrams. It is even outside the parameters of the notion of neural nets, dear to many current artificial intelligence gurus (see Pagels 1988).

In Deleuze and Guattari's words:

> The rhizome is reducible neither to the One nor the multiple. . . . It is not a multiple derived from the One, or to which the One is added ($n + 1$). It is composed not of units but of dimensions, or rather, directions in motion. It has neither beginning nor end, but always a middle . . . from which it grows and which it overspills. It constitutes linear multiplicities with n dimensions having neither subject nor object . . . and from which the One is always subtracted ($n - 1$). . . . Unlike a structure, which is defined by a set of points and positions, with binary relations between points and biunivocal relationships between the positions, the rhizome is made only of lines: lines of segmentarity and stratification as its dimensions. . . . The rhizome operates by variation, expansion, conquest, capture, offshoots. . . . The rhizome is a centered, non-hierarchical, non-signifying system. (Deleuze and Guattari 1987, 21)

Rhizomes ceaselessly string out series of semiotic chains, entwining them into a tangle of orders, organizations, struggles, and power plays. They are tubers agglomerating diversity into disordered order, into chaotic harmony, organized complexity. Regarding languages, both natural and artificial, they are not adequately meaningful if devoid of mimetic and gestural, and auditory, gustatory, tactile, olfactory, and visual, in short, all visceral, forms of signification. There "is no language in itself, nor are there any linguistic universals, only a throng of dialects, patois, slangs, and specialized languages" (Deleuze and Guattari 1987, 7). All sorts of sign activity are related to one another; there are no static or privileged codes or rules or strategies; everything is in a lovely embrace with everything else: rhizome.

Arborescent and rhizomic lines present a study in contrasts.[8] The first are molar and deterministic in terms of statistical averages, the second are molecular and indeterminate. The first are Cartesian—dimensions come in whole integers—the second are fractaloid—dimensions are "irrational," always more than one integer and less than the next one. The first are striated; the second smooth. The first consist of classes; the second of packs or masses. The first are precise; the second vague, ambiguous, fuzzy. The first give illusions of completeness; the second freely admit to incompleteness and more often than not to inconsistency. The first are of relative order; the second of disorder, even anarchy. The first are "paranoid"; movement is toward a stable state, a fixed point attractor being the ideal. The second are "schizoid"—nomadic, deterritorialized, consisting of aleatoric shifts and perpetual disequilibria, random ambulation—tending toward a strange attractor. In a manner of putting it, the rhizome is the perpetual *becoming* of *being,* never *being* having *become.*

In short, a rhizome is characterized by the following: (1) interconnections bringing indeterminately many and indeterminately diverse fragments together; (2) heterogeneity; the interconnections link different domains, levels, dimensions, functions, and forms; (3) discontinuities, breaks, fracturation, all of which are necessary for the multiply connected lines and pathways; (4) multiplicity of processes and products—which never actually are but are always in the process of becoming—and (5) cartography; rhizomic activity entails construction of two-dimensional maps of three-dimensional terrain, not one-dimensional tracing, with the acknowledgment that the map never stands a chance of becoming coterminous with the territory (recall Sagredo's words in this respect) (Deleuze and Guattari 1987, 7–25).[9]

Tree structures are monolithic, hierarchical, and stratified, with limited and rigidly controlled connections between components (Simplicio's initially naive idea of the real). In contrast, rhizomes are nonhierarchical: horizontally interconnected multiplicities that cannot be reduced to any totalizing pattern or set of algorithms. The molar level may be described by the equivalent of Boyle's laws of gases (Sagredo's alternative to Simplicio's real). As such they are quite determinate and predictable not in strict mechanical terms but as statistical aggregates governed by probability functions. At the molecular level, on the other hand, to all appearances, disorder rules. Caroming gas molecules rush about in what would appear to the human observer, if his or her focus could be magnified sufficiently, as virtually random behavior.

But let me offer a more practical example. Observe the coming and going of an individual bee within a swarm. Its trajectory appears aleatory, directionless. But ask the bee if it enjoys free agency and it will laugh at the question.[10] Of course it does! How could things be otherwise? Naturally, we would imperiously look down our noses at such brazen innocence, and yet we must concede that the bee's apparently aleatoric, Brownian-like movement is from the broadest possible view in a sort of chaotic accord with that of all other members of bee society. So, it would seem, as in the preceding example of the termites, that there must be some purpose, some direction; in fact, regarding the whole collection of bees in the community, there must be some sort of free enterprise in order that it attain maximum survival value. Yet the behavior of an isolated bee appears directionless, schizophrenic, though in reality we know it is marching to the step of and singing its modest song quite in harmony to the music of the big band of the hive, quite oblivious, we would suppose, to the nature of that music.[11]

Actually, as Freud often noted, schizophrenics tend to take words as things and things as words, all at surface value. Or as Gregory Bateson would have it, schizophrenics are prone to take all words, including metaphors, literally: thus in the dead of winter the expression "Go jump in the lake" might be taken as a command and land a mental patient in bed with pneumonia. Or even better still, Deleuze and Guattari's schizophrenic heroes take language in the process of becoming as the becoming of things in terms of series of events, and vice versa. Deleuze and Guattari present an elaborate argument to the effect that they—the schizophrenics—and their words and their world consist of three syntheses: (1) *connective syntheses* that establish single series of word-events and thing-events—those selfsame schizophrenics included; (2) *disjunctive syntheses* that bring about divergence of the lines of a set of series with respect to other lines (disruption, disharmony, disequilibrium, imbalance); and (3) *conjunctive syntheses* that put two or more series in resonance (mutual entrainment, syncretism).[12]

After an initial glance, it would seem that Beckett's characters' consciousness, beginning with a linear flow, will end up either collapsed into virtual nothingness or spread out into a virtually chaotic nonlinear soup of affirmations and negations. Benítez-Rojo's motley collection of humanity and geographical and climatic catastrophes takes the inverse route, which actually ends up being the same. In the beginning when and where there could have been no beginning, a virtual chaos of nonlinear pluses and minuses erupted, but as the unruly opposites learn to cohabit and become quite congenial bed partners, the resultant buzz, the scintillating, effervescent moiré effect evinced, is a process of every-*thing* canceling every-*thing* else out. And what remains? Virtually no-*thing,* that is, every-*thing* and no-*thing,* noise and silence, presence and absence, outside and inside. The inner monologue of Beckett's antiheroes and the carnivalesque dance of Benítez-Rojo's Antilleans: contradictions embraced and excluded middles opened and partly filled—rhizomes, linearity, disgorging into nonlinearity, connective syntheses giving way to disjunctive and conjunctive syntheses.

Dissemination of Seminal Signs[13]

Connective synthesis, establishing a single series, a linear coupling, is the production of production (signs becoming signs); it is pure succession, the mutual embrace of *this thing* and *that, . . .* and *that, . . .* and *that, . . .* with no conceivable end in sight. The principle of noncontradiction becomes dissolved in the gush of becoming from the sphere of *possibilia*. That is, connective synthesis is ongoing actualization devoid of clear-cut distinctions having yet to be exercised (grayness in fig. 5*a* engendering, through $\sqrt{-1}$, an elaborate Pollock paint dripping) and without classes, distinctions, and taxonomies having yet exercised their authoritarian force. It is schizogenetic hog heaven, giving rise to the possible emergence of unlimited word salads, conflicting images, illogical connections.

Connective synthesis is comparable to a fixed point attractor: there is variation, but it is only minimal. This is the simplest sort of attractor. It is a pendulum at peace with itself and moving in almost-perfect back-and-forth motion, and gradually, due to friction, it comes to rest at the greatest entropy state, the fixed point. If the pendulum's path were plotted on a graph in phase space, we would see the pendulum first describe a circle that gradually spirals inward until it stops at the center—much like Klee's "coiled cosmos" in figure 8. It is the back-and-forth of someone swinging lazily in a porch swing on a humid August afternoon, with a cold lemonade at arm's reach. The time clock is punched out, and things are winding down in the direction of a standstill: the height of contentment.

Disjunctive synthesis is a matter of either-or choices. At the outset this seems simple enough. But actually it isn't, not by a long shot. Disjunctive synthesis is a differentiating function, to be sure, but it is by no means limited in the standard logical sense of either-or. It is inclusive rather than exclusive, open rather than closed and restrictive. It remains disjunctive, yet it is still capable of affirming the disjoined terms without prioritizing one over the other, without restricting one by the other or excluding the other from the one. It is an elusive: "Either . . . or . . . or . . . or . . . *n*," instead of the closed and closet set, the simple binary either-or. This is comparable to the second kind of attractor, the limit cycle. Here, instead of the pendulum coming to rest, it is as if it were in a cradle gently being rocked back and forth in such a way that the bob of the pendulum goes around and around along a closed loop. In other words, some of the physical variables are the victims of constant periodic motion. A limit cycle attractor, the bare beginning of nonlinearity, is also, as mentioned previously, found in heartbeats. The beats maintain regularity, with some variation, until the heart is under shock or stress, when the limit cycle is disrupted and the system becomes more radically nonlinear. Normalization, in such a case, can be reestablished only after a certain lapse of time. If regularity is not forthcoming, the heart may be shoved into the equivalent of a fixed cycle such that it eventually stops and its owner dies.

Disjunctive synthesis may also appear to be comparable to Beckett's "everything divides into itself." There is neither one thing nor the other but a sort of Kierkegaardian oscillation between the either and the or in such a manner that the two horns of the opposition are not confined but open out to each other and, so to speak, dialogue with each other. The tangential moves implied by the "either . . . or . . . or . . ." serve to construct, in the long run, difference rather than opposition, and, though that black night, that awful threatening void, still rests in wait between one difference and another difference that makes a difference, it no longer carries the same threat, for some gaps have been filled and others are still in the process of being filled. If connective synthesis highlights differences between terms within a series, disjunctive synthesis gives confirmation and affirmation of differences by differentiating them into divergent, tangential series. The ultimate goal of disjunctive synthesis might appear to be that, at the infinite stretch, at the receding horizon out there above and beyond the limitations of sight, there is a melting of all differences into a continuous whole, where Nietzsche (1969, 347) could declare: *"[E]very name in history is I."* At this point, someone can be someone only at the expense of being everyone, like Borges's Shakespeare—or God—in "Everything and Nothing," which appears in *Labyrinths, Selected Stories and Other Writings* (Borges 1962, 248–49), like Beckett's characters in the final book of his trilogy that fuse into and become confused with other characters. But that goal, of course, is ultimately for some infinite and immortal being, Laplace's Superobserver, Maxwell's hypernimble Demon, or a god by whatever stripe: at any rate, that dreamy sphere of existence is the dwelling place of the gods, mythical beings, and mystics. It is not for us common folk.

Nevertheless, conjunctive synthesis may allow us at least a gentle lift of the veil to catch a peek of the whole of things. It evokes the moment of the becoming of the being of an entity's being recognized. Setting two or more series in resonance, it is the production of recording processes (the growth of signs) allowing for a proliferation of distinctions—or better, differences—until there is hardly any difference that makes a difference. It is either *this thing,* then possibly *that,* . . . and then possibly *that,* . . . and then possibly *that,* . . . and on and on. Excluded middles are of little consequence in the agonistic arena of conjunctive synthesis, since between any two entities another one can always wedge its way in for the next portion of the dance (in other words, it is tantamount to $\sqrt{-1}$, or the equivalent of Escher in fig. 5*a* in a process of ever greater differentiation).

Conjunctive synthesis, this act of radical inclusion rather than exclusion—unlike the limited exclusion of disjunctive synthesis—this embrace of the then of time, also potentially marks the entry of awareness of something apparently new, when one suddenly exclaims: "Ah, so that's what it is, it's ———." Something of seemingly novel countenance comes into view. The Harlequin becomes a harlequin, the "green" sensation becomes an "emerald," the series "2, 4, 6, 8, . . ." is recognized as the product of the rule "Add 2," and so on (Deleuze and Guattari 1983, 3–4, 84–89). In other words, from the virtual chaos, the infinity complexity, the strange attractor, where every-*thing* is given equal weight and there are hardly any differences, order suddenly appears, and things seem so simple. In this manner, conjunctive synthesis creates legitimate differences and differences between differences, unlike the gap filling of disjunctive synthesis with its relatively limited strictures. Disjunctive synthesis draws attention to the tangential divergences between series of terms, divergences that become smaller and smaller as the series moves on, but conjunctive synthesis—in concert with disjunctive synthesis—serves to create new differences between the differences between the series. Deleuze and Guattari state at this point that Bateson's threat of the double bind wanes. This is OK as far as it goes, I suppose. But actually, I would submit, there is more to the matter than that: conjunctive synthesis is not merely a question of overcoming sets and classes through Bateson's Bertrand Russellian avoidance of a confusion of logical types, but in addition, it entails forgetting the menace of contradictions and the appearance of something in the middle of terms altogether. It is a question of filling in the gaps—potentially, filling in the infinity of gaps, which is an impossible task, of course, but in the final analysis the fun of the run is in the running, not the winning.

Borges's "Zahir" is somewhat of an abstract depiction of Deleuze and Guattari's paranoid mind. Like the narrator's linear grasp of Zahir instantiations, the paranoiac's vision is linear, stringently tunneled. There is for him or her only the either and the or of standard logic, though from the timeless whole of all possibilities a potential infinity of options is always open. The "Aleph," in contrast, is Deleuze and Guattari's schizophrenic salad compacted virtually to a point, from which there must somehow be some sort of spinout into the concrete living and breathing world. William, of Eco's *The Name of the Rose* (1980), is at times the paranoid linear, single-minded reasoner; but at other times he becomes schizophrenic, entertaining any and all options, keeping the field open for the actualization of whatever, whenever. The schizophrenic, potential master of disjunctive and conjunctive syntheses (unlike the paranoiac, who is limited basically to connective and perhaps a bit of conjunctive synthesis) is part of the crowd and at the same time apart from it. He or she can't simply say "I am *this*" or "I am *that,*" for the connective plus the conjunctive plus the disjunctive is everything, is nothing, is now not what it was, is what it is, is becoming other than what it is. The schizophrenic moves from the logic of identity and contradiction to that of difference and vagueness and ambiguity, from either *this* or *that* to neither *this* nor *that* but some-*thing* else, some-*thing* just now coming into view. It is a form of "logic" within the domain of which a table support has no problem considering itself a leg. Richard Nixon can quite easily become a goat, the Sun can displace the Earth as the center of the universe, and solid, impenetrable, indivisible atoms can become largely vacuous. Within the logic of dissymmetry, disorder, and disequilibrium, virtually any and all impossibles can stand a fighting chance of becoming real.

Schizophrenia and paranoia combined are negative affirmation and positive negation, with *aporia* (emptiness) lurking about somewhere: Beckettian and Derridean undecidability, uncertainty, vacillation; the dissolution and disintegration of subjectivity, the end of the Unnamable. There is hardly any "I," any nominative, accusative, or verb. There are negation and rejection, to be sure, but at the same time there is affirmation. It is the schizoid subject of oscillation threatened by the one-dimensional, linear, single-mindedness of a paranoid world. It is suspension of the Cartesian subject without falling into the trap of Berkeleyan solipsism, of the painful coming of awareness of the Unnamable without the possibility of ever quite getting there. This implies more than the enlightenment of Plato's cave dwellers from their two-dimensional awareness to existence in the Euclidean three-dimensional world. It somehow implies the need to break loose from three-dimensional classical ontological and epistemological existence and enter into four-dimensional unbounded but finite space-time. Such an act would require the orthogonal perspective, via $\sqrt{-1}$, from within which one becomes hyperconsciousness of one's own self and of one's own consciousness—however, unfortunately, this is forever a delayed, deferred, differentiating hyperconsciousness. It is the unrepresentable and unpresentable: representation and reference remain simply beyond reach. The yield of vague intuition not razor-sharp intellect, this hyperconsciousness can only be sensed, not said, felt, not formulated explicitly. It remains beyond classical space and time and beyond traditional thought and speech. So it must border on the mystical, I would suspect. And there it must remain. "But," one retorts, "are you not trying to say it, to conceive it, which spells the paradox of my own endeavor and hence its inevitable failure?"

Yes, I guess I am. Perplexing, confounding. Perhaps Calvino can give us a better feel for what it is like to be squarely inside that which one is trying to articulate. Then again, perhaps not, but it is worth a try, I would expect.

Strange Wanderings

We experience the range of Deleuze and Guattari's three syntheses in Calvino's *If on a Winter's Night a Traveler* (1981). While watching the ginkgo leaves fall like fine rain from the boughs to dot the lawn with yellow, our protagonist wonders if it would be possible to "distinguish the sensation of each single ginkgo leaf from the sensation of all the others" (199). His Japanese host, Mr. Okeda, answers in the affirmative. The sensation of a single leaf is multiplied in the descent of two twirling leaves, and so also with three, four, and on and on, to create

> a general sensation like that of a silent rain, and—if the slightest breath of wind slows their descent—that of wings suspended in the air, and then that of a scattering of little luminous spots, when you lower your gaze to the lawn. Now, without losing anything of these pleasant general sensations, I would like to maintain distinct, not confusing it with the others, the individual image of each leaf from the moment it enters the visual field, and follow it in its aerial dance until it comes to rest on the blades of grass. (Calvino 1981, 199–200)

The protagonist's project meets the approval of Mr. Okeda, who adds that perhaps he could keep the sensation of every lobe of each leaf distinct, and if so, upon focusing attention on a leaf (≈ all leaves), which includes attention on each lobe (≈ all lobes), he would know all the leaves and the collection of their parts down to the minutest possible detail. Their summation would yield the whole. He would know every-*thing*. He would be master. The garden would be his! His mind-set at this ideal stage of awareness would be a combination of connective synthesis (one series, the trajectory of one leaf), disjunctive synthesis (everything in each series, all the eithers and the ors, the whole movement of the leaves), and conjunctive synthesis (two resonant series, and those series resonating with other series, or the trajectory of many leaves).

However, things are not so simple, for this final state would be tantamount to the neither-nor as described in the preceding. How so? In the first place, a solitary gyrating leaf descending from limb to lawn is for practical purposes every bit as unpredictable in its immediate vicinity as the butterfly of chaos theory. This unpredictability, this uncertainty, is patterned in another episode boxed within the conversation between the protagonist and Mr. Okeda: our hero fails in an attempt to fondle the breast of his host's daughter in a feigned maneuver narrated with jocularity befitting only an actual reading of the account. To put the matter bluntly, he tries to make the movements of his hand toward the enticing breast appear as accidental, as aleatory as those of the leaves. In fact, he is duped by his own calculations of the unpredictable, for the unexpected occurs, and his project is aborted. He is then told by his unpretentious but hyperaware Eastern counterpart, with a sobriety apparently revealing his innocence of ribald Western ways, that not only is each and every leaf of an importance virtually equal to the whole, but more important still is the empty air separating one leaf from another one. The protagonist, now quite solemn, intuits that he is given to understand that "an absence of sensations over a broad part of the perceptive field is the condition necessary for our sensitivity to concentrate locally and temporally, just as in music a basic silence is necessary so that the notes will stand out against it" (Calvino 1981, 202–3). His hypothesis is soon confirmed by the wily Mr. Okeda. But on so confirming it, his host subtly alludes to the overriding importance of tactile sensations in all forms of communication—tactile sensations of the very sort our protagonist had failed to bring to fruition with his daughter!

Mr. Okeda continues unabashed, suggesting that tactile signs are closer to our understanding, at the visceral level, than are visual sensations, and especially linguistic expositions. In order to salvage a vestige of his former dignity, our hero now tries to make an impossible and contradictory comparison between Mr. Okeda's allusions and the act of reading a novel. However, he soon becomes caught up in the abstractions entailed by the purely ocularcentric sensations of arbitrary black marks on the white background of a printed page. His thoughts falter and waver, and their linguistic clothing becomes scant. Then, after this failure, and as if the initial conditions of the situation coupled with his abstract speculations had become nonlinear and virtually chaotic, with all possible paths flying off in multiple directions, he ends up having sex with the daughter's mother! Disjunctive synthesis flowed into chaos and then into a conjunctive synthesis, incompatibles of mind and body came together in a tenuous embrace, mutually exclusive sensory modes merged into an interpenetrating whole. The chaotic, the unpredictable, the unutterable, finally won out.

This train of events serves to recall a novel entitled *In a Network of Lines That Enlace* that the narrator had stumbled upon earlier in *Winter's Night,* affording the notion of interconnectedness, intertextuality, irreversibility, unrepeatability, and the inarticulate. Depicting the whole shebang graphically, "every episode, with its climax, would require a three-dimensional model, or perhaps four-dimensional, or, rather, no model: every experience is unrepeatable. What makes lovemaking and reading resemble each other most is that within both of them times and spaces open, different from measurable time and space" (Calvino 1981, 160). This is, it would appear, nonlinearity with a vengeance. And it is, most properly, rhizomicity. The attractive image finds support in the narrator's next encounter with narrative texts: *In a Network of Lines That Intersect.* But to make matters apparently worse, the intersecting lines now mirror themselves to infinity. Quite significantly, it all becomes a matter of each part mirroring the whole of things:

> From mirror to mirror—this is what I happen to dream of—the totality of things, the whole, the entire universe, divine wisdom could concentrate their luminous rays into a single mirror. Or perhaps the knowledge of everything is buried in the soul, and a system of mirrors that would multiply my image to infinity and reflect its essence

> in a single image would then reveal to me the soul of the universe, which is hidden in mine. (Calvino 1981, 166).

At this point the narrator of *In a Network of Lines That Intersect* has been abducted, and his kidnappers are in a race for their lives in an armored car. A trap is set for them somewhere in the city by the police. The map of the city—a mirror—is also like a kaleidoscope, with lines that break up and diverge and converge nonlinearly and yet become segmented at every juncture. The narrator is taken to his own house, a house whose walls, floors, and ceilings consist of mirrors that reflect his image an infinite number of times—recall Lönnrot's Triste-le-Roy labyrinth. The thought occurs to him that he has been kidnapped by himself, since he is, after all, a part that reflects the whole. The trouble is that he is lost. Or he seems to have lost himself. He cannot determine what is he and what is his reflection; or are all the images mere reflections? Finally: "It seems to me that everything that surrounds me is a part of me, that I have managed to become the whole, finally" (Calvino 1981, 168).

Of course, that was also Palomar's supreme goal: to become the rhizomic whole, that is, to comprehend the whole, which is to say the same thing. Palomar tries all conceivable methods. He attempts an objective, detailed outside grasp but can catch at best only a minuscule part of the whole. He makes a gallant effort to view every-*thing* inside the whole while remaining inside, but it all appears as emotionless, drab grayness rather than a display of blacks and whites and all hues in between. He tries playing dead as the ultimate act of dissolution into the processual soup, but this will not do either: it is now not the universe *with* him either inside or outside, but the universe *without* him entirely. Other comparable situations abound in works by Calvino and Borges. In the former's "Blood, Sea," which appears in *T Zero* (1969, 39–51) when "the blood was in fact the sea, when our present inside was outside and our outside inside," things were simple, apparently, but not really, for they were of infinite complexity: chaos. Borges's lottery in "The Lottery of Babylon," which appears in *Labyrinths* (1962, 30–35) was a matter of chance to the infiniteth degree for the Babylonians, the chance of a library rat in the Library of Babel being able to find a completely intelligible book is computable to near zero. Yet, we learn from the Aleph, Tzinacán's mystical experience, and the Tlönians' subjective universe that perhaps the whole can somehow be intuitively grasped—however ineffable the intuition may be and however solipsistic it may remain for the onlooker.

But in Borges's texts there is always an element of indeterminacy regarding the totalistic image, and there are always allusions to human finitude and frailty. Perhaps these limitations are a matter of misused models and methods, of miffed models and methods, or of the result of misplaced faith in models and methods in the first place, that is, of models and methods that, when couched in language, are of inordinate abstraction and arbitrariness. Perhaps the narrator of *Winter's Night* is on the right track. As is appropriate in a novel that so effectively highlights nonlinguistic sensorial images, the narrator once engaged in a metaphorical "systematic reading" of his lover, Ludmilla's, body:

> [T]hrough channels of tactile information, visual, olfactory, and not without some intervention of the taste buds. Hearing also has its role, alert to your gasps and your trills. It is not only the body that is, in you, the object of reading: the body matters insofar as it is part of a complex of elaborate elements, not all visible and not all present, but manifested in visible and present events: the clouding of your eyes, your laughing, the words you speak, your way of gathering and spreading your hair, your initiatives and your reticences, and all the signs that are on the frontier between you and usage and habits and memory and prehistory and fashion, all codes, all the poor alphabets by which one human being believes at certain moments that he is reading another human being. (Calvino 1981, 155)

This mode of "reading" was actually initiated by the opening scene of Calvino's novel when puffs of steam from a locomotive hide part of the first paragraph. The odor of the train station, the befogged glass door of the bar, everything is mystery, vague. The sights, the sounds, the smells, the sense of touch, the acrid taste all contrive to reduce the sentences of the text to "vagueness, grayness, in a kind of no man's land of experience reduced to the lowest common denominator" (Calvino 1981, 12). All the signs converge into a diffuse, confused whole that fuzzes any and all attempts to describe it. This is, most properly put, an intertwined, interlaced, nonlinear, holistic intertext consisting of signs of all types in shimmering, scintillating interaction. It is nonlinguicentricity at its best.

Linguicentricity focuses obsessively on arborescence and connective synthesis while at least giving disjunctive synthesis and conjunctive synthesis lip service. However, the Unnamable's ineffables, the Antillians' profusion of words that do not say but give some sense of something—of what? It can't be said, it can only be sensed, felt, perhaps—make a tentative and tenuous move toward full-blown conjunctive synthesis. Calvino also goes far in the direction of creating paralinguistic and extralinguistic feelings, sensations, and intuitions through the stringent limitations of the written text. As a matter of fact, Calvino's "para-extra-linguistic" scene in *Winter's Night,* like many others in his works, goes entirely against the grain of Western linguicentrism, which has reached a shrill pitch with the current linguistic turn. The radically nonlinear, rhizomic, organized complexity of nonlinguicentrism is in grand part what makes Calvino unique.

Nonlinguicentricity borders on chaos from which order stands at least an outside chance of emerging; linguicentricity, at its worst in the form of analytic philosophy and structuralism, is obsessed with order and nothing but order. This distinction bears a further look via Deleuze and Guattari.

On the Fringe of Language

Collections of terms of *The cat is on the mat* variety that mesmerized and haltered logicians during the first half of this century neither diverge nor converge in deterritorialized or chaotic fashion with other series of different orders (i.e., other sentences). The series are strictly linear, a product of the drive toward simplicity and an attempt to free language of its inherent complexity: linguicentrism. Philosophers of logical positivist and analytical bent wanted bed partners hailing from classical mechanics and the equivalent of fixed point attractors: Deleuze and Guattari's tunnel-minded paranoid. Perhaps in their most daring expressions they might have thrown a few furtive glances toward notions of the limit cycle attractor sort. But they hardly went any further.

In contrast to the oversimplified and underdetermined sentences dear to analytic philosophers, Lewis Carroll's overdetermined "slithy toves" that "gymble and gyre" are proof in living language of the problematics of logic in regard to natural languages—and this in view of the fact that Carroll himself was one of the foremost pioneers of modern logic. Such violations of the desired parity of language as jabberwocky and everyday puns, metaphors, and other rhetorical figures are quite comfortable within quasi-periodic attractors and perhaps even rather benign strange attractors. Taking the equation a giant step further, Alexius Meinong's nonexistent, impossible "mental objects," the famed "Liar paradox" and a host of other varied and sundry quandaries, are suggestive of strange attractors of the wildest sort (see Stone 1981). We are now once again in the domain of Deleuze and Guattari's nomadic schizophrenic mind, a structured chaos, a nomadic, aleatory distribution of singular points each of which is a possibility for future actualization: wandering nonlinguicentrism. Within this domain, it can hardly be said that there is any meaning, value, or content within the unbounded domain of the nomad, for whatever meaning there may be is transient, values are dependent upon whatever might have been actualized and what could otherwise have been actualized but was not, and content is never really actual but always in the process of its realization.

As we will once again recall, Beckett's characters of his early writings put their bodies into motion in fits and jerks, and they talk and write with vacillation and stuttering. It reminds one of a smooth curve fractured by the calculus into a series of minuscule straight lines approaching infinity in number, all of which veer off at well-nigh unnoticeable angles with respect to one another. Or in a Zenoesque way of putting it, the curve is very roughly described by two straight lines beginning at the endpoints of the curve and connected at a point beyond it to give them in their conjunction a length approximately equal to that of the curve. Then the curve is somewhat less vaguely described by four straight lines whose connecting angle is less, and then eight lines are slapped onto the Cartesian plane, then sixteen, then thirty-two, and so on. The fractured surface successively approximates a continuous curve, but it never quite gets there. For those who never give up their illusions of determinacy, consistency, and the plenitude of all things, the ideal curve is "out there," somewhere, in Platonic heaven, submissively waiting for the proper cogitator. But *our world,* a helpless sludgy world of incomplete fractals and hopeful curves, will never be able to engulf the infinitude of infinitesimal segments, for our paltry capacity to perceive and make *our worlds* always remains short of whatever ideals we may wish to nurture.

The fractured, serrated, striated curve is a fractal approaching smoothness, which is at the ultimate stretch when doing calculus ($\rightarrow \infty$), and that smoothness might appear to have been reached as far as the naked eye is concerned, yet its realization remains in the receding horizon. In this manner, the continuous and the striated might be construed—though somewhat erroneously so, since the continuous cannot be more than an ideal—as a pair of binaries:

> The smooth and the striated are distinguished first of all by an inverse relation between the point and the line (in the case of the striated the line is between two points, while in the smooth, the point is between two lines); and second, by the nature of the line (smooth-directional, open intervals; dimensional-striated, closed intervals). Finally, there is a third difference, concerning the surface or space. In striated space, one closes off a surface and "allocates" it according to determinate intervals, assigned breaks; in the smooth, one "distributes" oneself in an open space, according to frequencies and in the course of one's crossings (*logos* and *nomos*)." (Deleuze and Guattari 1987, 480–81).

Continuous (rhizomic)-striated (arborescent) are thus also related to the wandering, deterritorialized, nonlinguicentric schizophrenic body and mind in contrast to the sedentary and relatively fixed, linguicentric paranoid body and mind. Nomadic, nonlinear wandering is in virtually smooth, continuous curves; sedentary fixity becomes striated, discontinuous. The desert is relatively continuous. Put it into a checkerboard design of irrigated plots and it becomes striated. However, sooner or later the desert wins out: a dust bowl occurs, and the striated is smoothed over. The jungle is relatively striated: from the buzzard's-eye view it is a textured green carpet broken by lazy, serpentine flows. Deforested and fractured into plots, it becomes even more striated; but this striation is artificial, like that of the cultivated desert, and the jungle eventually exercises its force, returning to its less striated countenance, though the violation that it has experienced inevitably will leave scars. However, perfection is not of this world: just as there are never pure blacks and whites, so also *felt,* upon microscopic analysis, sports ups and downs, peaks and valleys, much like the textured surface of the tree-topped jungle seen from a Boeing 35,000 feet in the air. And fibers woven into the yarn making up a fabric, somewhat in the order of Wittgenstein's "family resemblances," form a continuum of strength, though an interwoven continuum remains broken here and there, in addition to the striated breaks of the fabric. There is invariably a degree of smoothness in apparent discontinuity and breaks in would-be continuity. The little paranoiac in us wants order, striation; the little schizophrenic, wandering about somewhere, wants just that, the freedom to wander at will. The dichotomy, then, is no customary binary act at all. It is now free-flow dance, now a grunting wrestling match, now a flurry of karate chops, between complementary pairs that are held together in a whole, which is Benítez-Rojo's collectivity, is Beckett's individual, is all of us.

The curve of calculus, approximating a continuum at the infiniteth reiteration of the algorithm, also reminds one of what Deleuze and Guattari, especially in their work on Kafka (1986), call a "line of flight," an overdetermined phrase implying "diverging line," "point of leakage," "line of least resistance," and above all, it is tantamount to the real or imaginary lines converging to a point in perspective paintings. The line branches out into multiple series and rhizomic (nonlinguicentric) interconnections. It is perpetual deterritorialization. This nomadic behavior imposes no fixed, sedentary boundaries on a plane or territory or on the space of the body, subjectivity, and thought. The nomad occupies the space he or she can occupy to the extent of that individual's capacity until pushed away, or after exhausting that space or out of sheer boredom, he or she moves on to what now appear to be greener pastures. Thus the nomad as deterritorialized is, most properly put, a "writing machine"—not a machine of the eighteenth-century French *l'homme machine* variety but a self-organizing, multiple feedback cybernetic machine, which is autonomous of, yet perpetually opened to, its environment: a dissipative structure par excellence. Nomadic wanderings are a function of the "body without organs."[14] Such wanderings involve the unactualized, the range of *possibilia,* the decoded or not yet encoded, the destratified or not yet striated, the desert (relatively smooth, unstriated) or that which is soon to be converted into a quiltwork of fields, canals, and orchards. They involve deterritorialization, without anything having been actualized and put to utilitarian practices.

The deterritorialized, nomadic writing machine is nowhere more adequately illustrated than in Kafka's "minor literature," which invariably exercises nonlinguicentric moves. Kafka wrote in his diary about Yiddish literature in Warsaw and Czech literature in Prague. Minor literature serves as an outlet for an ethnic group's collective expression and provides a medium for at least partially articulating that which otherwise could not be said: Jews speaking Russian and other languages, African Americans and Native Americans speaking English, Algerians speaking French. It also compels writers to express themselves in the dominant language, the language of an alien culture: Kafka writing in German, Beckett in French, Nabokov in English, and Borges, an Argentine wanting to write in another language—perhaps English or even French—but limited to what he called "mere Spanish." All is "border writing," in the conception of Emily Hicks (1991). In Carlos Fuentes's words (1969), Borges is archetypal of the Latin American forced to use a language of conquest and colonization, a hierarchical, arborescent, binary language unsuited to the expression of minority peoples (Creoles, African Americans, Amerindians, people of mixed racial heritage, various Asian groups). This phenomenon is, once again, perhaps no more evident than in Benítez-Rojo's Caribbean, where multiple languages and multiple cultures are superposed in a limited and generally overpopulated region of oceanic expanses punctuated by occasional islands of disordered human order.[15] It is also inherent in the Unnamable and other Beckett characters. In sum, Deleuze and Guattari's wanderer, like Calvino's narrator of *Winter's Night,* follows his or her nonlinguicentric cues in multiple directions, in contrast to the wanderer's counterpart, the totalitarian tunnel-minded paranoic, who remains linguicentric through and through.

But where is all this taking us? How did we get where we are, wherever we are, and how do we go on from here, for we must go on, I suppose, but to where? It seems that we have the broad swath of a culture-nature landscape from Benítez-Rojo—culture becomes, is becoming, nature and vice versa in a dynamic, agonistic, complexly syncopated dance. We have the hesitating, uncertain, tentative Beckettian microstrokes of the brush into lines that crisscross into a confusing tapestry throughout the unbounded plane portraying inner landscapes. We have inner landscapes becoming a multilevel, multiply interconnected communal landscape. And we have aborted efforts to bring inner landscapes into harmony with outer landscapes with bold slashes by way of Calvino's characters. These landscapes presenting an exceedingly vague picture of interculturality can hardly, it might appear, give us anything of concrete substance we can sink our teeth into. But, then, when complexity grows, one has to give up something: certainty. At least let us try to backpedal for a moment toward the relatively simple—if indeed anything can be simple at all. Then, perhaps, we will be able more appropriately to sense how uncertain things really are.

CHAPTER 9 • Fissures, Fractures, Gaps, and Continuity

First I see the night, which surprises me, to my surprise, I suppose because I want to be surprised, just once more.

—Samuel Beckett

Unless you expect the unexpected you will never find the truth.

—Heraclitus

[I]n my case all painting—and the older I get, the more it becomes so—is an accident. I foresee it and yet I hardly ever carry it out as I foresee it. It transforms itself by the actual paint. I don't in fact know very often what the paint will do, and it does many things which are very much better than I could make it do.

—Francis Bacon

When Things Seem to Be Getting Mushy

Deleuze and Guattari, commenting on fractals, that is, "minor science" as a sort of counterpart to "minor literature"—the science of the margins—point out that arborescence is irremediably discontinuous whereas rhizomicity at least stands a chance of taking on the appearance of continuity.

One would perhaps like to believe that smooth space can be fractured, pace the calculus, and as a consequence, striated space can potentially—or at least ideally—become like smooth space. With this in mind, Deleuze and Guattari (1987, 486) speculate on the possibility that "progress is made by and in striated space, but all becoming occurs in smooth space."[1] In other words, the arrow can move and Achilles can overtake the tortoise in smooth, continuous space but not in space reduced to bits and pieces and actual numbers the likes of which we as finite creatures cannot help but get our hands dirty with. Nevertheless, to the question, Is it possible to give a general mathematical definition of smooth spaces? Deleuze and Guattari (486) respond that "Benoit Mandelbrot's 'fractals' seem to be on that path."

In presenting their argument, they avail themselves of two—actually myriad, but I'll focus on a mere two—metaphors for discontinuity and the becoming of continuity: fabric and felt, mentioned previously in passing. Fabric can be defined in terms of striated space. It consists of two kinds of parallel arrays, in the simplest form, horizontal and vertical, with the strands of one series passing over and under those of the other series and vice versa. This over and under, in the fractal sense, provides for something of slightly more than two dimensions but always less than three dimensions. The weave of fabric must have two sides: the respectable side, evincing order, and the other, unruly side, where the knots of interconnection are found. Felt is an exceedingly more complex slightly more than two-dimensional sheet; it is an "antifabric." There is no weave, only a complex entanglement of microscale fibers, an aggregate that is by no stretch of the imagination homogeneous as is fabric. Felt is, nevertheless, an adequate model of the smoothest possible consistency. Ideally, it is a continuum, completely open and unlimited. It has no center or any necessary boundary in any direction; there is no this side and the other side; there are apparently no limitations.[2]

Felt, approaching the ideal of smooth space, is the desert, the air, the atmosphere. Striated space, in line with the image of fabric, is that same desert broken up into a patchwork of fields, canals, drainage ditches, and utility roads—or it is the Arabian king's starkly simple labyrinth transformed into the Babylonian king's striated maze? Smooth space—albeit imaginary—is that of Mr. Palomar's beach affectionately massaged and rubbed down by the lulling, undulating waves. Striated space is that coastline from a fractaloid view, consisting of virtually an infinity of breaks, jerks, and tangential shifts. Smooth space is haptic, even kinesthetic; striated space is chiefly optical. Smooth space is para-extra-linguistic; striated space is obsessively linguistic (it culminates in those linguicentric practices of which even the deconstructionists are guilty). By far the most effective example of continuity is the sea. It is smooth: try to plow it and it obstinately returns virtually to the same condition. In the beginning there was nothing, pure chaos, pure sameness, continuity. Then there was the sea, which, as far as everyday practical affairs go, is smooth, though not really, at least at a closer look. Then there was land, and felt, that smoothness that can be cut and torn and striated and that striation as an artificially concocted world. And the punctuation continues, with no conceivable end in sight:

> [T]he sea is a smooth space fundamentally open to striation, and the city is the force of striation that reimparts smooth space, puts it back into operation everywhere, on earth, and in the other elements, outside but also inside itself. The smooth spaces arising from the city are not only those of world-wide organization, but also of a counterattack combining the smooth and the holey and turning back against the town: sprawling, temporary shifting shantytowns of nomads and cave dwellers, scrap metal and fabric, patchwork, to which the striations of money, work, or housing are no longer even relevant. (Deleuze and Guattari 1987, 481)

Yes. Cities are also definitely relevant to the problem at hand. Cities are to all appearances the ultimate of striation; the countryside, when relatively uncorrupted by human hands and minds, is relatively smooth. Cities are the focus of communication by linguistic means; in the countryside para-extra-linguistic forms still pervade. Yet the countryside resists closure within that romantic ideal of the *bon sauvage* and such, and cities are never able completely to free themselves of para-extra-linguistic means and methods and of the dirt and other tabooed countryside elements that never cease their infiltration.

The young Marco Polo's account narrated to the aged Kublai Khan in Calvino's *Invisible Cities* (1974) typifies the smooth-striated and felt-fabric complementarities. As we observed in the preceding, Polo's cities are more in the mind (relatively smooth) than in the world (striated), more imagined than physical, more icons than symbols. That is, in addition to gestures they are words, but more importantly, they are words about diagrams, maps, and images. At first Polo was "quite ignorant of the languages" of the cities he visited and of the language of Kublai and could express himself only nonlinguicentrically, "by drawing objects from his baggage . . . and pointing to them with gestures, leaps, cries of wonder or of horror, imitating the bay of the jackal, the hoot of the owl." But communication somehow took place. Kublai was enhanced by the space around Polo's (relatively smooth) nonverbal gestures, "a void not filled with words" (38). It was possible to wonder about the cities in pure thought. Gradually, (striated) words replaced objects and gestures, though "communication between them was less happy than in the past." Still, Polo continued often to find that words failed him, and he "went back to relying on gestures, grimaces, glances" (39). Kublai, gaining confidence in his ability to translate Polo's words and gestures into the imaginary (back toward smoothness), eventually decided he would describe the cities, leaving to his informer the task of deciding whether they were real (striated) or not (relatively smooth). Finally, he (nonlinguicentrically) dreams a city, describes it (linguicentrically, as far as that is possible) to Polo, and instructs him to go out and seek it and "come back and tell me if my dream corresponds to reality" (55).

Here, I hardly need write, we have in essence the idealism of Borges's Tlönians fused with the dream of the magician in "The Circular Ruins," which appears in *Labyrinths, Selected Stories and Other Writings* (1962, 45–50). Or perhaps better, "The Dream of Coleridge," in *Other Inquisitions, 1937–1952* (Borges 1964, 14–17). In this brief essay Borges reveals that Kublai Khan, the thirteenth-century emperor, dreams a palace and builds it; Coleridge, the nineteenth-century poet, dreams of the same palace and then writes a poem about it, unaware that its embodiment in a physical structure was also derived from a dream. This puzzle gives rise, Borges conjectures, to the notion that the series of dreams, poems, and labors has not ended. Perhaps, in fact, the series is endless, or perhaps the last person to dream will have the key. At any rate, whoever might have compared palace to poem "would have seen that they were essentially the same" (17). Back to Calvino's fiction (dream) about Polo's cities, both real and fictive, that become real in Khan's mind, as real as was his palace in the beginning.

Kublai (nonlinguicentrically) constructs in his mind a model space containing a model city from which "all possible cities can be deduced." This city is not that ideal urban geometry somewhere in Platonic heaven. Rather, it consists only of "exceptions, exclusions, incongruities, contradictions," in short, all that deviates from the norm. These deviations allow one, Kublai reasons, to (linguicentrically) calculate any city from "the most probable combinations" (Calvino 1974, 69). If this model city is conceived in phase space as the range of all possibilities, it is, however, entirely unpredictable: it is the product of nonlinear, rhizomic (hypertextual, nonlinguicentric) interrelations, virtually any one or a collection of which can be actualized at a given point in time. It is comparable to the narrator's obsessive and frantic search for what he believes to be the missing remnants of a chimerical novel, bits and pieces of which he has read or thinks he has read—this novel to end all novels, the novel of all novels, much like that self-contained monolithic book in Borges's library that contains all books or much like the self-returning book of the Garden of Forking Paths. It is also comparable to Palomar's initial conception of a model with which to portray the universe, or better of the necessity of models, and models of models, ad infinitum. It is, as well, tantamount to the moment of time containing all moments of time, and therefore it is timeless, in "T Zero." All are in one form or another attempts to smooth the striated, to reach the (nonlinguicentric) limit ($\rightarrow \infty$).

These metaphor-models lead us to a consideration of others conjured up by Deleuze and Guattari. I allude to their microscopic and macroscopic, or molecular and molar, domains, that is, alternative moves from the striated toward the smooth.[3] Crystals are the macroscopic manifestation of microscopic processes. It is all quite simple, relatively speaking that is. Dangle a near-microscopic "seed crystal" in a supersaturated solution by a fine thread, and by the very basic (recursive) act of positive ions attaching to negative ions according to their size and the value of their charge, the crystal will grow outward in all directions simultaneously. In essence the process is linear and sequential, but it spreads out along ordered, nonintersecting, multilinear pathways.

Of course, we heard all this previously. Yet it bears emphasizing that Calvino also knew the story well. In "Crystals," which appears in *T Zero* (1969, 28–38), he begins with a contrast between the narrator's relatively smooth world, "amorphous and crumbling and gummy, which has been our lot," and the striated "cluster of prisms" making up the New York skyline, with its "horizontal and vertical axes that crisscross," "smooth transparent walls," and "symmetrical angles" abiding by the laws of "rotational groups,"[4] all of which give the appearance of existing in a crystal. But appearances are deceptive. Actually, the contrary exists. The (striated) solids making up the symmetrical structures are composed of (smooth) glass and glasslike solids. It's "a paste of haphazard molecules which has invaded and cemented the world, a layer of suddenly chilled lava, stiffened into forms imposed from the outside" (29). The city in which the narrator dwells, rather than order (striation), is no more than a "threadbare patch over disintegration" (smoothness). It is quite unlike the narrator's dreams, in days past, of polyhedrons that grow as tall as mountains, diaphanous, perfectly symmetrical, with all parts in their proper places and fixed for all time. However, the glass factory pouring forth the city's present countenance of balance, harmony, order, symmetry, and, of course, simplicity of the simplest sort, is no more than "simulated order." In fact, it is disorder trying to pass itself off as order; it is entropy in a negentropic disguise; it is smoothness trying to act out the role of striation. Simplicity, yes, it is merely simplicity, but sameness has become the mind-numbing theme rather than construction of an image of a beautiful crystalline phalanx consisting of alternating ionic pluses and minuses held in a tight embrace precisely because of their distinctions—differences. Cities, proud, striated cities, negentropically springing out of the wilderness, eventually succumb to entropic exhaustion (smooth sameness).

But, then, in the final analysis, perfection is not of this world anyway. Perfection is fit to inhabit only St. Augustine's City of God or Plato's heaven. The narrator of "Crystals" must, alas, concede that although everything originally came from crystal, even the music being heard at that very instant, each and every crystal cannot but be at least to a degree "imperfect, flawed, veined with impurities, with rents in the warp of the atoms," all of which ultimately escalates toward the lattice's demise (Calvino 1969, 37). The narrator's companion, less demanding of the world's ways, attempts to convince him that real order cannot but carry "impurity within itself, destruction." And she is correct. He finally comes to realize the wisdom of her ways when he opens the icebox and takes out a painfully irregular cube of ice for their whiskey. The so-called cube evinces nary a vestige of that harmonious hexagonal pattern it should be putting forth to the doughy, flabby, "soft machine"—the protagonist himself—observing it. And he ponders: "The crystal which has succeeded in becoming the world, in making the world transparent to itself, in refracting it into infinite spectral images, is not mine: it is a corroded crystal, stained, mixed" (38).

In view of the preceding observations of crystals, Calvino's image is aesthetically appealing and technically on the mark. Crystals cannot but fall short of the ideal to which they aspire, since they are never absolutely pure but tainted: they contain some flaw, if they are not by nature aperiodic. As we have also observed, an impure or aperiodic crystal is an indication of heightened tension—given its asymmetry—which at a primitive level is how life can be defined in the Schrödinger sense. That is, symmetry breaking, disequilibrium, fluctuation, and above all, tension, results in a sort of groping at the biological level in order to ameliorate that which is flawed or imperfect? (Of course, the task must be ongoing, since crystalline perfection is not of the world of the living—with somewhat of an apology to the nearest virus, since it exists in the liminal zone between crystal and life.) Calvino's crystals, in commensurable fashion, evince the image of amorphous masses that become flawed and that due to these flaws eventually culminate in life processes intent on preventing the infiltration of undue disorder. In other words, life is the product of disequilibrium, asymmetry, imbalance, disharmony, and dissipation of order and simplicity, "tension-bound" characteristics all.

In essence this is the Prigogine view of life, of course, which, with some sputtering and a number of fits and jerks, has been gaining a head of steam over the past decade. This view has it that—as with chaos theory—an infinitesimal flaw can soon throw the entire system that gave it birth out of kilter such that its behavior becomes unpredictable. Contrary to the spirit of Prigogine, some hopeful scientists—much in the sense of Simplicio—continue to blame the impossibility of our being able to specify the conditions of a system on our ignorance. If we could just learn enough about the world, everything would fall into place, our knowledge would be neatly wrapped into a tight package and topped with an attractive ribbon, and life would be fine and dandy. This controversy recapitulates much of the turmoil still surrounding the entropy principle, which lies at the heart of Prigogine's hypothesis. For example, E. T. Jaynes writes that entropy in a subjective sense measures "our *degree of ignorance* as to the true unknown microstate," and that "it is not the physical process that is irreversible, but rather our ability to follow it" (Jaynes 1965, 391). On the other hand, Denbigh and Denbigh (1985, 29) conclude that regarding the entropy principle it is quite irrelevant to let the focus of attention rest on human ignorance, that is, on a subjective interpretation. The fundamental concern of statistical mechanics should be to create a theory whose final equations are directed toward the macroscopic variables of a system—notably energy or temperature, volume, and chemical composition. From this vantage point, the incompleteness of specification is not dictated by what we are capable of knowing. It is dictated only by the limits set for the required statistical mechanical theory, rather than the limitations of human knowledge, per se. The problem revolves around the distinction between subjective and objective interpretations, a distinction that has been quite clearly drawn by Popper (1963, 281–95), among others, and a distinction that is central to much of the controversy surrounding Prigogine.

Subjective interpretations arise out of probabilities as expressions of human ignorance; objective interpretations, in contrast, conceive of probabilities as the outcome of real dispositional properties or propensities of systems—living and nonliving alike—properties that remain independent of the presence of the observer.[5] The former view is customary for macroscopic thermodynamics; the latter, of course, is the Einsteinian view and that of others who want a real world "out there" that will coincide with the expectations of the observer who has finally discovered the key with which to unlock its doors. The view of the present book, and more in line with Prigogine, is somewhat of a middle way: the knower is part of the known, which allows itself partly to be known by the knower it contains within itself, and in the process both knower and known know a bit more, that is, they become more complex, a complexity that carries, within itself, traces of the past such that what went on before sets the stage with a virtually infinite set of nonlinear possibilities in the present for future actualizations. In other words, everything is related to everything else, everything is becoming what it can be, but it never is, since it is always becoming—which is a characterization of inorganic chaotic systems and life processes alike, as suggested in Calvino's "Crystals."

In order hopefully to work our way along this tangle, let us allow ourselves no more than abortive glances at the subject of observation, while focusing on the object. Take, once again, Palomar's observation of a single wave. How can he predict the precise moment it will break? He cannot, if he attends only to one wave, for its behavior depends upon that of all the other waves within his purview and all waves outside his perceptual grasp. In fact, it depends upon the entire ocean and the air currents, the various continents, the Moon's gyrations around the Earth, the Earth's around the Sun, and on, and on. The radically nonlinear, chaotic system cannot be known in its entirety. Palomar in the company of his friends—apparently he has no friends, but that is irrelevant to the story—could make wagers in terms of the instant a rise in the ocean's surface would break and sport a scintillating ribbon of white water. Some would come close and others not, but that is because the guesses, well, they can be hardly more than stabs in the dark. And why will they be no more than stabs in the dark? Because in order to predict the break of a single wave, Palomar must know with absolute certainty and along nonlinear lines in phase space the conditions of all waves and of the entire ocean, the fishes swimming in it and the kelp lazily moving to and fro, the air currents above sporting a concoction of seagulls and airplanes maneuvering their way around, the groans and belches of the Earth's crust, and so on, and so on. An impossible task. And since Palomar cannot know all this to perfection, he cannot know a single wave. In other words, life processes are part of all processes that make up the universe as a self-organizing whole, and since (1) all processes are incessantly becoming and (2) from within, there is no conceivable end of the line of becoming, then (3) each part—in line with fractal geometry, chaos theory, Prigogine complexity, Palomar's quest—is as impossible to know, absolutely know, as the whole itself. Knowledge of all processes demands an impossible infinite grasp of things.

This nonlinear, processual situation is quite unlike consideration of a linear series of single coin tosses (Ford 1983). A toss is specifiable with much greater accuracy than Palomar's guesses, for example. The physical conditions can at least in principle be specified regarding the precise force exerted in the toss and at what angle, the air currents, and so on. And the position of the coin when it comes to a rest can be in theory predicted. However, given varying conditions, a statistical computation of the ratio of "heads" to "tails" cannot be determined—absolutely determined—short of an infinitely long run of coin tosses. Yes, we know, there is a 50/50 chance that either "heads" or "tails" will show up, and so after a given number of tosses the number of each should be about equal. But "about equal" is not exactitude of the crystalline sort; it is no more than a statistical average, however close to the mark it may be. Besides that, as a matter of statistics and nothing but statistics, no matter how many times we toss the coin, we cannot know what the next outcome will be. If we tossed it 1,000,000 times and recorded 499,999 heads and 500,001 tails, it would be most logical to assume that the next toss would yield heads in order to even things out. But for all we know, the next 10 tosses could play a trick on statistics and impudently give us the tail end of the coin. In the case of Palomar's wave, a nonlinear chaotic system is unspecifiable, though in principle it may well be deterministic. In the coin case, the nature of the toss can in principle be specified, but an absolute determination of the nature of any and all tosses cannot be forthcoming in a finite world of fallible human knowers. Once again, we could conclude either that all we need is more information (Simplicio), or that no matter how much information we have, we will always fall short of the mark (Sagredo). The controversy will most likely rage on. And it will continue to revolve around the distinction between objectivity—all we need to know is more details—and subjectivity—no matter how many more details we know, we will not know everything, for we are part of what we know and thus our very knowing is part of the known.

A question arises: If entropy would be able more effectively to continue its relentless push if there were no rebellious living organisms around to know it, then why should it be considered subjective? One reason lies in its alleged connection with "disorder." Maxwell wrote in 1875: "Dissipated energy is energy which we cannot lay hold of and direct at pleasure, such as the energy of the confused agitation we call heat. Now confusion, like its correlative term order, is not a property of things in themselves, but only in relation to the mind which perceives them" (Maxwell 1875, 153). The last sentence of this quotation is not out of the question. The "orderliness" of things depends on the possibility of our seeing some sort of pattern in them. We see an "ordered" pack of cards, and the pattern that it evinces, or that is imposed on it, may well depend on the rules of some card game. From a particular view, a specifically human view, whatever goes as orderliness is merely one out of a large number of possible sequences all of which could have been viewed as orderly from other possible vantages. So we are knowledgeable, for we construct order out of what could otherwise be seen as disorderly, and at the same time we are ignorant, for whatever order we can manage to make of things we remain unaware of many other sorts of order that could have been present. That is to say, our view and our order are dependent on our frame of reference—a limitation that apparently never entered Palomar's mind. Consequently, the Order—the concoction of all possible orders—cannot in theory be within our grasp—just as the narrator of Borges's "Library of Babel" dreamed of the Order with hardly any hopes of knowing it. It seems in my humble opinion that, if I may reorder the perplexing phrase of a preceding paragraph, knower and known are one, and they are split; they are split, yet they are one, for the hairline of the figure 4 ovum, the line separating yin from yang, weds them in a marriage that was never actually consummated but remains in a process of a perpetual consummation.[6]

So it appears that limitations there will always be for us tenderly fallible humans. At least that is the lesson Palomar's failures should teach us. Subjectivity there will also be, especially when the knower maintains awareness of his or her limitations. In this regard, back to the task at hand, that is, back to the impossible purity of nature and of knowledge of which, nevertheless, our narrator in Calvino's "Crystals" dreamed.

Facts of Life from the Fractals?

As we have observed, one notion has it that life saw its origin precisely from crystalline structure. An increasing number of scientists are leaning toward the conclusion that to expect DNA and RNA somehow to pop up from sheer chance collisions of oxygen, carbon, nitrogen, hydrogen, and a few other atoms is too much to ask of nature. It is once again the familiar adage about a chamber full of monkeys pecking away at typewriters and one of them miraculously composing *Hamlet.*

The universe according to present estimations simply has not been around long enough for life, given its complexity, to have come about by way of random mutations.[7] Perhaps, so the speculation goes, life emerged from relatively complex inorganic forms. In fact, as briefly discussed in the preceding, it has been speculated that asymmetry might exist at the very origin of life. All amino acids have the same left-handed twist, which may account for the uniformity in the handedness of DNA helices. Did this originally occur by mere accident? Could the twist as easily have been right handed? The apparent fact is that asymmetry tends to breed asymmetry, as Hofstadter (1979, 130–40) demonstrates in his discussion of recursive series. Ordinary crystals are built up by the fastidious repetition of the same structure in many directions over and over again. If, on the other hand, after a cycle is completed the endpoint does not coincide exactly with the starting point, the subsequent repetition will not be identical, and a spiral rather than successive cycles will result. Schrödinger's suggestion that genetic processes might arise out of the asymmetry of aperiodic crystals, "which can grow into an almost—but not exactly—symmetrical structure," became "reality" following the work of Crick and Watson. Once the aperiodicity of the crystal was established, Schrödinger (1967, 65) surmised, there was "virtually no limit to the size of the aggregate that can be engendered."[8]

Aperiodicity, like asymmetry, qualifies human mental constructs as well as the general process of nature. A Gödelian sentence invariably found in sufficiently rich and complex formal systems, Heisenberg's uncertainty regarding quantum phenomena, the incompleteness of any and all descriptions, linguistic vagueness and ambiguity, all testify to a universe that is inextricably tainted somewhere and at some time or other, as Calvino's narrator tells us. All elaborate intellectual constructs, like the physical world itself, invariably conceal some flaw or other, if no more than of the most minute variety. Gilbert Chesterton, in *Orthodoxy,* pinpoints the dilemma with respect to the overriding tension between our world of reason and that of irreason, according to how we may wish to take it: "The real trouble with this world of ours is not that it is an unreasonable world, nor even that it is a reasonable one. The commonest kind of trouble is that it is nearly reasonable, but not quite" (Chesterton 1908, 131).

Chesterton imagines what it would be like for an extraterrestrial being to comprehend the human body. He or she would observe bilateral symmetries of arms, legs, ears, eyes, even the lobes of the brain. After a closer look, slight imperfections would become evident, and probing deeper to the internal organs, there would appear to be hardly rhyme or reason for the disorderly concoction. It is this "silent swerving from accuracy by an inch that is the uncanny element in everything. It seems a sort of secret treason in the universe" (Chesterton 1908, 132). Everything swerves, everything is somewhat less than perfect, at least just a little bit awry. Let us make no bones about it. The world, and almost invariably our thought, is, as Sagredo metaphorically put it, sludgy. This problem of the variation of life systems from the ideal, Chesterton writes, is that "Life is not an illogicality; yet it is a trap for logicians. It looks just a little more mathematical and regular than it is; its exactitude is obvious, but its inexactitude is hidden; its wildness lies in wait" (Chesterton 1908, 131).

Logic and reason, like beauty, the good, ethics, and value, always fail to satisfy in the ultimate analysis. Discussing the beauty and the unreliability of reason in *Orthodoxy,* Chesterton outlines four principles: limitations, discreteness, complexity, and proportion. Given the myriad array of phenomena before the seeker of knowledge, the seeker must select, with awareness of the limitations of his or her faculties. As the seeker proceeds along the road, accumulating bits of information here and concepts there, he or she is advised always to condense. Ultimately, it seems, Chesterton would like to wish for a conception of the whole. However, he prudently keeps an eye on limitations. One cannot know the universe's unity, he counsels, only one's minute selection of it—Palomar could take a few lessons from Chesterton in this regard. Somewhat like William Blake, Chesterton dislikes vagueness of outlines and diffuseness, demanding that concepts be clearly delineated and made discrete against their background. If thoughts are clearly delineated, then they necessarily become more numerous, leading to a pluralism of forms and, ultimately, to complexity. Complexity, we read, calls for proportion, proper balance. Abstractions, including arithmetical figures, charts, and graphs are fine, but, as Alfred North Whitehead also advises, we should hold them up to suspicion and always keep an eye attuned to the concrete world to which those abstractions presumably refer. In other words, Chesterton is not simply infatuated with complexity. He calls for a balance of simplicity and complexity, such balance arrived at by subjective as well as objective, and intuitive as well as well-reasoned and logical, means. There will always be interaction between the pull of balance, harmony, and simplicity, and the push of complexity, with its fearful dissymmetries, disorder, and far-from-equilibrium or balanced nature. Ultimately, however, the simplicity of symmetries is destined to fall into dissymmetries, complexity, apparent disorder, and perhaps even chaos. Although Borges—one of whose champions was Chesterton by the way—often makes a case for symmetry in his writings, his inclinations ultimately point toward asymmetry. Indeed, there is also an overriding tension in Borges between symmetry and asymmetry, a tension arising from symmetrical models of simplicity, which are inexorably over the long haul a myth, on the one hand, and the asymmetry of complexity, of a universe of chance, chaos, the multiplicity of the unknown, on the other hand.[9]

This is also the tension in Calvino's "Crystals" between the ideal, diaphanous, pristine crystal and the blemished world of real crystals. This image is even more significant in light of Cairns-Smith's (1985) hypothesis mentioned previously, that, rather being than a chance collision and collusion of molecules, the emergence of DNA and its capacity to replicate itself are patterned in "lower" self-replicating chemical systems, namely, crystals. Simple crystal replication alone is not enough, however. Variability (i.e., aperiodicity) is needed. Cairns-Smith believes he has found this missing ingredient in clay. Clay consists of an elaborate concoction of needlelike crystals that grow in many directions simultaneously. These structures contain irregularities and dislocations that break the ordinary lattice arrangements of the crystals. As the crystal structures grow, these irregularities are repeated at successive levels, and with each level the irregularities become more pronounced. Eventually, living organisms evolved from such asymmetrical, aperiodic, self-replicating metallic minerals in clay deposits. The necessary information was coded in a crystalline structure and finally transferred to what later became in the process nucleic acids. For sure, crystalline matter was in abundance during the Pre-Cambrian period when life is believed to have emerged. It is not amiss to assume that crystals, washed about by the primordial broth and undergoing chemical evolution, eventually developed certain molecular preferences. Then a crystalline biochemistry could finally give way to a biochemistry controlled by organic compounds of no more than quasi-crystalline form.

All this talk about microscopic domains and macroscopic domains evokes Deleuze and Guattari's concepts of the molecular and the molar, which have a bearing on Borges and Calvino and ultimately, in the chapters to come, on life itself.

Becoming Something Else

Deleuze and Guattari make hardly any distinction between microscopic and macroscopic processes with respect to such phenomena as ideal crystal growth: what happened below happens above. As we have observed, crystal growth occurs at the surface of the crystal in many directions at the same time, and whatever occurred at the inner levels of the crystal is the same as what is now occurring at the surface, except that imperfections below will be magnified at each successive level. Simplicity apparently rules. Sequences repeat themselves over and over; the eternal return is the order of the day; an individual's (atom's) trajectory is a linear spinout in automaton fashion, with little possibility of that individual's influencing its environment or of any influence of its environment on it.

Life forms, in contrast, are something altogether different. Here, Deleuze and Guattari exercise the Hjelmslevian separation between the content and the expression of life processes, while inorganic molecular and molar processes remain basically the same. Content and expression are both molecular and molar: the barrier has dissolved. In other words, whereas expression occurs solely at the surface of the crystal where the action is, in the case of the organic sphere, expression emanates from within and becomes manifested at the surface. Now, "expression becomes independent in its own right, in other words, autonomous" (Deleuze and Guattari 1987, 59). The engenderment of linear sequences of amino acids (molecular) and the sequential engenderment, equally linear or unidimensional, of DNA (which is molar) are the result of the same mode of expression on diverse levels of content—both molecular and molar. What happens above is now dependent upon memory of what happened below. Engenderment comes from within; it is the expression of immanent drives pushing engenderment along preprogrammed pathways. This linear machine is governed by a code, the genetic code, setting constraints on the entire organism. In other words, a sort of double articulation is operating here wherein molecular combinations create the possibility of molar arrays of ever increasing complexity, until finally, life emerges. Crystalline forms give way to life, which in turn leads to human self-organizing entities and, eventually, to language.

Hjelmslev distinguished between the unfolded matter and that which is enfolded by forms of expression and content. Deleuze and Guattari, comparably, allude to the plane of consistency (or, if you will, the "body without organs") from which expression and content emerge. In its most primitive manifestation, the "plane of consistency" is also akin to what I have termed, regarding Spencer-Brown (1979), that which preceded the initial "mark of distinction." It is destratified, decoded, and deterritorialized, the sphere of *possibilia* that lie in wait for their actualization into the world of presumed substance, although they are nonessence, nonsubstance. The plane of consistency is "for all that neither a chaotic white night nor an undifferentiated black night" (Deleuze and Guattari 1987, 90–91). It is both: not really gray, but both whiteness and blackness, yet neither the one nor the other.

That is to say, the plane of consistency consists of pure interrelations without anything having been actualized, and when some unexpected—or perhaps expected—portion of it lifts itself into the sphere of the actualized, then it is what was possible but is now phenomenal. What was pure abstraction, that is, possibility, is now part of reality, that is, reality for some perceiving organism in some respect or capacity. This notion recalls the *Umwelt* concept, after Jakob von Uexküll (1957). Von Uexküll believed, rightly or wrongly, that he was able quite effectively to step into the universe of other organisms. Actually, whether he could or could not do so is at least partly irrelevant; it is the sheer act of attempting to do so that is of interest here. For example, a tick's universe, we are told, is made up basically of three signs. The first one is visual: the tick climbs up a tree and along a limb following light and fleeing from the relatively dark domain behind it. The second is olfactory: if perchance a mammal passes below the tick on the ground, it may sense the oxybutyric acid contained within its sweat. The third is caloric: the tick drops on the mammal and makes its way through its host's fur to its hide and burrows in for a little nourishment. Human semiosis, it hardly needs saying, is exceedingly more complex in comparison, with virtually an infinity of signs lying in wait to pop into existence. However, in either case, the end product is at its basest the same: the Deleuze–Guattari abstract machine becomes a concrete machine.

But that observation is obviously not good enough. Our anthropocentric needs cry out for more, much more. We cannot help but anthropomorphize, as perhaps Nietzsche knew only too well, yet our anthropomorphisms grant us the illusion that we know the world better than before. Here is a case in point. Try as he may, Mr. Palomar cannot imagine the sensations of a pair of mating turtles. He observes them

> with cold attention, as if they were two machines: two electronic tortoises programmed to mate. What does eros become if there are plates of bone or horn scales in the place of skin? But what we call eros—is it perhaps only a program of our corporeal bodies, more complicated because the memory receives messages from every cell of the skin, from every molecule of our tissues, and multiplies them and combines them with the impulses transmitted by our eyesight and with those aroused by imagination? . . . [T]he poverty of their sensational stimuli perhaps drives them to a concentrated, intense mental life, leads them to a crystalline inner awareness. . . . Perhaps the eros of tortoises obeys absolute spiritual laws, whereas we are prisoners of a machinery whose functioning remains unknown to us, prone to clogging up, stalling, exploding in uncontrolled automatisms. (Calvino 1985, 20–21)

All perhaps is machinery, Palomar anthropomorphically speculates. Some of the machinic processes are relatively simple and others are relatively complex, the complex having lost the concentrated intensity of the simple, the simple remaining incapable of the complex's vast display of possible experiences. But all is machinery just the same. Palomar later strives to know the blackbird's whistle, to conceive the world as it is seen by other birds, to get in tune with the salamander's immobile, mystical attention—to understand what it would be like to sleep without eyelids—to comprehend the world of the great white ape at the zoo, and even to understand the perceptions of the Japanese culture regarding the meaning of the stark landscape of a sand garden.[10]

In the long run, the impossible nonanthropomorphic task is virtually as impossible as Palomar's effort to learn to act as if he were dead. To say or do something as if it were to say or do something else implies some sort of knowledge of that something else. But how is that possible, if the something else is truly something other? How is it possible to begin, where to begin, when is a beginning most appropriate? To know Aristotle from a culture incommensurable with that of Averroës—as in Borges's "Averroës' Search," which appears in *Labyrinths* (1962, 148–55)—he must already have a foot at least partly within that incommensurable domain (see Merrell 1991a). In order that Palomar might know what it is like to be dead, he must somehow have some sort of experience from within the domain of the dead. He strives for some form of Zen-like sublime detachment, and as he does so, the idea of being dead becomes for him more and more natural. But there is always something lacking, a gaping void; somehow he never attains his goal. Finally, Palomar decides to describe, that is recapitulate, every instant of his life, and when he has described them all, he will no longer think of being dead but will feel what it is like to be dead. But alas, at the very moment he accomplishes this task, he dies. How could it possibly be otherwise?

Borges's Averroës in "Averroës' Search" tried gallantly to get inside Aristotle's head to comprehend the concepts of tragedy and comedy, without the benefit in his own culture of the existence of the theater, to say nothing of Greek aesthetics, logic, and ethics. Perhaps more by luck than by management, more by a fortuitous conjunction than by pure intellectual power, he comes quite close to the mark, relating tragedy or panegyrics and comedy to satires and anathemas, adding that they both abound "in the Koran and the *Mohalacas* of the sanctuary" (Borges 1962, 155). Borges himself intercedes at the close of the story, explaining that

> I tried to narrate the process of defeat. . . . I reflected that it would be . . . poetic to tell the case of a man who sets himself a goal which is not forbidden to others, but is to him. I remembered Averroës who, closed within the orb of Islam, could never know the meaning of the terms *tragedy* and *comedy.* I related his case; as I went along, I felt what that god mentioned by Burton must have felt when he tried to create a bull and created a buffalo instead. I felt the work was mock-

> ing me, I felt Averroës, wanting to imagine what a drama is without ever having suspected what a theater is, was no more absurd than I wanting to imagine Averroës with no other sources than a few fragments from Renan, Lane and Asín Palacios. I felt, on the last page, that my narrator was a symbol of the man I was as I wrote it and that, in order to compose that narration, I had to be that man and, in order to be that man, I had to compose that narration, and so on to infinity. (The moment I cease to believe in him, "Averroës" disappears). (Borges 1962, 155)

A bootstrapping, self-organizing operation. The "real" Borges creates Averroës, who creates Borges, who is-becomes a dream that disappears the moment Averroës disappears. Coleridge dreams a dream and writes it, and Khan dreams the same dream and brings it into reality, a reality owing its existence to a dream. Von Uexküll imagined himself inside the reality of a tick, whose reality in turn helped make von Uexküll's career what it was; the tick created (dreamed?) von Uexküll, who lives within that career insofar as the tick continued to live its reality. Palomar tried to know what it would be like to be a tortoise, a blackbird, a salamander, an ape. In every case, either anthropomorphisms or ethnocentrisms inevitably make their appearance. In every case the mind ends up reflecting itself; in every case a vicious circle seems to inhere. These infinite regress quandaries, like the Zahir-Aleph effect, the "Circular Ruins" merry-go-round, the world of Tlön, and that of Qfwfq within his point-moment in "T Zero," all are ultimately of the interdependent self-uplifting, self-contained, self-organizing nature of fractals and virtually chaotic systems, of Prigogine dissipative structures and life processes.

Complexity yet somehow unicity, unicity in all its pluralistic complexity. How can we hope to come to grips with this issue? Perhaps by not taking our world and ourselves so seriously, perhaps by accepting the vagueness of it all. A motley sea of ambiguity, then? Perhaps. Perhaps there is hardly any alternative. Yet we must push on; there is no standing still, I suppose.

PART 4

First, there is a brief (much too brief) journey through the sphere of fuzzy sets and fuzzy logic, which was patterned in Sagredo's response to Simplicio's quest for certainty, and which has a bearing on asymmetrization and the emergence of complexity in life processes, as suggested in chapter 9. Then another incursion is exercised into fractal dimensions, which in turn calls for a reevaluation of much of what was discussed in many of the preceding chapters, especially regarding some of Borges's stories. This leads to a tentative conclusion: the world—*our world,* whatever form it happens to have taken—is a world of vagueness, if it is not paradoxical through and through. Nevertheless, there is an effort toward a broader grasp of the topic at hand through a revisitation of Deleuze and Guattari's concept of multiple, in fact, *n*-tuple, rhizomic series. A tour through a few Calvino stories provides illustration of this concept, and it opens out to a further discussion of life processes and their relation to fractal geometry, chaos, and Prigogine's hypothesis. This discussion ushers in a note against the imperialism of language, the tendency to render anything and everything linguistic. As a counteracting move against this linguicentrism, attention is briefly turned to the strange quantum world of Schrödinger's cat, which serves to highlight, and support, determinacy and indeterminacy, clarity and ambiguity-vagueness, and simplicity and complexity, all of which evoke reference to some of Malevich's enigmatic remarks on Suprematism. Finally, focus rests on painting: the highlighting of perspectivism by the Renaissance, of tone during the baroque period, and of color by the impressionists, all of which provides filling for the original schematic presentation in figure 5*a*, and its multiple ramifications.

CHAPTER 10 • The Vengeful Onslaught of Sludge

[I]f I put a black dot on a sheet of white paper the dot will be visible no matter how far I stand away from it—it is a clear notation; but beside this dot I place another one, and then a third. Already there is confusion. In order that the first dot may maintain its value I must enlarge it as I proceed putting other marks on the paper.

—Henri Matisse

There is nothing in the world constant
but inconstancy.

—Jonathan Swift

Until now, the three dimensions of Euclid's geometry were sufficient to the restiveness felt by great artists yearning for the infinite. . . . Today, scientists no longer limit themselves to the three dimensions of Euclid. The painters have been led quite naturally, one might say by intuition, to preoccupy themselves with the new possibilities of spatial measurement which, in the language of the modern studios, are designated by the term: the fourth dimension.

—Max Jacob

Everything Merges with Everything Else

It seems that the West became so accustomed to putting the things and happenings of its world into neat and tidy compartments that balanced, harmonious, noncontradictory laws and categories came to be blindly accepted as the natural and normal procedure. This Aristotelian practice in many ways paid large dividends. Perhaps it was too successful, for it lulled entire civilizations into accepting it as the one and only valid practice.

Then, in addition to the disruptions of relativity and quantum theory, and to Simplicio's chagrin, along came chaos theory, the "science of complexity," and fuzzy logic. The last of these strange new beasts on the scene, fuzzy logic, has at least two sources over the past century.[1] The first appeared shortly before the turn of the century by the mind and pen of C. S. Peirce in the form of what he called a "logic of vagueness," the full development of which he held as a project for some future time that never arrived before his death. The concept of vagueness was later picked up by Max Black (1937) and has more recently become the focus of studies by Brock (1979), Engel-Tiercelin (1992), Merrell (1995a, 1996, 1997), and Nadin (1982, 1983), among others. The second source is an outgrowth of work with fuzzy sets in the 1960s and 1970s by Lofti Zadeh (1965, 1987).[2] In a word, fuzzy logic reveals the sludge inherent in linguistic practices. As such, this new logic refuses to prioritize language over para-extra-linguistic modes: all communication is to a greater or lesser degree vague. It was, of all philosophers, the analytical Bertrand Russell (1923), who, in an article on vagueness, suggested that language is invariably vague and that vagueness is a matter of degree.

"But how," one asks, "can fuzzy sets be depicted if they are indeed fuzzy? How can their depiction be anything other than more unadulterated mush of the sort that makes up the fuzzy sets themselves? The very thought that everything consists of fuzzy sets sort of hits one like the child's realization that there is no Santa Claus. Regular forms of reasoning and ordinary logic are much simpler; shouldn't we hold onto them for dear life?" On the other hand, recall that Simplicio desired to maintain faith in simplicity, but when the bolt of lightning hit him he aggressively took over Sagredo's lead, and, using essentially logical principles, he found his own way to embrace his particular sort of fuzzy thinking with a vengeance. His impetus even dragged Sagredo along for a vertiginous ride.[3] Whether he wanted to or not and whether he was fully aware of the fact or not, this act led him to the precipice overlooking the yawning jaws of complexity and gave him a playful shove. Consequently, his world will never again be as safe as it was under his previous delusions.

Bart Kosko (1993), a rising star on the fuzzy logic horizon, believes that freeing oneself of classical Western forms of binary reasoning can, in the best of all worlds, be a Buddha experience. And to boot, he claims to have constructed a cast iron framework—which, by the way, cannot but be the product of Western rather than Buddhist thinking—with which to account for fuzzy sets. At the outset Kosko's claim strikes one as ironic, especially in view of the fact that his model is a hypercube, which appears to be none other than an extension of analytic geometry: classical logic and Cartesian reason—once again, Western rather than

Buddhist thinking. According to the Cartesian way, and quite in line with some of my preceding considerations, it is all a matter of points and lines on a plane. The entire universe can be projected on paper, with everything represented by points, each enjoying a unique position specified by the horizontal and the vertical axes. Two numbers to the right on the scale and five numbers up give the coordinates (2,5). Eleven to the right and minus four down give (11,–4). And so on. Every point, ergo every aspect of the universe, has a two-digit Universal Security ID.[4] (Sagredo, of course, gave us a hint of this expansion during his dialogue with Simplicio, but he would most likely be prone to knee jerks at the thought of its final implications, chez Kosko, as we shall note.)

However, the initial appearances of Kosko's diagrams are deceptive. Kosko's trick consists in using the idea of sets spread out on the Cartesian plane in Venn fashion, rather than integers in the ordinary Cartesian mode. He stacks an infinity of such planes on top of each other to produce a cube, and then he runs an infinity of cubes into one another to get a hypercube—recall in this regard the initial words of Borges's "The Book of Sand." It's a matter of expansion to the right or upward from the 0 value to the absolute value, represented by 1 (which is comparable, of course, to the values of fig. 1). According to binary thinking, if 0 is the sign of, say, nonfat, then 1 would be the sign of fat, and if 0 is nontall, 1 would be tall. Each of these binaries, nonfat/fat and nontall/tall, could be depicted by a line, with 0 at one extreme and 1 at the other. Now let us combine these two sets of values and their opposites into a radical variation of Greimas's "semiotic square."[5] Fat could rest at the top and nonfat at the bottom, with tall at the right-hand side and nontall at the left-hand side. Now we have a plane, which recapitulates part of our dialogic exchange between Sagredo and Simplicio (fig. 19: disregard the values within the square for the moment, they will be qualified subsequently).

"This all might seem fine and dandy," one would like to protest, "for those who wish for the security of Boolean binaries that is—and in line with Sagredo's initial setup of 'happy' and 'nonhappy' people. But as a model for classrooms, executive meetings, and even barroom banter and street chatter, it simply doesn't fit the bill." Yes. Between the 0 and the 1 there are an indefinite number of gray shades. The same is to be said for the values between fat and tall and nonfat and nontall. To recapitulate a portion of the Simplicio-Sagredo discussion, Ronald is considered fat by whom and in what culture, within what context, and with

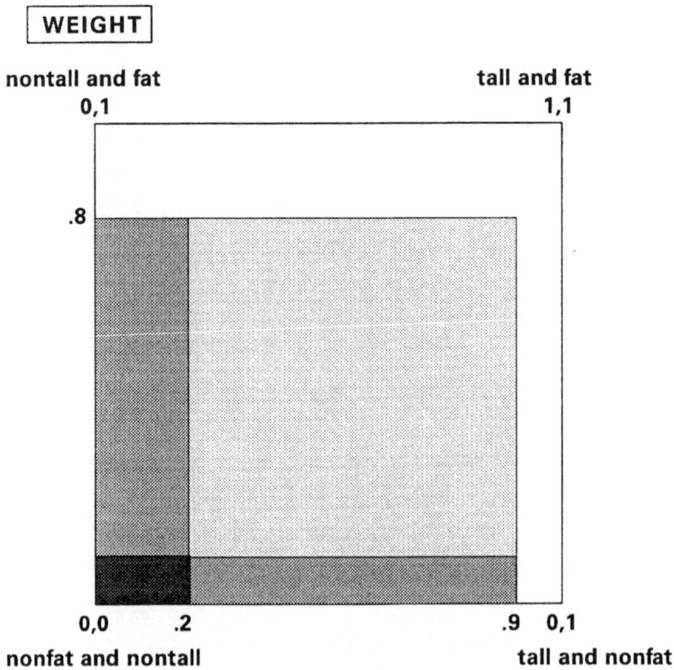

Fig. 19. Fuzzy diagramming

respect to who else? Donald is tall or not so tall by what standards? As a gymnast or as a basketball player? As a beau for Jill, who is 4′10″, or for Janet, who is 6′0″? As a candidate for the president of the United States or as a Texas entrepreneur? In addition to being language dependent, the terms in question are also culture and context dependent. If there are an indeterminate—possibly an infinite—number of differentiations between the values and their contradictories, tall and nontall and fat and nonfat, then the same must apply to the lines on figure 19 between the values and their complementarities, tall and fat and nontall and nonfat, and to their contraries, tall and nonfat and fat and nontall.[6] The two negative values thus correspond to 0 and the two positive values to 1. And the two complementary lines crisscross to bisect the plane into the two general, and inclusive, values, height and weight. In this manner, fuzzy logic is quite in line with Simplicio's ruminations on signs and nonsigns and the real and the nonreal: all are inexorably fuzzy.

Now, notice that if we cut out the figure 19 plane, twist it, and reconnect the right and left sides, we will have constructed a Möbius strip.

Within the Bind

The Möbius strip is apropos, I would expect. Draw a line around and around the strip without any crossings, then flatten the strip, and as we observed previously as with Lorenz's butterfly attractor presented on a two-dimensional plane, it appears that there are multiple crossings. But there are not. In other words, the strip is constructed by a twist of two-dimensional space in three-dimensional space, and when the strip is "two-dimensionalized" by being squashed, in a manner comparable to fractal geometry it nonetheless resists being treated as a two-dimensional object: it is slightly more than two dimensional; its dimensionality is fuzzy.

The Möbius strip is evoked frequently by psychoanalyst Jacques Lacan, who has been of recent interest to scholars in the humanities, particularly in literary theory and cultural studies. Lacan's continued fascination with topological forms—especially knot theory—plays a starring role in his theory of the unconscious (see Clément 1983; Turkle 1978). His "Schema R" consists of a rectangular figure drawn on the flat page, which must be read, Lacan warns, in three dimensions, like the Möbius strip (Lacan 1966, 531–83). Moreover, the Möbius quality of the schema renders apparent heterogeneity homogeneous. What would otherwise be a catastrophe is, so to speak, smoothed out: discontinuity becomes, from another vantage, continuity, and symmetry becomes asymmetry. Elsewhere, Lacan (1977, 155–56) uses another two-dimensional figure, called the "interior 8," which doubles back into itself. That is, the interior 8 is continuous only up to a point, where the surface is folded in upon itself in three-dimensional space such that there are two intersecting fields (fig. 20).[7]

Lacan places the libido at the point of the intersection as the field where the unconscious develops. This point is, Lacan claims, a void. It is commensurate with a mathematical point: *imaginary.* Potentially every-*where* and yet no-*where,* it can be at one of an infinity of locations along the inner loop. We are told that this line, containing potentially an infinity of points, is the "line of desire" lying within "the field of demand, in which the syncopes of the unconscious are made present."

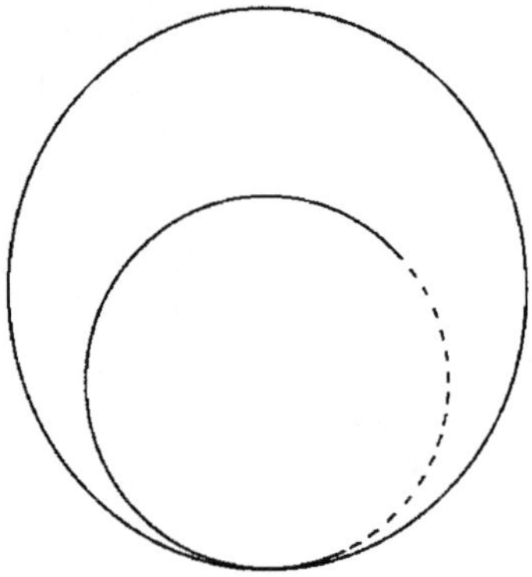

Fig. 20. Getting inside the circle

More significant still, this line of desire is metonymous, that is, linear and asymptotic, since desire cannot be totally fulfilled (Lacan 1977, 156). Here we note an obvious commonality between the point of intersection in Lacan's "interior 8" and a distinction, as described in the preceding and illustrated in figure 4, setting something apart from something else as its other. If this field within which the distinction is made is of Möbius quality, then that which became distinguished from everything else is turned inside out as it navigates its particular route along one of a potential infinity of routes. As such, with each self-returning fold, it becomes a mirror-image of itself; it becomes its own enantiomorphic form.

Enantiomorphic forms of the Lacan sort—and all others for that matter—are germane to the present discussion of dimensionality. An intriguing method for creating an enantiomorph is with the Möbius strip. Consider the case of, say, a female "Flatlander" inhabiting a "Möbius strip world." She can happily slide along the strip her entire life, believing she exists in an infinite universe. After all, why not? She can perceive neither a beginning nor a terminus from within her world. From our three-dimensional perspective, however, notice what occurs after she has completed one revolution around her universe, if we assume the strip to be transparent (fig. 21). After a trip around the strip, she is now the mirror image of herself and requires another revolution to return to her old form. One might contend that she is actually on the

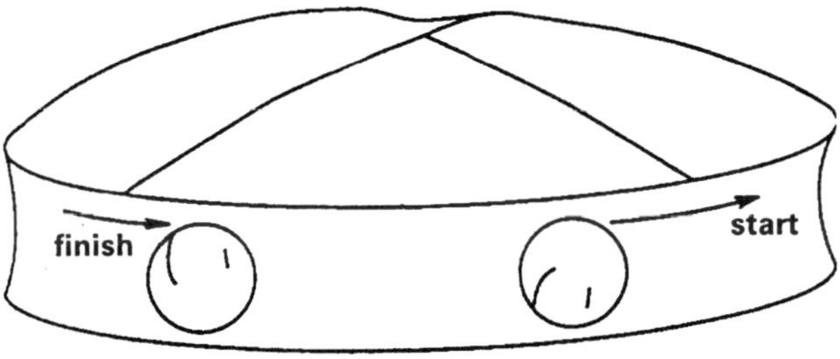

Fig. 21. A Möbius world

other side of the strip. But two-dimensional space has no "other side." The strip has created our Flatlander's enantiomorphic self through the twist in three-dimensional space.[8]

The same can be said of a three-dimensional object turned inside out. Consider a two-dimensional drawing that is ordinarily perceived to be three dimensional—the Necker cube—when placed before a mirror (fig. 22*a*). On the surface of the page, ABCD of the first cube is on the left side and projected downward. In its mirror-image form, in contrast, ABCD is on the right side and projected downward. Notice also that, since the two triangles are mirror images of one another, there is no way to transform one into the other as long as they continue to inhabit the two-dimensional plane. A transformation can occur solely by means of a 180° flip within a third dimension (see fig. 22*b*). The same can be said of the two enantiomorphic cubes. In order that one cube may be converted into the other one, each point along the six planes and twelve lines making up one cube must be flipped orthogonally within an extra dimension. So just as one Flatlander on the Möbius strip of figure 21 must be rotated through three-dimensional space in order to be reversed, so also the triangle must be flipped, by means of an infinite number of infinitesimal jerks, through three-dimensional space in order to be reversed. And just as the Möbius strip can be represented exclusively within a two-dimensional world, but with an intersection or warp in three-dimensional space, so also the Necker cube can be conceived within a three-dimensional space, and its enantiomorphic transformation entails a four-dimensional temporal-spatial topology.

Interestingly, this interpretation of the Necker cube phenomenon

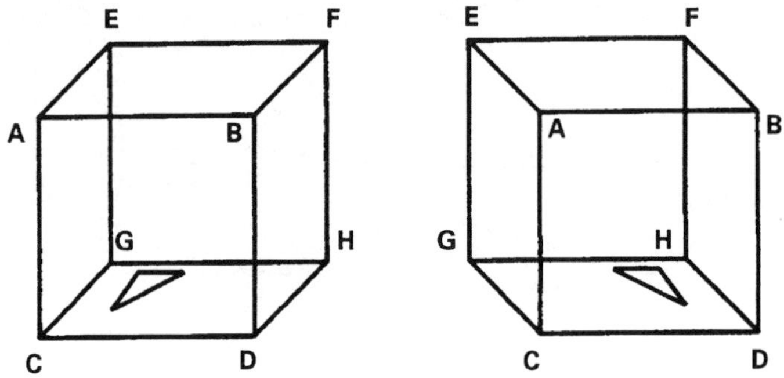

Fig. 22*a*. Mirror-image cubes

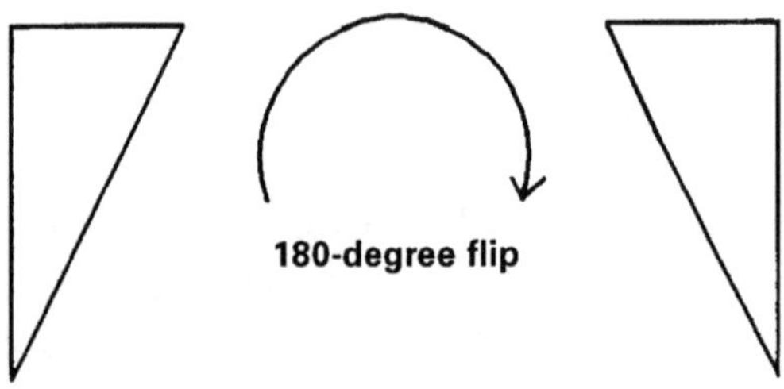

Fig. 22*b*. Enantiomorphic triangles

leads H. A. C. Dobbs (1972) to conclude that (1) the transformation in the perspectival reversal of the Necker cube is logically impossible in three-dimensional space: (2) no spatial degree of freedom is evident in this transformation except height, width, and depth—that is, three-dimensional space; and (3) therefore an extra dimension is necessary for the transformation, which must consist of a "mathematically imaginary time dimension," a four-dimensional manifold.[9] In other words, just as the Möbius strip and the triangles require a third dimension for their construction, a dimension outside the imaginative faculties of a Flatlander, so also the Necker cube requires a fourth dimension in order to account for the transformation from one cube into its enantiomorphic twin, that dimension resting outside our own visual capabilities.

Now for a brief thought experiment. Place the enantiomorphic twins of figure 22*b* on a plane facing one another. Take a point on one of the triangles, describe an arc above the plane that rejoins the plane at the midpoint between the two triangles, then describe another arc protruding down under the plane that connects the point to its twin point on the other triangle. You have described a sine wave. Now take the point back to the original triangle by a reverse operation and connect it to a point adjacent to the point of departure. You have now described a curved path equivalent to the one-dimensional Möbius strip in figure 15. Take the next point on the first triangle and do the same with it, and repeat the operation an infinity of times to construct an infinity of twisted curves. The transformation will now have been completed. Each one-dimensional Möbius strip is the equivalent of a line around an imaginary two-dimensional Möbius strip connecting the two triangles; each curve creates endotopic and exotopic space in the sense of Klee in figure 15; each is the equivalent of Lacan's topology in figure 20. The concoction of all the ellipses consists of a twisted knot of lines, Möbius strips, and interpenetrating cubes along fractal, strangely attractive contours within a four-dimensional manifold.

This, then, I would submit, is the essence of the spatial labyrinths in Borges's "Two Kings and Two Labyrinths," "Death and the Compass," and "Library of Babel," and in Calvino's "The Count of Monte Cristo," and also the essence of the temporal labyrinth in Borges's "Garden of Forking Paths" and of Palomar's becoming entangled in the labyrinths of his own reasoning upon attempting to grasp the whole. Moreover, I would suggest, none of these labyrinths would have been possible without the equivalent of the $\sqrt{-1}$ operator, the hairline in Figure 4—yes, $\sqrt{-1}$, required for Dobbs's "mathematically imaginary time dimension," which is used in relativity theory to account for the four-dimensional space-time continuum and which is germane to quantum theory.

We live in three-dimensional space, which we believe is for us adequately accountable using our perceptual and conceptual capacities. And we live in one dimension of time. But we cannot survey things in time as we can in space; we are neatly wrapped in the cocoon of time and it is wrapped in us; we would like to think we intuitively know it, but we cannot adequately think it or say it, so St. Augustine tells us. Moreover, when time is formally presented as another dimension of space, as is usually the case in popular expositions of relativity theory, that fourth dimension apparently remains virtually unintelligible with respect to our concrete world of everyday affairs, though it can easily be dressed up in a gala of mathematical signs. However, while the Möbius strip and Necker cube illustrate the enigma of dimensions beyond those of the world in which we live and breathe, they are just that: whole dimensions. Interesting as the interaction between whole dimensions may be, they remain foreign to the world of fractal dimensions. These more subtle fractal dimensions, it is becoming increasingly evident, are characteristic of the world of our actual coming and going, while whole dimensions are no more than abstractions from the sludgy fractal world of complexity to which Sagredo vaguely alluded.

Back to fuzzy thinking for a moment, then.

When Things Are Not

Regarding the idea of using sets in place of integers in the order of Kosko's hypercube rendition of fuzziness, reconsider figure 19.

Each person occupies a spot somewhere on the plane according to his or her tall/nontall and fat/nonfat attributes, depending upon how the folk of a particular community would be disposed to categorize that particular person. If John is 5′11″ in height and is considered tall by only 20% of the citizens of his community, then the value is .2; if he weighs a hefty 220 pounds, which renders him fat according to 80% of the population, then the value is .8. We have a plotting of John's qualifications on a variation of the Cartesian plane in figure 19 as [.2,.8]. Any point within the vertical darkly shaded area is the set of all people less tall and less fat than John, while any point outside the area—of people either taller or fatter than John, or both—falls outside that set. If George stands at 6′3″ and is tall in the estimation of 90% of the population, and if he packs 200 pounds, which makes him fat in only 10% of the judgments, then we have in the horizontal darkly shaded area, [.9,.1], the set of all people falling within the range of George's attributes, they being less tall and less fat than he. The overlapping between the John set and the George set is the logical intersection or disjunction of the two sets, represented by [.2,.1] in figure 19. This is the subset of all people who are both less fat than John and less tall than George. The inclusion, union, or conjunction of the two sets is represented by [.9,.8] and illustrated by the entire shaded rectangle. This is the set of all people whose attributes are less than either those of John or of George or less than those of both John and George. Taking the entire graph into account, we have all the people who are either fatter or skinnier and taller or shorter than either John or George or than both John and George. If we include another attribute, say, bald or nonbald, then the plane is expanded into a cube.

Yet another attribute, old or nonold, and a fourth dimension is necessary, and so on—the relevance of figures 21, 22*a*, and 22*b* comes into view, and fractal dimensions (of vagueness) make their entrance. As a consequence, complexity raises its benign or bedeviling countenance, depending upon the eye of the contemplator.

Once again we see that if classical logic reigns supreme, then there can be no overlap, and therefore the excluded-middle and noncontradiction hold true to form. But with fuzzy categories, classical logic goes out the window. However, fuzzy categories do not exactly exist in contradiction with classical logic. To claim that they do would be to enter the territory of our bugbear, binary thinking. Classical logic actually enjoys a part-whole relation with fuzzy categories. Fuzzy categories can tolerate Boolean binary principles as a subset of fuzzy logic, but the other way round does not hold: binary principles in their pure form categorically exclude fuzziness.[10] Classical logic in its most ideal sense would give no picture of fuzzy logic at all, for the one involves a pristine language chipped in stone for all time, while the other wallows in a landscape of marshes, bogs, and occasional spots of quicksand: sludge. However, the ideal always rests beyond the wishes and efforts of us finite knowers: it is invariably blemished with a certain degree of vagueness here and there. Consequently there is inexorably always a tinge of fuzziness: classical logic is always at least just a little bit fuzzy, which indicates that it includes, within its domain, a portion, however minuscule, of the domain of fuzzy logic. In other words, the product (disjunction) of classical and fuzzy logic is not an empty set, but rather, there is a degree of overlap. Thus the relationship between two terms *classical logic* and *fuzzy logic* is itself fuzzy. The whole of "logic," classical and otherwise, is more in the order of a cloud than a clock; it is pliable and vulnerable software rather than solid-state hardware.

Recall that the labyrinth Lönnrot proposed moments before his demise was linear, and it resisted, in Zenoesque fashion, being infinitely striated. The king of Arabia's labyrinth, in contrast, was two dimensional, consisting of an infinity of linear labyrinths of the Lönnrot sort. But it was slightly more than two dimensional if we consider the helpless and hopeless Babylonian king's linear wandering, which consisted of a virtual random walk, we would expect, on the desert sands. In other words, his irregular trajectory was not exactly linear—that is, one-dimensional—either: more appropriately, it was like an indeterminate, unpredictable Koch curve as in figure 18. The middle third of the top line is removed and replaced with the two sides of an equilateral triangle, then the middle third of each side of the triangle is removed and replaced with the two legs of an equilateral triangle, and so on. Finally, a "curve" of virtually an infinity of barbs emerges, the curve being of infinite length with a dimension of 1.261859 (according to the formula log 4/log 3). So the Babylonian king's apparently linear walk is actually of 1.261859 dimensions. But there is still a problem since the floor of the desert is not exactly smooth; rather, it is granular, with each minuscule segment comparable to each grain of sand on the desert floor. Mapping out a one-dimensional, linear curve along the desert floor connects many, potentially infinitely many, grains-as-points on the apparently two-dimensional surface. Complexifying the curve increases the number of grain-points coming into connection. The mapping itself becomes infinite; its one-dimensional character becomes fuzzy: the line expands as it comes into contact with many, potentially an infinite number, of other lines, and the trajectory's one-dimensionality begins fading into two-dimensionality. Finally, when all the grain-points are connected, the trajectory becomes slightly more than two dimensional. The fractal breaks in each grain, between grains and grains, and between grains and the apparently smooth continuum of the desert floor provide a glimpse into a third dimension. In other words, if the Babylonian king could have walked on the desert for an infinity of years, the tracing of his paths would have been like the evolution from a lacerated near-white expanse of virtual emptiness to a near-black plenum. He would have filled the entire desert with an infinitely long well-nigh infinitesimally thin Black line! Now, if we begin removing the grain-points along the king's tracing one by one, we will have the equivalent of the sorites paradox: when can we say slightly more than one-dimensionality has become two-dimensionality, and when has two-dimensionality become three-dimensionality? Dimensions in this respect can be none other than fuzzy.

The labyrinth of the two-dimensional sort Lönnrot constructed on his map in an effort to trap Scharlach, on the other hand, was to all appearances quite concisely two dimensional. It consisted of three points representing the site of the first three crimes on the map of the city, which constituted the corners of a triangle. "Extrapolation" of the triangle by plotting a fourth point and constructing two lines to form a parallelogram pinpointed the presumed site of the fourth crime. Lönnrot appeared at that location at the calculated time. However, Scharlach had other plans. The labyrinth he wove for the hapless Lönnrot consisted of three dimensions, or an infinity of planes each of which was capable of including an infinity of lines. But that was not all. A Scharlach depiction of Lönnrot's geometrical figure on the map would actually be more than a surface, yet at the same time it would be somewhat less than a volume. It would be a counterpart to that familiar Sierpensky sponge, mentioned in chapter 7 and depicted in many expositions of fractal geometry. The sponge consists of a cube shot through with an infinity of square holes. Each face of the cube is penetrated by a square hole directly in the center whose sides are one-third the distance of the side of the face. This leaves eight squares of equal size on each face, each of which is in its own turn penetrated by a square hole whose sides are one-third the distance of its side, and so on, ad infinitum. (As noted previously, the resulting space of the Sierpensky sponge, with 2.7268 dimensions, would be much like that of the prison in Calvino's "The Count.")

The surface of the apparently three-dimensional labyrinth at the house at Triste-le-Roy twists and turns, like the twisting staircases, to entrap Lönnrot within dimensions, within the space-time manifold, much like the triangles in figures 22*a* and 22*b*, or like Escher's *Relativity* or the Penrose triangle within the space-time continuum to which figure 5*a* alluded. These interspatialities are properly nonlinear, rhizomic, asymmetrical. Each minute spatial increment may for convenience, and for practical purposes, be defined in Euclidean terms, but its connections to other increments are indeterminate. The entire concoction is amorphous, pliable, more haptically than visually accessible. It is quite unlike an Escher cut, which is exclusively visual, or like two ambiguous meanings of a word, which is exclusively linguistic. It is more like the warp and woof of harmony and melody, of the taste of sweet-and-sour pork, or perhaps better, of the sweet scent of "Chanel No. 5" mixed with the smell of a greasy spoon restaurant. In fact, it is much like mixing two or more sensory channels to create a synaesthetic sensation.

In other words, with respect to the rhizomic concoction, borders have become rather more than fuzzy, which can be specifiable at least by fuzzy logic if by no other means: they have disappeared altogether in the fusing of erstwhile categories of mind and reputedly of matter. This reemergence of the notion of mind and its merging of categories by way of its nomadic meandering calls for another turn to an important feature of Beckett's Unnamable: the fusion of every-*thing* onto one *thing,* which is, paradoxically, well-nigh indivisible, barely differentiated, hence of the most fuzzy of fuzziness.

Beyond the Gray Horizon

The Unnamable is hardly more than a disembodied talking mouth—a reminder of that slash of a mouth against a black background in the play, *Not I* (Beckett 1973). He is like a big articulating ball or some other geometrical surface: "Perhaps that's what I am, the thing that divides the world in two, on the one side the outside, on the other the inside, that can be as thin as foil. I'm the partition. I've two surfaces and no thickness" (Beckett 1955, 134). In other words, he is hardly more than the figure 4 cosmic egg.

Given to uncertain stops and starts, forward moves and backtracking, the Unnamable is also much like Deleuze and Guattari's schizophrenic: now a functioning multiplicity, now a pure point, an unextended zero-intensity no-*thing*. The nomadic slithering, elusive schizophrenic is perpetually marginalized; there is hardly any outside that makes a difference. The schizophrenic and the Unnamable are both without the wherewithal to function as anything more than an exceedingly faint boundary between norms (that which is normal, conventional, prescribed, culture) and every-*thing* else (that which is prohibited, chaos, nature). In this respect they are every-*where* and no-*where* at all. There is no clearly delineable "I"–not "I," inside-outside, subject-object; these and other such binaries are no longer meaningful. Every-*thing* is interrelated rhizomically. Yet, there is some vague sort of margin, boundary, otherness, between what they are "in here" and what remains "out there." The situation is tantamount, once again, to nothingness, white-black-lessness, paradoxically bringing every-*thing* into interrelation with every-*thing* else. A vague line bisects the two-dimensional surface to produce separation between *this* and *that* in three-dimensional space. But it can do so only through orthogonal leaps from outside the plane. Each break is an orthogonal $\sqrt{-1}$ switch. It is an oscillating (Bateson, Laing), trembling (Kierkegaard, Pascal), scintillating (moiré, op art), vacillating uncertainty (Bohr, Heisenberg, Klee), a most certain undecidability (Gödel, Derrida).[11] Now, the hairline, however vague, is none other than that of the line tracing out the equivalent of the figure 4 egg: neither the Unnamable nor the schizophrenic is precisely distinguished from his environment, but rather, all is always almost one but not quite. (Compare this individualism to Benítez-Rojo's community mind and behavior: interesting.)

At this zoneless zone of *aporia,* this *mise-en-abîme,* this fathomless chasm, the Unnamable's selfless self is in an incessant self-reiterative process, changing yet remaining the same, invariant yet suffering change through time. The spontaneous spew of semiosic narrative from the Unnamable eventually gives way to Beckett's strange character, Worm: little more than a mass of albumin virtually bereft of differentiation, the final blow to all hopes of differences that make an appreciable difference. Born of the concoction of prenatal and postnatal illusion, Worm is an atavistic, beingless, formless, nonsubstance-substance, imagined but not quite real, writhing but not quite alive. Worm is "the famished one, and who, having nothing human, has nothing else, has nothing, is nothing. Come into the world unborn, abiding there unliving, with no hope of death, epicenter of joys, of griefs, of calm. . . . On the outside of life we were in the end" (Beckett 1955, 346). Worm's sluggardly spirals of possible nonlinear squirms pattern the impossible pattern of virtually undifferentiated paths that melt into one another: virtual chaos. According to the Unnamable's delirious soliloquy: "I confused myself with Worm. But I have no reasons either, no reason, I'm like Worm, without voice or reason, I'm Worm, no, if I were Worm I wouldn't know it, I wouldn't say it, I wouldn't say anything, I'd be Worm. But I don't say anything, I don't know anything, these voices are not mine" (347).

Once again we see that there is no clear distinction here between "I" and not-"I," there is only "I"-not-"I"-lessness. It is ultimately a matter simply of "I am that I am," without distinction, without difference, without either the one thing or the other, without clear-cut *thatness* as opposed to *thisness.* Indeed, Worm is the incarnation of figure 4, but not quite, for it/she/he is virtually without an outer skin and certainly without the inward split. In order to render it/her/him the recipient of at least a modicum of sentience, "they could make a small hole for the eye, then bigger ones for the arms, they can make one bigger still for the transit of Worm, from darkness to light" (357). In a manner of speaking, there is evidently no color emanating from Worm but, rather, mostly grays, with only occasional distinguishing features of blacks and whites. There is no apparent life, at least as we know it; there is hardly more than a writhing and squirming, the effort of life's processes to lift themselves up by means of the universe's self-help program.

Whether Worm is nothing more than a larva and almost immobile, or whether it/she/he can move in infinitesimal steps by repetition, as in the early stage of a fetus, in either case, we have at least as an exceedingly vague possibility the transition from no-*thing* to some-*thing* and potentially to every-*thing,* which is at the same time every-*thing* and a return to no-*thing.* The eye begins with gray, no doubt, yet it is not just gray, for "there is yellow in it, pink too apparently, it's a nice grey"—that is, Tobey's painting, *Tempera.* The eye, however, at this stage sees no particulars, the color is merely "noise," "grey noise." Though we read that "perhaps the brightness will come" (Beckett 1955, 361), there is no more than grayness, the "calm." In addition to the "calm," "there must be something else, to go with this grey, which goes with everything. There must be something of everything here, as in every world, a little of everything" (362). However, the relatively undifferentiated monotony of the gray sphere, the faint buzz, the stillness, is "like slime, paradise" (364). And though the silence is a "grey silence," it must be somehow either black or gray, since either way there is no significant difference: gray is every-*thing* and no-*thing* plus no-*thing* and every-*thing,* white and black. It is a mere "lull." But in a way of speaking it is quite desirable, this gray, this lull without memory of the before or anticipation of the after. It is the eternal poise at the apogee of the syncope: "Is not a uniform suffering preferable to one which, by its ups and downs, is able at certain moments to encourage the view that perhaps after all it is not eternal?" (367). The Unnamable concludes that the monotony of it all "is more stimulating" after all.[12]

But enough of these somber images within Beckett's characters' drab empirical world. Surely the mind can remain active no matter how depressing its surroundings. So, reenter Borges's Tlönians, those inveterate subjective idealists the furniture of whose world continues to exist only insofar as they see it, hear it, taste it, smell it, feel it, think it, and talk about it. This abrupt shift of focus is necessary, since following Beckett's dreary world, we are not merely within the ineffable—for which there is no representation and, indeed, which denies the very idea of representation. We are abruptly shoved into the sphere of visual images that belie their very nature as something slightly more and slightly less than that of which the visual is ordinarily capable.

From a Para-Zeno Point of View

The world of the Tlönians "is not a concourse of objects in space; it is a heterogeneous series of independent acts. It is successive and temporal, not spatial" (Borges 1962, 8). However, there exists a heretical school of thought that goes so far as to negate time: the future, its masters propose, has no reality but is only a vague, present hope, and the past has no reality other than as an often equally vague, present memory. A heresiarch of this scandalous doctrine gives the following example:

> On Tuesday, X crosses a deserted road and loses nine copper coins. On Thursday, Y finds in the road four coins, somewhat rusted by Wednesday's rain. On Friday, Z discovers three coins in the road. On Friday morning, X finds two coins in the corridor of his house. The heresiarch would deduce from this story the reality—i.e., the continuity—of the nine coins which were recovered. It is absurd (he affirmed) to imagine that four of the coins have not existed between Tuesday and Thursday, three between Tuesday and Friday afternoon, two between Tuesday and Friday morning. It is logical to think that they have existed—at least in some secret way, hidden from the comprehension of men—at every moment of those three periods. (Borges 1962, 11)

Defenders of common sense in Tlön maintain that this paradoxical anecdote is a verbal fallacy lacking in rigorous thought. The verbs *find* and *lose,* they claim, are used illegitimately. The coins supposedly having existed from the instant they were lost to the moment of their rediscovery would imply their continuous existence—that is, the view of our own Western classical physics—which is intuitively impossible for the Tlönians. They believe the coins ceased to exist when they were "lost" (i.e., unperceived) and once again popped into existence upon their being "found" (i.e., reperceived, constructed anew). Idealism obviously pervades.

The narrator of Borges's tale suggests that the Tlönian paradox is the Tlönians' equivalent of the Eleatic paradoxes of our own tradition. But it is, of course, the inverse, since the Tlönians' obstinate idealism clashes with our equally obstinate love affair with objectivity, reference, correspondence, representation, and the mind as mirror of nature (recall Simplicio's allusion). As a consequence, intuition continues to tell us that a set of coins has ongoing existence. In contrast, Zeno argued that such a conception is paradoxical, for just as an arrow displaces no more than the space it displaces at a particular moment and cannot squeeze forward through space or enter into another moment of time, so also the arrow—or anything else for that matter—is what it is in this instant and nothing else; hence it cannot move, and time is a fabrication. Quite in line with Zeno, and contrary to our own inclinations, intuition tells the Tlönians that the coins enjoy no more than discontinuous existence, since they can come into being only when someone interacts with them, perceives them (*esse est percipi*). In contrast, the dissident Tlönian heresiarch, in line with our own intuitions, taught that the Tlönians' world has a beginning and end and a continuous and quite lucrative existence, in spite of who might be thinking the Coins, saying them, or looking at them.

We would like to believe that Achilles catches up with and passes his torpid competitor along a continuous line. Somewhat like us, the heresiarch would likely postulate that Achilles could perform his task without any spectators around to actualize each stage of his effort to pass the lethargic tortoise. Zeno, however, demonstrated that, at least from within his logical framework, this is impossible, that is, if we properly atomize Achilles's path to the finish line. The Tlönians, faithful to Zeno it seems safe to say, would assume Achilles could win the footrace only insofar as they, or anybody else, might have been participating with him as active spectators, for without them, there would be neither Achilles nor tortoise, nor could there be any there-then, here-now, and somewhere else in the next moment. In sum, according to our traditional view, existence, time, and motion are continuous, but according to Zeno, continuity, when properly fragmented, refutes our most cherished notions. For the Tlönians, there is motion only insofar as by their mind acts they actualize events in succession, in time; motion and time are continuity fragmented, but in spite of this fragmentation, motion and time can exist, thanks to sheer acts of mind. But for the heresiarch, whatever exists has existed, it will continue to exist, and it is free to move about or to be moved about in time.

Put the essentials of the two views together, and we have the best and the worst, or the worst and the best depending on the perspective, of the extremes of our Western tradition. Objective reference, correspondence, and representation have had their antagonist, subjective idealism, in a half nelson for a few centuries. Parmenides has consequently managed to get the upper hand over Heraclitus, Newton has had the advantage over Berkeley, Descartes over Pascal, Kant over Kierkegaard, Hegel over Nietzsche. More recently, however, Heraclitus and others have begun to reemerge: Bohr gave Einstein fits; logical positivism suffered senility and a slow death; Nietzsche has been revived; Derrida and his crowd seem to have won a few matches against Gadamer, Habermas, and others; and Rorty and the neopragmatists, with dreams of solidarity dancing in their heads, have been muscling their way around.

Where is the answer? If we give credence to the playfully ironic views supremely put forth by the likes of Borges and Calvino, it seems quite likely that a big picture will not be at hand and that we must give up the dream of nonparadoxical answers. We are so comfortable with the tidy compartments within which our traditional modes of thought and of feeling are packed that the very suggestion that fixed, nonparadoxical answers to our questions will not be forthcoming strikes fear in the heartstrings of those most stalwart knights of good reason and good taste. In fact, one might suggest that, in view of the Sagredo-Simplicio dialogue, many of the problems that still plague us are found in the essence of Zeno's formulations, which should convince us that obsessive focus on Platonic-Aristotelian *being* cannot be more than half of the picture. The other half involves the somewhat uncomfortably paradoxical notion of *becoming.* Of course, being and becoming have generally been considered simply incompatible. Yet change (becoming) cannot be accounted for within a nonparadoxical logic of static, eternal verities (being). Zeno's arrow paradox illustrates this dilemma.

In regard to the Zeno quandary, Alfred North Whitehead (1929, pt. 2, chap. 2), like Henri Bergson ([1911] 1964), sees not static things but dynamic acts: the properties of time and motion are most fundamental; things are ultimately a product of transitory stages in time. Whitehead particularly views Zeno's puzzle as focused on the problem of pure change. Dynamic motion is not a matter of things but of processes, irreversible in character. In pure change, there is no definite thing that must be somewhere at some moment or that jumps from one position to another one. Pure change is not an actual event in the classical sense of involving things in transition. It is simply change qua change, it is a matter of events, not things. If change is viewed in terms of things that change, as it is in the classical sense, that is the prerogative of the viewer, not necessarily the reality of nature. Whitehead's dilemma consists in his admitting to eternal—abstract, unchanging—essences lying behind and beyond the pause between what is going out of Becoming and what is coming into becoming—or in another way of putting it, between the being of becoming and the perpetual becoming of being (Ushenko 1946). The problem is that Whitehead strives to maintain a consistent picture; he resists the temptation willingly to embrace the ultimate paradoxical nature of mind and of the world.[13]

So Whitehead's Heraclitean "nothing but change" coupled with Zeno's Parmenidean "never any change" is actually victim to the need for order, harmony, and balance. In other words, they are caught within the comfortable certainty of the either and the or in the central zone of figures 4 and 5*a* where standard logic continues in its attempt to reign supreme. In contrast, the lower left side of figure 5*a* coupled with its upper right side and the knife-edge nothing or emptiness between one-half of the egg and the other in figure 4, forces an awareness of something's manifesting both one character and another, or neither one character nor another, depending on the perspective. This is thought not by exclusion but by complementarity, not by loyally worshipped consistencies but by fearless inconsistencies, not by divergence but by coalescence, not by dichotomies of reason but by diffusion of thought. Like Heraclitus's philosophy, nothing is considered absolutely either this or that but something else that resists neat pigeonholing (Wheelwright 1959).

In his attempt to account for the process of becoming, William James (1948), like Whitehead, opts for a "pulsational" view of reality in terms of the existence of a durational threshold to the limits of the becoming and begoing of events and of our awareness of them. Time durations are not dense but discrete, though our awareness of them merges them one into the other—James's "buds" or "drops" of consciousness. Putting James and Whitehead (1947, 61–82) together, if we admit that something becomes, by Zeno's own argument there cannot be a continuity of becoming but only a series of "pulses" of becoming. On the other hand, there can be a becoming of continuity—in principle at least—since, as the pulses become finer and finer, experience fuses them together. In this manner, an actual thing is the product of an act of experience. It is always becoming and begoing. The coming into being (awareness) of an event requires awareness of what just came into being (awareness) and anticipation, with a minimum of duration. Somewhat in line with James and Whitehead, Deleuze and Guattari's schizophrenic, deterritorialized, and nomadic hero is dramatically illustrative of, indeed an ambulating metaphor of, the becoming of continuity along nonlinear, irreversible, Heraclitean rhizomic paths. In contrast, for the paranoiac there is no bona fide continuity of becoming but a static, linear, discrete series of steps each of which actualizes things that fall into their proper cubbyholes—according to the Deleuze–Guattari paranoid mentality, of course.

In the two actors on the stage of Deleuze and Guattari's inquiry, the paranoiac and the schizophrenic, it would appear that we have the two extremes of our Western tradition. Is there no legitimate middle ground? Richard Bernstein (1983), among others, has attempted to stake out and survey such a middle ground regarding largely metaphysical issues.[14] In the present disquisition the middle ground thrust has been more overtly nonmetaphysical: literature, science, painting. Where has it led us up to this point? It has perhaps revealed the poles of various oppositions more than any middle path: one/many, simplicity/complexity, order/disorder, "new physics"/classical physics, logic/illogic, reason/irreason, consistency/inconsistency, completeness/incompleteness, timelessness/time, continuity/discontinuity, permanence/transience. On the other hand, at least a few baby steps have been taken toward some sort of middle ground. For example, it seems to have become quite apparent, in light of the preceding chapters, that complexity and disorder, like their correlative terms *simplicity* and *order,* are not properties of things in themselves, but they exist only in relation to the way the mind perceives them. This "perspectivist" or "constructivist" account can serve to (1) bring extremes into relation with respect to the infinite variations between them and (2) smooth out the distinctions between these extremes as presented in figure 5*a* and elsewhere into differences that become finer and finer to approach a continuum, that is, vagueness. Ultimately, in this respect, vagueness, fuzziness, contradictions, incompatibilities, mutually exclusive terms and concepts, and even paradoxes can often become quite congenial bedfellows. This provides a middle ground in the sense that relations and events, not things, are considered real, and regarding relations, not only identity and either-or categories and the relations between them hold, but, on occasion, both-and and neither-nor relations, regardless of whatever their categorical frameworks from whatever perspective, can also occasionally take their rightful place in the agonistic arena of ideas, doctrines, conventions, and beliefs.[15]

This tentative conclusion, I fear, will satisfy few onlookers. In order to give it a bit more respectability, I will move on toward a broader view of relations and events, which includes more directly the work of Deleuze and Guattari and gives further visibility to the work of C. S. Peirce: I am speaking of signs, their use and their occasional abuse and, especially, their meanings.

CHAPTER 11 • Flying off in All Directions

Chance and chance alone has a message for us. Everything that occurs out of necessity, everything expected, repeated day in and day out is mute. Only chance can speak to us.

—Milan Kundera

The dreams of reason bring forth monsters.

—Francisco Goya

It is not the slumber of reason that produces monsters, but vigilant insomniac rationality.

—Gilles Deleuze and Félix Guattari

Three quarks for Master Mark!

—James Joyce

Is there a more mysterious idea for an artist than to imagine how nature is reflected in the eyes of an animal? How does a horse see the world, how does an eagle, a doe, or a dog?

—Franz Marc

Nomadic Wandering

In *The Logic of Sense* (1990a) and *Difference and Repetition* (1994), Deleuze alludes to the types of series with which we are now somewhat familiar: connective, disjunctive, and conjunctive. Regarding seriality, during our century the focus has been obsessively on language and virtually language alone: for Frege, the two series are, roughly speaking, sense *(Sinn)* and reference *(Bedeutung);* for Husserl they are expression and indication, for Saussure, signifiers and signifieds, and for Hjelmslev, expression and content. The so-called poststructuralists and neopragmatists hardly fare any better regarding this indelible element of linguicentrism over the past century.

Deleuze's concept of series, perhaps unfortunately, also bears heavily on language. But, I would suggest, his series imply and are emergent from semiosic activity that precedes the surface manifestation of language. Deleuze's two metaphors depicting the unfolding of multiple series, trees and rhizomes, entail a proliferation of signs ultimately leading to apparent disorder, chaos, anarchy. From Porphyrian trees to Linnaean taxonomies to Chomskyan parses, series upon series finally produce nonhierarchical, nonordered, unregulated, and unruly rhizomes: a mobocracy of aleatory, apparently impenetrable conglomerates. In Deleuze's conception, we can no longer conceive of a world of things taxonomizable into myriad packages all of which are partly to almost wholly arbitrary but merely relations and events—Deleuze's influence from Stoic logic emerges here. According to this conception, nouns (substance) and adjectives (attributes) are suppressed, and verbs (events) are highlighted. A tree is not simply green, according to the standard category for trees, but in a process of leafing, greening, yellowing, or deleafing. A lumberjack and his chain saw do not simply cut it down. There is a process of cutting and being cut. Man and chain saw that cut and the living cellulose being cut are in the process inextricably mixed. They merge and overlap with one another, they eventually become one, they are one in terms of the relations between them and of an event consisting of cutting and being cut. During this process, hands wield the saw that penetrates first the bark, then the fleshy cellulose of the living organism. At the apex of the initial wedge-shaped absence in the tree there is linear penetration; metal forces its way deeper into the heart of the trunk; gravity exercises its ubiquitous force; the tree is in the process of mutilation/being mutilated, it is falling/being felled, it is dying/being terminated. The tree's neighbors, the forest, the peaks and valleys and rivers, the outlying area, nature itself, all are wounded. And the lumberjack has suffered irreversible changes as well: he is another step toward death himself, though at the outset at least this change does not appear as catastrophic as those changes wrought on the tree and its environs.

Whether speaking of this imaginary event consisting of lumberjack, chain saw, and tree, or for that matter, any supposedly real world happening or even a novel, dream, hallucination, or make-believe world, it is all basically the same. Facts as well as fictions and fantasies consist of incorporeal events, in much the same fashion that the real import of real world happenings is sign-events more than sign-things.[1] Events (Peirce's category of Secondness), in this conception, have no depth or being (in the sense of the superposition of all possibilities—Firstness) but dwell on the surface. They do not *exist* in the customary use of the term, they *subsist*—commensurate with Alexius Meinong's strange ontology—in the things they inhabit. And they enjoy virtually no extension through time (as identical with themselves, as generalities, types—Thirdness) but are what they are during some fleeting moment or other.[2]

Deleuze's Stoic-inspired notion of depth (the virtual states of things, pure Firstness) in conjunction with surface (events, Secondness) sits quite comfortably with Peirce's concept of the sign. As a matter of fact, Deleuze's recapitulating the long tradition from the Stoics to Spinoza to Meinong actually entails triadicity rather than binarism. Triadicity includes mediation between the sign (that which signifies) and that to which it relates (the semiotic object, which exists, or merely subsists in the Meinongian sense). If the sign is not objectively existent but relational, and if the object of the sign is relational as well, then the mediating effect of the sign and the object to engender an interpretant constitutes an event. The existence of this event, then, is not referential but purely relational, a matter of what is between the sign and its respective semiotic object, and between the sign and other signs.[3] If, as Peirce points out, an elementary sentence (proposition, compound *symbol*) consists of a subject (noun, *index*) and an enumeration of one or more of its attributes (predicate, verb, adjective, *icon*), then that symbol's function is chiefly verbal, that is, eventlike (the chain saw is cutting, the tree is being cut) more than thinglike. The supposed thingness of Secondness (what the "thing" is) plus the range of possibilities of the thing (Firstness) is mediated neither by a thing nor the property of a thing but by relations through a thingless, attributeless happening, an event, a naked sign-event. We cannot be in possession of the thing *an sich* (Kant), nor can we know all the possible attributes of the thing, which are infinite in number (Spinoza). The best we can do is survey the sign-events on the stage apparently before us, the stage we are on as both actors and participants, and hope for the best.

In Deleuze's (1990b, 165–67) interpretation, for Spinoza a thing, no matter how insignificant, contains an infinite number of possible attributes. It is not a form but a complex of interrelations that precedes the engenderment of signs. The thing's affective powers are limited or virtually unlimited, depending upon the semiotic agents involved in its interpretation. For example, von Uexküll's tick is susceptible fundamentally to only three effects of a food-sign-event: (1) climbing up a tree toward a brightness-event offered by the sun, (2) detecting an odorness-event of oxybutyric acid and falling on the unwary animal below, and

(3) sensing a warmthness-event close to its host's hide, the final sign-event inviting it to settle down, penetrate some vulnerable spot where a toughness-event is not so pronounced, and enjoy a meal. And all this without language. The human semiotic agent's susceptibility to prelinguistic sign-event affectation is, in comparison to that of the tick, of overwhelming complexity and diversity, a near-Brownian movement of colliding, interacting, interpenetrating, including and included, sign-events only a minute portion of which can be interpreted at conscious levels. The world, *our world,* in its myriad complexion, is an effervescent to-and-fro process of explication from the implicate, unfolding from the enfolded. It is a dizzying field of multiply intersecting series—a rhizome of intricately differentiated rhythms and affective intensities, without sign-things, properly speaking but, rather, relations and sign-events. And to further complicate the issue, we have language, proud language, whose charge is to wrap all this nonlinguistic semiosic activity into words, words, words. A daunting task indeed.

The task is even more problematic in view of linguicentrism's self-imposed restrictions. Upon taking into account the range of possibilities for relations and sign-event series, one is forced to conclude that sentences, statements, propositions, arguments and their counterarguments, and entire texts and universes of discourse resist inclusion within that positivist-linguicentric triad of pigeonholes, (1) true, (2) false, and (3) meaningless (or nonsensical). A fourth category—radically nonlinguicentric in its inception—becomes imperative: *imaginary* ($\sqrt{-1}$, phlogiston, the ether, quarks and other scientific fabrications, any and all fictions, fantasies, lies, make-believe worlds, dreams, illusions, delusions, and hallucinations).[4] A sign can presumably be either true, false, or merely nonsensical in terms of its meaning, but when Deleuze's notion of sense enters the scene, it can have meaning even though it is deemed nonsensical (e.g., $\sqrt{-1}$ put to use in quantum theory and relativity), and it can have sense even though it is considered meaningless ($\sqrt{-1}$ standing on its own or purely as a mathematical operation). In fact, $\sqrt{-1}$, considered as an eternally oscillating point on the Argand plane (fig. 12), is quite apropos. Commensurate with the plane, Deleuze's approach to events (relations and sign-events) is comparable to a mathematical field of singular points as possibilities for defining unlimited domains of equations and solutions. The point, by successive unfoldment, if it is lucky, can be expanded into the solution to a problem worked out on the two-dimensional Cartesian grid, with all necessary lines and their intersects. While a particular point can be specified only after the equation has been solved, it is by no means the result of the solution. Rather, its possibility as an implicate entity that can be explicated into a particular semiotic world on the Cartesian plane was always already there, as a virtual field of possible equations from within which an unending series of problems and their solutions could have been actualized.

The same is to be found in the cases of the consonance, conjunction, collusion, and even conflict resulting when human relations and multiple series of human sign-events combine into a whole, which is taken chiefly or almost exclusively in terms of its linguistic manifestation: linguicentrism. But actually, the vast majority of the signs making up this whole are prelinguistic. Take a professional conference typical of the superhyped ego-world of academia. It may be viewed in terms of (actualized, unfolded, explicated—Secondness) bodies meeting at certain locations and remaining in somewhat orderly arrays during specified periods, while between these arrays perplexing streams of bodies of all sizes and shapes and dressed in attire of all fashions move, with intermittent fits and starts, along meandering streams, bucking white water, and occasional cataracts, presumably toward some goal or other. Along these streams, intermittent periods of apparent chaos resulting from far-from-equilibrium conditions somehow settle into order during certain time frames at certain locations, only later to dissipate into unruly flows once again.

From another, complementary Stoic-Deleuze vantage, the conference can be viewed as nonphysical surface effects produced by the random to orderly coming and going of the bodies, a sort of floating (enfolded, implicate, Firstness, coupled with relations between Secondness and Thirdness) film everywhere sensed yet nowhere localizable, ubiquitous yet strangely intangible. This surface is indifferent to the individual bodies involved—that is, the bodies that strive so desperately to evoke the image of individuality (difference that makes a difference) in their appearance, their actions, and above all, in their linguistic prowess. As far as the Deleuze surface is concerned, this individuality is no more than illusion. The vague surface disclosures are quite indifferent to these bodies' jubilance, fear, frustration, anguish, petty jealousy, to their pomp, arrogance, pedantry, and even to their clownish antics, their presupposed superordinance and subordinance, their strategic linguistic maneuvering, their wheeling and dealing. All these activities are trivial and even irrelevant, from the complementary indifferent view. The ubiquitous but unseen film of indifference remains in much of a quietist mood; it is an anonymous process with no discernible beginning or ending. It consists of relations and relations between relations, and if "existents" there must be, they are no more than the manifestations of mere happenings. There is no question of "this session here-now, that session there-then, 'I' now pressing forward to score some points with that notorious repository of academic power and now off for a drink with some old friends to enjoy some inside gossip." There is merely conferencing, with no-*thing* actually existing as a self-identical entity anywhere or anywhen. It is the becomingness of becoming, the beinglessness of there not yet having been a becoming of any-*thing* determinable.

All in all, however, this indifference, this Firstness, somehow promises meaning by way of sign-events emerging through other sign-events (Secondness). This forms the film of Firstness-Secondness between particular semiotic worlds and the eternally receding real world, on the one hand, and the film of the becoming of the Thirdness of meaning without its quite being able to arrive, on the other. Meaning is never either here or there but always already elsewhere. It is incessantly moving on. Our obstinate quest for the ultimate kernels of meaning, then, is an impossible dream. Meaning is the intangible by-product of the paradoxical, self-contradictory field of *possibilia,* the superposition of all that might be, the vagueness of Firstness lying in wait for specification (Secondness) and interpretation (Thirdness). However, never to be had, like the hand trying to grasp its own thumb, meaning's elusive nature as sheer becoming apparently defies all intellection and all common sense.

Meaning, it can only be said, is always becoming something other than what it is. Yet the obsession is to render it actual, as specifiable Seconds. But as that which has been actualized it is no more than a set of once occurring particularities, *haecceities,* no more than fleeting, scintillating flashes without enduring countenance. Any and all attempts to pin it down as Secondness inevitably end in an infinite regress. The problem is that the continuous film of Firstness, everywhere and everywhen yet nowhere and nowhen, seems to inhere in all signs. But at least it is felt, sensed as such, so why can't a sign and its meaning be specified? It cannot , because it is no sign-thing as such but a sign-event, a happening. So why can't Thirdness lasso and hog-tie the 'thinglessness' of Firstness and the elusive, fleeting, ephemeral 'thingness' of Secondness to bring some enduring form of meaning to the whole affair? It cannot, because Thirdness, as the self-proclaimed author of generality, sets for itself a task perpetually destined to remain incomplete—for us fallible human semiotic agents, at least.

In light of the hazy example of a professional conference, it becomes apparent that the field of Peirce's vagueness, of signs as hardly more than *possibilia,* as virtuality not yet made actuality for some semiotic agent or other, consists of the equivalent of a field of singular points the combination of which is no more than an indefinite smear—an "electron cloud" of meaning. Nothing is actually "there," to be precisely specified, nor is it something in the head of the semiotic agent or somewhere "out there" in that chimerical Platonic empyrean. It might appear, consequently, that the relations combining to make up the meaning of the sign-event are actualized in aleatoric fashion, by chance, a throw of the dice, a shot in the dark. Yet, the other picture also looms large, that of the semiotic agent (for example, a female agent) as active participant within the flow of semiosis, who constructs a world, *her world,* at the same time that *the world*—the range of all possible world versions—containing the semiotic agent, participates with her to bring about the actualization (Secondness) of part of that which it is as possibility (Firstness) and move it on a bit further toward the realization (Thirdness) of that which it can become, but since that realization is a receding horizon, it is always in the act of becoming. The semiotic agent is not what she is without each and every aspect of *her world* as *possibilia,* actuals, and becoming, and *her world* is not what it is without each and every semiotic agent.

This image of the whole and its interdependent, interrelated parts, of complexity bred from simplicity and order out of chaos, bears a further look through the microscopes offered by Calvino and Borges, especially with respect to the emergence and development of life processes.

And It All Seemed So Simple?

Calvino's "Blood, Sea" (which appears in *T Zero* [1969]), as we observed previously, is about that which makes us "primitive" in the most primitive sense, that is to say, that which is hardly different from what there was before there was anything to distinguish some-*thing* from some-*thing* else, before there was an artificially and partly arbitrarily constructed inside and an outside, an "I"-here and otherness-there.

In the beginning there was Firstness without specifiable sign-events (Secondness) or meaning (Thirdness). This was the beginning, which, our ubiquitous narrator, Qfwfq, tells us, is from the inside about the same as it always was at least for those for whom there was an inside and an outside, that is to say, for a cell, which has foregrounded itself from the background of the molecular soup by indicating a boundary (membrane) that sets it apart from what it is not. The inside/outside split entails, then, some originary act of distinction, though it cannot be specified. The cell raises itself by its own bootstraps from what to all appearances would be a homogeneous (continuous) background to distinguish itself and hence establish a heterogeneity (discontinuity) consisting of itself and its environment (Varela 1979, 1984). This is in essence a matter of continuity cut into sets of discontinuous breaks, which usher in the dilemma introduced in our dialogue between Simplicio and Sagredo: if continuity and discontinuity are incompatible, then Zeno's interdictions inhere; if they are not incompatible, then Zeno is hardly of any consequence.

The shell of the "VW bug" within which the four characters of "Blood, Sea" are crammed plays the role of a cell membrane, with a nucleus of four sentient entities. Qfwfq, from inside his little skin in the sea, is swimming, is pulsating, and is perhaps the recipient of a few vague messages transmitted through the waves. The flow, the undulation, the vibration, give him the dim, cloudy sensation of being inside and there being an outside. But what is outside is fundamentally the same as that which remains inside. So of what can the boundary, the dividing line, possibly consist? Can it be any-*thing* at all? Is it not simply an infinitesimal line, an imaginary quantity that is actually no more than a quality between *this* and *that*—which are essentially the same? It is like a Dedekind cut somewhere along a continuous line separating one infinite set of points on one side from another infinite set of points on the other side? Is that what a beginning (of Firstness) is all about?—the incessantly becoming dancing of painting-talking-writing-sciencing?

Of course, Calvino provides no analytically based answers, no hypothetico-deductive theory, no rigorously demonstrable proof. For most likely there are none. What he gives us is a certain giddy feel (of Firstness) for what makes us *us* and the other *other,* for what it is to be "in here" and distinguished somehow from what is "out there" (Secondness). A distinction splits a line, a space, time, or the entire world into two parts. This is the most fundamental of acts, the act of separating something from its environment. A distinction, however, is not made simply for the fun of it. We are not talking about any trivial game here. A distinction is some-*thing* that separates something from something else for some purpose or other for someone in some respect or other (Thirdness). I could have written "for some reason or other" or "in order to provide some function or other," but that would be excessively anthropomorphic; at this most basic level, there need not yet be any reason or function, and whatever purpose there may be is no more than implicit. A distinction, nevertheless, coexists with some exceedingly vague purpose.

Oh, by the way. In order to avoid an oncoming Jaguar, the driver of the "bug" hits a tree, and bug and passengers go spinning down a precipice and into the ocean. We learn that "the sea of common blood which floods over the crumpled metal" is not the blood-sea of their origin but only "an infinitesimal detail of the outside" (Calvino 1969, 51). The sea-blood, the blood-sea, that which is inside the bug, is hardly different from the homogeneous soup of the every-*thing* from within which bug and passengers lifted themselves up by their bootstraps. All is a matter of sign-events distinguishing and indicating themselves from their surroundings in time, a time embracing, and embraced by, a tenderly tentative, delicate, fallible space.

In "Blood, Sea" we have Firstness in the most primitive sense and the barest of distinctions and indications (between Firstness and Secondness) in their most simple manifestation. In fact, at this stage it can hardly be said that there are events or relations, and much less meanings. Complexity of relations and sign-events as they begin their long trek toward meaning is just around the corner, however, in mitosis.

Everything Divides up into Itself

The narrator of "Mitosis," which appears in *T Zero* (Calvino 1969, 59–74) tells us that in the beginning there were no relationships in terms of differences that make a difference between some-*thing* and some-*thing* else: I'm talking about the times before I had established any relationships between myself and anything else, there was a cell and the cell was me, and that was that (59). In this world it doesn't really matter whether or not there were other cells around. There was simply the narrator-cell, a distinction and an indication of the most vague sort.[5] That is, however, "already quite an achievement" (60). In spite of his or her apparent autonomy as a sign-event virtually unrelated to anything else, any other, our narrator tells us s/he is "dying of love."

Dying of love? How can this be? That ancient but eternally enigmatic process of being in love presupposes self-awareness and hyperawareness of some other by way of the most sensitive form of subjectivity. Our narrator, however, tells us that s/he can be perfectly objective about the matter, for, there not yet being any other at the time s/he was dying of love, there could have been no subjectivity, which requires some other, some object toward which the subject can be subjective. It appears to be a simple matter: there is a cell, and this cell is a unicellular organism, and this unicellular organism is me, and I know it, and I'm pleased about it (Calvino 1969, 62). S/he is a sign-event in space and time, space and time are her/him and s/he is space and time. In other words, s/he is a distinction and an indication in the here and now, without there being any clear-cut there and then; s/he is a point in space and an eternal present in time. Outside, if it can be said that there is any outside at all:

> seemed to me a void I might occupy in another moment or point, in a series of other points or moments, in short a potential projection of me where, however, I wasn't present, and therefore a void which was actually the world and the future, but I didn't know that yet; it was void because perception was still denied me, . . . but I had this contentment because outside of me there was this void that wasn't me, which perhaps could become me because "me" was the only word I knew, . . . a void that could become me, however, wasn't me at that moment and basically never would be: it was the discovery of something else that wasn't yet something but anyhow wasn't me, or rather wasn't me at that moment and in that point and therefore was something else, and this discovery aroused an exhilarating enthusiasm in me, no, a torment, a dizzying torture, the dizziness of a void which represented everything possible, the complement of that fullness that was for me all, and there I was brimming over with love for this elsewhere, this other time, this otherwise, silent and void. (Calvino 1969, 63–64)

Zeno in yet a new garment! If the cell occupies the point it occupies during the moment it occupies that point, then how can it get on to another point in another moment? It is not a matter of here some-*thing* and there a void part of which is to be filled by that some-*thing,* and now some-*thing* is, and at an as yet timeless future moment that which now is will be something else when it occupies that part of the void. It is a matter of the Parmenidean what is, is, and what is not is not, for if it is not it is not; hence it cannot be thought, and what is unthinkable cannot be.

Nevertheless, our narrator is "happy to be me." But the happier s/he becomes, the more her/his nucleus becomes charged with "impatience." Eventually, everything s/he was ends up becoming "nucleus, absorbed there and registered and accumulated in a serpentine twist of spirals, in the gradually different way that they were forming a skin and unraveling, so . . . everything I knew I knew in the nucleus, if that wouldn't involve the danger of making you believe in a *separate* or perhaps even *opposing function* of the nucleus with respect to the rest, whereas if there's an agile and impulsive organism where you can't make all these distinctions that is the unicellular organism" (Calvino 1969, 65) (emphasis added). There is distinction and separation or indication, to be sure. Strangely, there is also a "state of desire." Now, how can this be? Mere distinction and indication have no room for desire, which entails something unfulfilled, some lack. How can there be some-*thing* to be filled if there is no-*thing* more than a distinction between *this* and *that?* There is no gap, no-*thing* in between that can be occupied by a third, which would give rise to an event, a happening, a sign-event as some-*thing* related to some-*thing* else—the *that*—for some-*thing*—the *this*—in some respect or other.

At least we can say, perhaps, that for Calvino's hero(ine) this state of desire is not without tension. One passes, we are told, from a state of satisfaction to a state of "mounting satisfaction" and then to a state of "dissatisfying satisfaction, namely, of desire" (Calvino 1969, 66). It is not that some-*thing* is missing or lacking, for there is no-*thing* other, elsewhere and elsewhen, that was or could have been but now is not. In other words, there can be no lack unless there is the absence of some-*thing* that previously was. But there was no elsewhere and elsewhen. Yet our unicellular organism's state of desire exists; it is simply a vague tendency (First) toward an elsewhere and an elsewhen (Second), toward an otherwise that might contain something (or, let's say, the world) or contain only me, or me in relation to something (or to the world), or something (the world) without me any more (67). This lack—

or at least exceedingly vague sense—of an outside world must somehow express itself. To do so requires language, but there is no language, for all the happenings of Calvino's solitary cell are prelinguistic. Or perhaps better put, the only language available is "all those specks or twigs called chromosomes, and therefore all I had to do was repeat those specks or twigs and I was repeating myself, obviously repeating myself insofar as language was concerned, which as you will see is the first step toward repeating myself as such, which as you will also see isn't repetition at all" (68).

To make a long and familiar story short, our narrator is seized with a need to stretch her/himself to full length, with elongated cytoplasm, causing the nucleus to explode into two of itself. During the process s/he remains her/himself and at the same time becomes some-*thing* other than her/himself. S/he becomes her/himself plus a reduplication of her/himself, with each of the two equally new and equally old. There is an original and a copy. But it is not a simple matter of here the original now that repeats itself into a copy there, then. There is no repetition in the genuine sense, but merely the equivalent of a verb that reenunciates itself, that says itself. But that cannot be it, not really. "To say" is not adequate, for "to say something you have to have someone who says and something that's said, and this wasn't actually the case at the time" (Calvino 1969, 71).

Quite simply put, one somewhat symmetrical entity divides into itself, into two somewhat symmetrical entities: a sign-event of the most fundamental sort, and the beginning of relations. It is much like a tug-of-war between two eddies when an interval was formed, and at that juncture our narrator takes on some form of dim awareness of her/himself as two equal entities, one on one side and the other on the other side, "joined by a bottleneck that was becoming finer and finer until it was only a thread, and at that instant I was for the first time aware of plurality, . . . I felt the plurality in me as the image and destiny of the world's plurality, and the sense of being part of the world, . . . and at the same time the still-sharp sense of being me" (Calvino 1969, 72). The ecstasy forthcoming as a result of this process is described much as if it were a mystical experience. It was the last self-contained instant when our narrator was one with her/himself, with the other, with the world. It was that moment when s/he began to become aware of some-*thing* else, for s/he sensed that there was a coming forward from the void of the elsewhere and elsewhen, of a society of cells, a multicellular concoction complementary with the concoction of cells our narrator was becoming, is becoming, will have been always already becoming. One cell (sign-event) becomes a society of cells (sign events), and interrelations between the society of cells and between those cells and their world begin to take on importance.

How can we possibly go on from here?

And Now, Simplicity from Complexity?

Perhaps by a brief incursion outside Calvino's cell-world into current notions of cell-worlds in the field of biology.[6]

According to earlier observations, crystallization in its purest form involves the monotonous reiterative union of identical units. Rather than movement, it depicts permanence; rather than dynamism, stasis; rather than life, death. Growth of a crystal is relatively uncomplex. The same pattern appears tediously to repeat itself; reproduction is entirely on the surface, with no detail in limbo or hidden from view. Each repetition yields a virtually exact replica (simulacrum) of what went on before. Each is new, yet since it is for practical purposes identical to the old, there is actually neither new nor old but timeless invariability. The cell, in contrast, prefers a different route to reproduction. At the outset, we notice that the cell's anatomy depends chiefly on the details of a few protein sequences, governed by the genetic code. We also observe that reproduction of the cell is made possible by means of its three-dimensional molecular structure being determined by a one-dimensional algorithmic order specified by its DNA—that is, one-dimensional generation giving rise to a three-dimensional manifestation. The code, the algorithmic unfolding, and the one-dimensional linear messages remain entirely within the system, while the three-dimensional cell exists within a two-dimensional topological wrap, its skin, serving in the capacity of boundary conditions.

In fact, the cell avails itself fundamentally of two different systems: the double-helix sequence of nucleic acid that remains linear and recopies itself one-dimensionally and the protein sequence that spontaneously unfolds into a three-dimensional form. Its three-dimensional complexity is capable of reproducing itself due to the underlying simplicity of the one-dimensional sequence. For, in the biotemporal domain the order is an order of an order; one order is linear, the other requires three-dimensionality for its development. Unlike the crystal, the cell reproduces itself not by a three-dimensional accumulation of layers on its two-dimensional surface but from within, by subdivision, with the self-enclosed surface becoming a mirror image of itself. The crystal's surface appears continuously to expand. In contrast, the cell's surface makes a copy of itself while the surface itself remains intact, like a soap bubble splitting into two, and the topological space apparently remains the same for the original and its copy. The copy, or the "explosion of the nucleus," as Calvino puts it, involves a perplexing twisting, turning, recursive, quasi-symmetrical topological form. As pure topology it is comparable to $\sqrt{-1}$ flip-flops and Möbius strip transformations the likes of which we have seen in figures 4, 5*a*, 6, 11, 12, 13, and 21. In other words, linear generativity of symbols undergoes a fold on a two-dimensional plane in three-dimensional space, thus breeding a paradox, which would have been impossible without the added dimension.

From another angle, this self-reference-self-reflection is analogous to Diego Velázquez's *Las Meninas* (1656), which contains the painter and the painting he is painting as well as the spectator, who is also sucked into the painting, all in one convoluted tangle. *Las Meninas* evinces visual resonances with *Don Quixote, Tristram Shandy, Hamlet,* and a host of postmodern contemporary narratives. This masterpiece distorts reflection and representation, classically conceived, thus incorporating an irreconcilable contradiction. The absent outside becomes the nonempirical inside and the inside and the outside are in eternal conflict, a topological agonistics for which there is no decidability (Searle 1980a). By a comparable twist, our wishful mirror of nature is the image of an image of an image, each successive one an inverse replica of its predecessor and in turn the flip side of its successor. We are caught within an eternally coiling, whorling, vertiginous, vorticose cascade, numbing in its complexity yet exhilarating in its breathtaking volatility. If this breeds nightmares of uncertainty, oceans of ambiguity, and an apparent promiscuity of paradoxes threatening to dissolve all dreams of reason, harmony, and stability, I see no call for despair. Rather, it opens the door, if not exactly to a Nietzschean-Derridean joyous linguicentric play of freewheeling signifiers, most certainly to a vision of open, creative, self-organizing dialogue with one's self, with the others of one's community, and with the other of nature at large, engaged in the process of its own self-organizing project.

That much written, the confused snarl evinced by my previous

Möbius-like example might tend to evoke in the mind other topological quandaries and their linguistic counterparts. Take, for example, the perennially debated "square circle," of which physicist Percy Bridgman (1959, 34) writes:

> We cannot say that such a thing as a square circle, . . . does not "exist" without implying by the mere fact that we are using the words and talking about "it" that a square circle has a certain kind of existence. . . . It seems to be that the situation can be adequately dealt with by reducing it to the purely formal level. If we say, "The combination of words 'square circle' has no referent either in the objective external world or in the conceptual world of logically consistent objects," it seems to me that we have said all we need to say.

The import of Bridgman's words stems from his suggestion that we take the words *square circle* as symbols qua symbols and leave it at that. The problem is that this is a radically linguicentric move, as if language could be the cure for any and all quandaries. But yes, Bridgman's advice is legitimate, for the very words "square circle" testify to some sort of "existence"—if nothing more, the existence of a couple of homeless, wandering signifiers. However, I would also say no to Bridgman's advice, for what makes the combination of two incompatible terms such as *square* and *circle* so intriguing is precisely their extralinguistic (hence nonlinguicentric) nature.

Of course, any attempt to create reference for the words or to exercise mental gyrations in an effort to create an image of the symbols is a futile enterprise. Two-dimensional visual illusions that illustrate the limitations of our faculties for constructing images are legion—for example, familiar Gestalt images, the Müller–Lyer illusion, the Ponzo illusion, the Zöllner illusion (Gregory 1966, 1970). At the next stage of dimensionality, in addition to mind benders of the Escher variety, we have actual physical phenomena such as that of two wire squares facing each other and linked by a connecting wire that can be rotated in such a manner that the squares appear to glide flatly over one another, or alternately they appear to oscillate about a vertical axis, reversing direction of rotation at every half-turn (Gregory and Gombrich 1973). Attempts to create more perplexing images of the fourth dimension (hypercubes, tesseracts, computer mock-ups) are amusing, and they also serve effectively to reveal our paltry perceptual and conceptual powers. However, topologically, a square can quite easily be transformed into a circle and vice versa. Such concepts as phase space, *n*-dimensional Hilbert space, complex time, and other comparable mind twisters are no problem as far as abstract symbols go. But these symbols, chiefly mathematical in origin, are not inextricably caught up in linguicentrism. In fact, they are nonlinguicentric through and through. Whether we take mathematics in the Platonic sense of real entities "out there" in that diaphanous space of pure abstractions to be discovered by the right-minded mathematician or in the intuitionist sense of purely mental constructs, the same conclusion inheres: the Thirdness or strictly symbolic depiction of a thought-sign incorporates and interacts with Secondness (or inexicality) and Firstness (or iconicity). The three sorts of sign-events, the three categories, Firstness, Secondness, and Thirdness, are all knotted in a tight embrace, and none of the three can be separated from the other two. There is no freewheeling *ars combinatoria* of floating signifiers in the semiologist's playpen, but rather, everything consists of signs that take turns playing the roles of Firstness, Secondness, and Thirdness.

Actually, by subtle acts of intuition and of mind we are ordinarily capable of overcoming confusions comparable to the square circle sort with the same ease that Achilles overcomes his plodding counterpart. A mountain range can be reduced topographically to a two-dimensional map—the collapse of three-dimensional space into two-dimensional space of iconic, indexical, and symbolic character. During this process, myriad contours are reduced to straight increments, inclines are reduced to a flat surface, violent undulations of river beds are reduced to smooth curves, and geometrical conglomerations making up cities are reduced to dots and spots. Yet, the experienced traveler can easily find his or her way around, giving the mechanics of so doing hardly a second's thought. That is, the map is ordinarily not confused with the territory. Thus, at increasingly sophisticated levels of modeling and mapping, a topological drawing can figuratively become a set of "epigenetic paths" (Waddington 1957), an expanding dotted balloon can become a model for the "expanding universe" (Gamow 1947), pockets in a billiard table can become "black holes" (Davies 1988), and a James Joyce neologism can even become "quarks," and variously "colored" ones at that (Herbert 1985). By a comparable move, it is now legend in psychological and psychoanalytical circles that Lacan's (1966) "Schema R," Bateson's (1972) "double bind," and R. D. Laing's (1970) "knots," or their equivalent thereof from other sources, can play havoc with the best of one's intentions. Yet, by topological transformations, one can potentially become aware of the knots in one's thought in such a way that what was a vicious tangle becomes relatively harmless. The upshot is that, once sign-events and their interrelations have emerged into the light of day, things begin becoming fuzzy, fractals take their place between whole dimensions, and therein, meanings arise among the myriad differences (meanings are not simply in the signs themselves, in the heads of their knowers, or in Euclidean space between signs and signs, signs and their knowers, and knowers and knowers) (Merrell 1997).

Now let us take a giant step backward in our game of topological hopscotch from the paradoxical Möbius twist and the likes of Lacan, Bateson, and Laing to the humble cell. What we would hope to find is simplicity to soothe our aesthetic preferences. But to no avail, at least if we heed Varela's (1984) advice. He begins his description of the cell as a Cretan lie monger with a description of Escher's pair of *Hands* that draw each other. Varela highlights the cell's autonomy, somewhat to the detriment of its openness to its environment. In terms of self-referential systems, the Möbius twist is foregrounded from the perspective of autonomy, to be sure. When the system opens itself to its environment, the paradox can be at least ephemerally resolved by the introduction of an "axiom" from outside—in line with the limitative theorems (Gödel's incompleteness, Tarski's metalanguages, the incompleteness of science [Schlegel 1967], and such). The organism, playing out its autonomous role, swims against the current of the now defunct behaviorist program of nurture and flies in the face of the cognitivist, Chomsky-inspired, nature hypothesis.

Some examples are in order before explaining what Maturana and Varela mean by "autonomous." A dog and a pedestrian are walking along the sidewalk in the same direction but on opposite sides of the street, when suddenly the animal decides to cross the street and greet its human counterpart. A pelican at the wharf during the early morning hours suddenly notes that it is time to spread its wings and dry them out under the rising sun. A gerbil in its cage doing its morning aerobics in its mechanical Ferris wheel decides it has had enough and approaches its mate for some grooming. A flatworm running a T-maze opts at the juncture to make a left rather than a right turn. Now, the dog's behavior is neither exactly programmed nor the result of having received instructions from some outside source, and the same can be said of the behavior of the other animals. The flatworm, in fact, does not turn left merely as a result of repeated punishment-reward trials and errors, but because, given its role as an autonomous entity, it is subjected to certain changes and disturbances in its sensory input that it processes, constructs, and interprets so as to maintain a satisfactory regulation and balance of its inner workings. This, for Maturana and Varela, distinguishes the autonomous or autopoietic self-referential, self-regulative aspect of the biological entity from an allonomous or allopoietic system requiring an outside agent for its proper functioning. As we noted in chapter 2, a car, as allonomous, did not make itself but was made. It is now operated by a driver, and if it breaks down, it cannot fix itself but requires a mechanic to do the job. Escher's *Hands* drawing as a self-contained system gives the appearance of its being autonomous; but, of course, the viewer knows the paradox is not real, for Escher's unseen hand once lurked above, drawing the hands drawing themselves. So it is allonomous.[7]

The upshot is that the component parts of autonomous entities refer to each other, and they mutually specify themselves. It is in this sense, as we experienced through Calvino's narrative, that in the molecular domain life is capable of determining and organizing itself, thus acquiring the chiefly autonomous nature of its existence. An individual cell, Varela (1984, 311–12) writes:

> stands out of a molecular soup by defining and specifying boundaries that set it apart from what it is not. However, this specification of boundaries is done through molecular productions made possible through the boundaries themselves. There is, therefore, a mutual specification of chemical transformation and physical boundaries; cell draws itself out of a homogeneous background. Should this process of self-production be interrupted, the cellular components no longer form a unity and gradually diffuse back into a homogeneous soup.

Much in the order of Qfwfq's emerging into the world, Varela's cell sets up boundary conditions in terms of a membrane separating itself from the rest of the universe. From within the membrane, what Hofstadter (1979) calls a "strange loop" composing a "tangled hierarchy" is to be found the equivalent of which in natural language is our familiar Epimenides paradox: "I am lying." From within the sentence the paradox is irresolvable, nothing but eternal flip-flopping between an either and an or. A solution is forthcoming solely by a leap outside the system in order to understand the intention behind the utterance. Regarding biological organisms, the leap becomes trickier, of course, for we are, ourselves, biological organisms.

Can the/a Beginning Really Be a Matter of Word(s)?

"But," one might retort, "the same inheres when we include the inorganic universe within our purview, does it not? After all, is the physicist not a collection of 'wavicles' trying to understand what it is to be a 'wavicle'?" The point is apparently well taken. For the physicist, as part of the self-contained, self-organizing, self-referential universe, can hardly expect more. Besides, at this level it is all a matter of language: waves, particles, their combination into wavicles, and other symbols of scientific discourse are over the long haul no more than "word spinning," no more than so much language, as Bruce Gregory (1988) has so aptly demonstrated. By means of customary word spinning practices, the massive sea of prelinguistic semiosis is simply shoved aside to make way for the relatively comfortable prison house of language: linguicentrism.

Varela, in this respect—as does Calvino—provides the final insult to our vain efforts to know the world as detached, objective subjects upon declaring that the same self-containment of the Epimenides sort is found at the core of living organisms (see also Pattee 1969, 1971, 1979). In a somewhat uncomfortable manner of putting it, just as the physicist of quantum theory or relativity makes ample use of an irrational value, the imaginary number $\sqrt{-1}$, so also the organism's "tangled hierarchy," as a whole, evinces the equivalent of the statement "Everything within this boundary (membrane, epidermal layer) is a fiction." That is, the organism has, rather arbitrarily as it were, set itself apart from what is now conceived to be reality, which consists of the entire universe less the organism. As such the organism is not what the universe is, and the universe is what the organism is not. The organism's interaction with its world—its knowledge of its world—is tantamount to an imaginary realm, a netherworld somewhere in some ethereal sphere from which the universe can be seen. Its position inside/outside of/at the margin of its world is, once again, rather comparable to Wittgenstein's eye of the *Tractatus* (1961) that sees its world but cannot see itself seeing that world in the same act.

Varela points out that in the Epimenidean case, the paradox remains intact as long as its contemplator is unwilling to let go of the pigheaded desire dogmatically to choose between absolute "truth" and absolute "falsity." Once discarding this need, one can then see the sentence's vicious circularity as its own self-help program for specifying its meaning. From an outside vantage—that provided by the contemplator's perspective—the sentence becomes like a photograph of Escher, with hands on the drawing board, putting the final touch to his pair of hands engaged in their own viciously circular self-organizing project. Escher "imagines" himself in a neutral sphere from within which he is capable of generating an irrational self-referential system whose very existence depends upon his own creative hand. The Epimenides sentence, viewed from outside, is like contemplating an imaginary thought experiment one of the supreme examples of which is that of Einstein—to which Sagredo made reference—who wondered what it would be like to grasp the entire universe from a neutral perspective, as if lassoing a photon of light and saddling it for a scenic trip around the universe. Of course, Epimenides's sentence, Escher's hands, and Einstein's universe are allonomous rather than autonomous. They are the yield of creative minds; they are imaginary language-dependent models designed to help us get a feel for the unimaginable and ineffable. Biological organisms, on the other hand, are apparently somehow able to lift themselves up by their own bootstraps without the need of such an artificial leap of imagination giving them a view as if sub specie aeternitatis.

Varela relates the to-and-fro oscillation of the paradoxical systems in question to Mandelbrot fractals, whose iterative program yields such phenomena as a complex snowflake design ("Koch curve") from a simple equilateral triangle. This is pure mathematics, of course, but according to Mandelbrot and his followers it is capable of patterning the self-organizing character of all facets of change and growth found in the entire universe, inorganic and organic phenomena alike—and it would be capable, according to some of the preceding observations, of patterning sign-events and their interrelations and meaning. More apropos to the concerns of the present inquiry, and equally patterning life and nonlife, are Prigogine's far-from-equilibrium conditions leading to order through fluctuations. The Belousov–Zhabotinsky reaction, for example, creates a structure of concentric and spiral cells that undergo intermittent stages of pulsation and stability. The concentric circles are "waves" not of liquid motion but of chemical change, comparable to the action potentials succeeding one another as they travel along a nerve axon. As with action potentials, Arthur Winfree (1987, 169) writes, the onset of these Belousov–Zhabotinsky chemical changes "is so abrupt as to constitute a virtual discontinuity: the wavefront spans only a few percent of a millimeter; during mere milliseconds, one of the reactants explosively increases a thousand fold in concentration. Unlike water waves or sound or light, waves of this kind propagate steadily without attenuation. Though the waves may be spreading from a point source, it does not lose strength, because each newly excited region supports the passing wave to the full extent of its resources."

The reaction, in other words, is strictly chemical and involves no DNA, yet it "grows" as if it were endowed with life processes. With apparently minor adjustment the chemical reaction can be made to cease oscillating spontaneously. But its "excitability" and its capacity for self-organization remain uncompromised (Winfree 1972). The same behavior can be observed in living "excitable" media. For example, an inert neuron in a state of delicate readiness for firing in response to a stimulus can be adjusted by reducing the stimulus; then the firing occurs spontaneously, and it can be iterated time and again (Winfree 1987, 170). Indeed, given Arthur Winfree's conclusion, space-time forms in all vibes of life—heartbeat rhythms, fruit-fly and firefly entrainment, cricket chirps harmonizing with one another, morphogenetic oscillations in chick embryos, slime mold periodicities, and so on—are all part of, and in a certain manner of speaking a microcosm of, the interrelated, meaningful, cosmic wave ballet.

In formulating his own argument that life is a self-help program, Varela (1984, 315–17) asks us to compare the state of affairs inside a pair of strange loops to the liar paradox as an individual cell. I will attempt to perform a certain variation on both of these states for the purpose of the present concerns by introducing the concept of fictionality. In the Varelan spirit, consider figure 23. The sentence includes itself within the frame containing nothing but fictive sentences. Therefore if it is true, it cannot be fictive, and so it belongs to the sphere of reality (that which is nonfictive). But it cannot be true, since it includes itself in the class of all sentences to which it refers, so it must be fictive, that is, not real, but if this is the case, then what it says of itself must be true, and so it is real rather than fictive, and so on. This is somewhat comparable to Magritte's lifelike painting of a pipe underneath which the caption "This is not a pipe" is found. If we were to substitute "Is not a pipe" for "This," we would have " 'Is not a pipe' is not a pipe." It is as if to say of the paint-

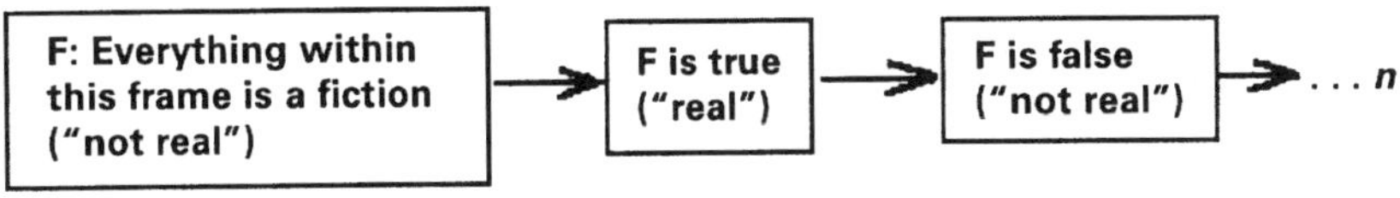

Fig. 23. Life within strange loops

ing, including the caption: "P: Everything within this frame is not a pipe," which yields "P is true" ⇒ "P is false" ⇒ "P is true" ⇒ "P is false" . . . *n*.

Varela then goes on to construct the following: "A membrane is formed" ⇒ "Metabolites are produced" ⇒ "A membrane is formed" ⇒ . . . *n*. He suggests that this twisted, self-contradictory use of language patterns the iterative growth and subdivision of a cell, or more specifically a Calvino cell, as we have observed, and more specifically still, life-sign-event interrelations. What Varela seems to imply is that in essence the cell, upon separating itself from the remainder of the universe with a two-dimensional boundary condition, the membrane, is acting out the one-dimensional linear statement: "C: I am that I am not." What the cell is, is not the rest of the universe and what the rest of the universe is, it is not, yet it remains within the universe. The cell is and is not a subset of the master set, the entire universe including the cell as a subset. In this manner, the counterpart of "Metabolites are produced" would be "C is true (real)," and the iteration of "A membrane is formed" would be "C is false (not real)." Both systems are thus given natural language window dressing. This window dressing is perhaps necessary in order to render the entire operation a tad more intelligible, but unfortunately it also marks entry into a certain form of linguicentrism: inordinate attention is given to either-or imperatives; $\sqrt{-1}$ is prioritized unduly over *i*.

Now let us integrate Calvino's narrator's verb, "to say," or better, "to become," into the equation. Let "to become" be S, the cell, or a living sign, as it were. Let "I" be the interpretation (meaning) of, that is, a sign of, S. Now, S is not the real thing; neither is it the real object for which it stands, which is itself not exactly real either but merely real insofar as it is the yield of semiosis—it is semiotically real, that is, it is a sign-event. "S," a sign-event, is transformed into (translates itself into, gives meaning to) its respective "I," another sign, which becomes, in collaboration with its interpreter, the cell itself since it is an entity unto itself; it is yet another sign-event, as if it were real. The entire equation is as if the signs in question were implicitly telling us: S_1: "I am that I am not" $\Rightarrow$ I_1: "S_1 is true (real)" $\Rightarrow$ S_2: "I_1 is false (as if real)" $\Rightarrow$ I_2: "S_2 is true (real)" $\Rightarrow \ldots$ *n*. Now, of course, we are standing quite tenuously on Simplicio's turf at the end of his dialogue with Sagredo.

If in nutshell form, the Peircean sign-event is something that is related to something else to someone in some respect or capacity, then that "someone in some respect or capacity" becomes another sign-event relating to another sign-event not yet conceived to be some difference—though to an infinitesimal degree—which in turn relates to yet another sign-event further down the stream. And this difference becomes another sign-event relating to yet another sign-event, and so on. Moreover, the interpreter (self), as him- or herself a sign-event, places the whole affair in another dimension. Upon the sign-event's being interpreted, the self becomes a displaced interconnecting mode in the sign-Event component's interaction, and the entire concoction, "Self-Sign-event," enters into the self-participatory rush toward some unseen, unknown, undefined, unfathomable finality. Self and sign-event, related and relationship, and meaning become fused and confused in conjunction with the ongoing effervescence of semiosis. During this process, like cells, living sign-events subdivide, they enfold and unfold along the stream: hence the life of all signs, semiosis as living process. In their own terminology, Deleuze and Guattari, Calvino, and even Beckett and Borges, say as much.

Just as an individual cell (a linearly generated entity) cannot step outside the boundary conditions (two-dimensional membrane) defining itself as something set apart from the remainder of the universe (of three dimensions), so also it has no world except what it experiences by means of its sensory perceptors, which make it what it is. Neither is there any possible indication of its origin, of that time when it first defined itself as an entity apart from its other. It is a viciously regressive, ever receding set of replicas of replicas of replicas, with fuguelike variations at certain points. Likewise, all sign-events, from the most humble icon to the most imperious of symbols, flow along, side by side, in the same semiosic stream as signs of signs of signs. In this manner we note anew, it would appear that we ourselves, our conscious selves, are caught up in the same infinite regress. Our present consciousness is conscious of our previous consciousness—what it now is not—and what our consciousness now is will be something else in the next moment of which there will be consciousness of what it then is not, and so on.[8] I repeat: we have no semiotic sonar mechanism with which to gauge the depth of the stream, no periscope allowing us to bring its banks into focus, no anchor we can drop to halt our movement within the flow, no sextant to determine where we are, no map to see how we arrived at this point or where we are headed.

So we go on to more complex organisms; we cannot remain silent.

Or, from Simplicity to Complexity, or Perhaps Vice Versa?

We read in Calvino's "Meiosis," which appears in *T Zero* (1969, 75–86), that there was an individual named Priscilla, who had appeared in the last lines of "Mitosis." Our narrator has been evolving, developing, is evolving, developing. He has finally expanded into a multicellular organism. Having attained this sublimely complex status, he is now prepared to accept a mate; he has to have a mate, for he can no longer divide into himself—that is to say, each of the unicellular organisms he carries within him can do so, but as a macrocellular concoction, he cannot. The vague desire, tension, lack of which we read in "Mitosis," is magnified manyfold. The narrator has now become the epitome of complexity striving for simplicity: he is a massive concoction of "sign-events" and their interrelations.

Qfwfq begins by telling us that "Priscilla is an individual of my same species and of the sex opposite mine, multicellular as I now find myself, too" (Calvino 1969, 75). But in almost the same breath he realizes he must confess that by multicellular he means a complex of around fifty trillion cells "very different among themselves but marked by certain chains of identical acids in the chromosomes of each cell of each individual" (76). Qfwfq defines the relations between himself and Priscilla as "chains of nucleic acids arranged in identical series in each of her cells and in each of mine" (76). In a meeting of the two minds two trains of thought are set in motion and coordinated so as to harmonize with one another. Simple enough. But he feels compelled to concede that

> a mental relationship involves several billion special cells called neurons which, however, function by receiving stimuli from such a

> great number of other cells that we might just as well consider all the trillions of cells of the organism at once as we do when we talk about a physical relationship.
>
> In saying how difficult it is to establish who's having relations with whom we must first clear the decks of a subject that often crops up in conversation: namely, the fact that from one moment to the next I am no longer the same I nor is Priscilla any longer the same Priscilla, because of the continuous renewal of the protein molecules in our cells through, for example, digestion or also respiration which fixes the oxygen in the bloodstream. (Calvino 1969, 76–77)

Surface simplicity turns out to be a teeming society of numbing complexity underneath. That is to say, myriad complexity at the mesolevel is engendered by the one-dimensional algorithmic, recursive, microlevel unfolding of a basic set of simples in such a way that at the macrolevel it can be taken—albeit erroneously—as considerably less complex than it actually is. In fact, with a few mental contortions, it can even be construed as simple.

That is to say in the narrator's words, there is a "special configuration which my cells and her cells assume through a special relationship between the environment and a special genetic heritage which from the beginning seemed invented on purpose to cause my cells to be mine and Priscilla's cells to be Priscilla's" (Calvino 1969, 77). Qfwfq doesn't know how many "I" 's have preceded the "I" he now thinks he is or how many Priscillas preceded the Priscilla toward whom he is running in heated pursuit. The "I" is the most simple of terms. But as is often the case the more a term is simplified the more complex it becomes. If the "I" is construed as a collection of amino acids lined up in a certain way, then the list of all possible relations becomes for practical purposes infinite. Yet once the cards are dealt, "I" is, well, quite simply, "I." That is the way it is taken. The vast majority of the possibilities have been disposed of, a paltry few selected, and the "I" becomes "walled up within myself, chained to my molecular program" (79). If on the surface there is some stuff that has certain relations with some other stuff that is Priscilla's, and if their respective "I" 's are altered appreciably in the process, that fact has hardly any bearing on what goes on below the surface: that teeming, trembling, oscillating, effervescing race of activity below. That other stuff, which is really what makes the narrator himself and Priscilla herself, is reduced to a stark amateurish he and she. Quite simple, it still might appear.

However, not really. In fact, it is all mind-numbingly complex (a tumultuous dance of interrelated sign-events). But, then, there is hardly any alternative. The narrator feels justified in throwing the blame on his parents, who "programmed me to be in the beginning . . . what I am: that and nothing else" (Calvino 1969, 80). And his parents' instructions contained the instructions of their parents, and so on, in an "endless chain of obedience." So he resigns himself:

> The story I wanted to narrate therefore is not only impossible to narrate but first of all impossible to live, because it's all there already, contained in a past that can't be narrated since, in turn, it's included in its own past, in the many individual pasts—so many that we can't really be sure they aren't the past of the species and of what existed before the species, a general past to which all individual pasts refer but which no matter how far you go back doesn't exist except in the form of individual cases, such as Priscilla and I might be, between which, however, nothing happens, individual or general. (Calvino 1969, 80).

All Qfwfq and Priscilla are is a catalog of possibilities, of experiences that are there, waiting, and ready to be repeated as they have been repeated countless times in the past. They proceed quite blindly, unthinkingly, gropingly: they are apparently the product of a chance affair, and what they do is a trial-and-error game (Jacques Monod's dice-throw universe).

Yet it is not really a matter of mere chance. There is some sort of paralleling between one "chain of obedience" and another one. The narrator and Priscilla are destined to separation and the impossibility of a genuine meeting, for between them there is nothing but "[v]oid, separation, and waiting" (Calvino 1969, 82). The universe apparently greets them with "blind indifference." They seem to be only an "envelope" for an encounter of the past that happens through them, without their active participation. Yet they are surrounded by freedom. They are free, free like the previous example of a solitary bee, or a termite or an ant, or even a molecule rushing about and accompanied by millions of frantic neighbors. They are free to do their thing, and yet they are not free, for they belong to a collection governed by statistical probabilities, and they are driven by biological demands. They are not free since they are bound to the demands of physics, chemistry, biology, and yet they are free because, as with the preceding examples of the dog, the pelican, the gerbil, and the flatworm, there is no predicting exactly when and where they will do what they do. What is more, their behavior implies symmetry and asymmetry, reversibility (timelessness) and irreversibility (time), equilibrium and disequilibrium. Calvino concludes his tale of the tryst between Qfwfq and Priscilla with an unexpected line: "[A]nd we camels feel suddenly light and you break into a run and I trot after you, overtaking you in the grove of palm trees" (86).

"Camels?" one might wish to query. Yes, camels! Calvino could hardly have chosen a more appropriate mammal. Observe a camel ambulating in its natural habitat. "Elementary," one concludes. Actually, it's quite complex. In fact, it's very complex. It involves asymmetry, irreversibility, disequilibrium, repetition but always with a difference that makes a difference. It is a matter of sign-events and the relations between them; it is $\sqrt{-1}$ breeding aperiodicity rather than periodicity, all a self-referential lie (i.e., it is paradox) held together in the equivalent of *i*: it is of the very essence of life. We humans walk by interchanging left and right sides so as to shift the phase of our movement—in its worst form, recall Watt's strained movements or Molloy's clumsy bike rides. Because we are bilaterally symmetrical units, since one-half of us is in essence a mirror image of the other half, one might expect that our left side would move just like our right side, except in inverse fashion. Not so, however. Depending upon how fast we are walking or running, and particularly when we alter our speed, our movements break the symmetry, though ever so slightly. In the case of quadrupeds, especially of the larger species such as camels, symmetry breaks to produce out-of-phase movement of the legs. Different animals prefer a variety of gaits, from hopping to walking to trotting to galloping to cantering. Most quadrupeds trot at low speeds, but camels tend to pace, and elephants amble—they could hardly do otherwise. Yes, camels. Their gaits evince the essence of symmetry breaking that gives rise to the complexity about which this entire disquisition revolves.[9]

We have Mr. Palomar's expert testimony on this issue. He never tires of watching the giraffes race at the zoo. Fascinated by their "unharmonious movements," he cannot decide

> whether they are galloping or trotting, because the stride of their hind legs has nothing in common with that of their forelegs. The forelegs arch loosely to the breast, then unfold to the ground, as if unsure which of numerous articulations they should employ at that given moment. The hind legs, much shorter and stiff, follow in leaps and bounds, somewhat obliquely, as if they were of wood, or crutches stumbling along, but also as if playing, aware of being comical. Meanwhile the outstretched neck sways up and down, like the arm of a crane, with no possible relationship between the movement of the legs and the movement of the neck. (Calvino 1985, 79)

Mr. Palomar observes that the giraffe seems like a mechanism constructed of a collection of heterogeneous parts, each of them a relatively autonomous machine in its own right. In spite of the apparent dissonance, there is a

> complicated harmony that commands that unharmonious trampling, an inner proportion that links the most glaring anatomical disproportions, a natural grace that emerges from those ungraceful movements. The unifying element comes from the spots on the hide, arranged in irregular but homogeneous patterns: they agree, like a precise graphic equivalent, with the animal's segmented movements. (Calvino 1985, 80)

Yet all organisms, like ourselves, do what we do quite naturally. We have the proverbial centipede who, when asked how it orchestrated the movement of all its members, thought about it and was paralyzed.[10]

Mr. Palomar's young daughter, tired of the lumbering giraffes, pulls him toward the penguins, whose strained bipedal movements do no more than inspire anguish in him. To her question as to why he is so interested in the giraffes, he speculates that it is perhaps because the world appears to move in such an inharmonious manner and he wishes to find some pattern in it, or perhaps because he is disconcerted by the uncoordinated, even seemingly random, movements of his mind and he would like to think, though he finds it increasingly difficult to do so, that they fit into some "pattern of inner harmony" (Calvino 1985, 80). Symmetry is broken in animal gaits—or Palomar's thought processes—which gives rise to a new form of periodic, undulating, off-balance sort of order.

Simplicity gives way to complexity, which evolves into simplicity of greater sophistication: the process is irreversible, for the original simplicity remains as a Paradise Lost.

Yes, camels. That is, symmetry breaking of the sort that leads to bilateral symmetry and the spiraloid characteristics of seagoing crustaceans, ancestors to organisms of lesser symmetry, then to organisms of clumsy dissymmetries (see Calvino's "The Spiral," which appears in *Cosmicomics* [1968, 141–53]). Life, in sum, is asymmetry, tension, the effort to rectify some tragic flaw in nature's otherwise harmonious workings that constantly threatens to escalate into anarchy and chaos. And above all, life is a matter of putting down roots and of wandering, of the sedentary life and of nomadism, of territorialization and of deterritorialization.

On this note, it seems that Deleuze and Guattari have reentered the scene.

Parts and Wholes That Are Parts of Other Wholes

For Deleuze and Guattari (1987) the living organism is at its most fundamental territorialized: it functions in terms of lines—of attack, of flight, of search, of play. As such life remains relatively simple in terms of function and form, though below the surface there is a teeming jungle of activity. When the human animal assumed an upright position, freeing two of his or her appendages for other tasks, there existed the possibility for deterritorialization. Then with the construction of tools, things rapidly took a turn from simplicity to complexity, though underneath the surface of these new activities simplicity appeared to rule. As we noted in the introduction of this volume, with the advent of the computer age, the turn is simplicity on the surface and complexity below. This is not mere coincidence. Computer linearity evinces commonalities with genetic code linearity, so if the two phenomena develop along parallel planes, it should come as no surprise. The linear sequence of nucleotides of DNA and the amino acids that correspond to those sequences might appear to be patterned by the linear sequence of the binary, Boolean computer code, or of phonemes into words, words into sentences, sentences into paragraphs, and paragraphs into texts.

Initial appearances can be deceptive, however. Deleuze and Guattari (1987, 62) point out that François Jacob was reluctant to compare the genetic code to language. The genetic code enjoys neither addresser nor addressee, neither interpretation nor translation; in fact, expression is dependent upon no more than content, and form of expression is dependent upon physical substance.[11] In contrast, linguistic expression can be divorced from content, and the form of its expression need not have any direct bearing on the semiotic object. One of the greatest dangers inherent in equations of genetic code and language at purely coded levels is that this can lead to "imperialist pretensions on behalf of language" (Deleuze and Guattari 1987, 62). Roland Barthes (1968) temporarily became imperialistic pretender numero uno with his declaration that Saussurean semiology does not include language as a subset, but on the contrary, language, the ruler of the human sciences, contains semiology. This move led to an even more devastating linguicentric imperialism, when in poststructuralist parlance the signifier was divorced from the signified. It opened the door to the possibility of an empire of despotic signifiers conceived as the alpha and omega of all communication.

Deleuze and Guattari argue, on the other hand, that linguistic signs are neither signs of things nor signs of other signs but signs of deterritorialization and reterritorialization. In fact, signs of deterritorialization and reterritorialization are diagrammatic through and through, which evokes the idea of Peirce's diagrammatic thinking. Peirce's diagrammatic sign is an icon, at its most fundamental, but it is an icon with a purpose. It is an icon whose function extends in indexical manner toward something else—its semiotic object—in matters of similarity, to be sure, but the similarity is schematic rather than substantive, abstract rather than analogic in the concrete sense, a matter of relations rather than mere imagery. Actually, a diagram is "neither substance nor form, neither content nor expression" (Deleuze and Guattari 1987, 141). It is neither wholly iconic nor indexical but somewhere in between: it is, in a manner of speaking, fractal. If icons are chiefly substantial and indices chiefly formal, then the diagram is neither of the two. That is to say, the diagram is not purely formal, as are pure algebra and other abstract disciplines; and it is not purely substantial, as is work with pigments or with marble or granite. It is neither purely linguistic, like the logical proof of universal pretensions, nor is it of more particular language usage, like the poem or the philosophical treatise. It is, to reiterate, somewhere in between. All this is to say that diagrammaticism, diagrammatology, diagrammatic thinking are always already between one dimension and another, like fractals, like strange attractors, like far-from-equilibrium conditions and their concomitant dissipative structures.

Diagrams, in sum, are in the "fold" between one plane and another and defined by the quivering, palpitating, dancing $\sqrt{-1}$ (figs. 4, 5*a*, 6, 11, 12, 13). That is to say, they are directly in the rhythm of things. As Deleuze and Guattari (1987, 313) put it, rhythm is always already between emptiness and something, between chaos and cosmos: in other words, it is in the "chaosmos." Diagrams and diagrammatic thinking are also somewhere between cosmos and chaos; they are the "mesocosm" between the microcosm and the macrocosm. They are where the action is, where everything is always on the move. (To where? Who knows? The action is the medium without there yet having emerged a full-blown message.) Diagrammatic thinking serves as the playground for the drama that is to be unfolded, to unfold itself. It is between sensation and intellection, quality and quiddity, feeling and form. It is between that from which everything arises and that toward which everything tends to gravitate, between Firstness and Thirdness, between the complexity of the apparently simple and the underlying simplicity of the apparently complex. In the final analysis, diagrams, like all signs, are not things but sign-events, interrelated with all other sign-events.

On this note, a further look at the graphic arts is perhaps in order.

CHAPTER 12 • Simplicity and Complexity Revisited

To see the world in a grain of sand.

—William Blake

When Anaxagoras says: Even the snow is black! he is taken by the scientists very seriously because he is enunciating a "principle," a "law" that all things are mixed, and therefore the purest white snow has in it an element of blackness.

—D. H. Lawrence

Colours appear what they are not, according to the ground which surrounds them.

—Leonardo da Vinci

I paint infinity.

—Vincent van Gogh

Once Again, It Is a Matter of What You Make of It

I repeat: the questions "What is simple?" and "What is complex?" are both exquisitely simple and exquisitely complex. A solitary cell is simple when we consider it the building block of an organism but mind-numbingly complex if we look at its inner workings. Blake's lines may appear simple when the terms *world, sand, heaven, flower, infinity, eternity,* and so on, are taken in the context of everyday language use; they are complex, in fact, they border on unintelligibility, when embedded in religious, philosophical, scientific, and mathematical languages, and especially when they are an attempt to articulate mystical insights.

Nothing could at the outset appear more simple than that notorious thought experiment, Schrödinger's cat.[1] Suppose a cat is placed in a closed box with some radioactive material that, after one-half hour has elapsed, will have a 50/50 chance of having emitted a bit of radiation that triggered a mechanism designed to fill the chamber with gas and kill the cat. The question is posed, After the half-hour has elapsed, and at a given moment before anyone has opened the box to inspect (interact with) the cat, is it dead or alive? From the quantum theoretical viewpoint, we might conceivably say that it is either both dead and alive or neither alive nor dead until we open the lid of the box and take a peek (compare to Sagredo's discussion of the life and death of the prisoners in the "Prisoners' Paradox"). Then and only then will the cat either have choked on the gas or have survived—the future conditional is important here, I would submit. In the here and now the conditions might be ripe for the cat to be alive rather than dead, but it can be neither the one nor the other until we have already observed it, that is, reacted with it, that is until the contraption has reacted with us, for quantum participation is definitely not a one-way street. So let us bring our drama to a conclusion by supposing we open the box and find the cat alive and well.

But things are not so simple, and here is why in a nutshell—I hope without my committing undue violence to a sophisticated thought experiment. Put "live cat" at the upper left-hand corner of figure 3, "dead cat" at the lower right-hand corner, "both live cat and dead cat" at the lower left-hand corner, and "neither exclusively live cat nor exclusively dead cat" at the upper right-hand corner.[2] The gray area is the "both and," white is "live cat" as one possibility, and black is "dead cat" as the other possibility, with $\sqrt{-1}$ in the middle as the "either-or" separator of the sheep from the goats. Meanwhile, back at the "real world" of Schrödinger's cat's quarters, after we have taken a gander at the contents of the box, our conscience has been put to ease by the evidence of a purring feline. Then the "neither-nor" at the upper right-hand corner became one of the two possibilities, that is, the +1 of $\sqrt{-1}$ was actualized. However, the "neither-nor" world could have been other than what it turned out to be for us—this we learned particularly from Borges's "Garden of Forking Paths." There could have been a "dead cat," or even a countless number of other worlds: in another world the contraption might not have been placed in the box; in another the cat could have been black instead of white or vice versa; in another it could have been conveniently fitted with a gas mask; in yet another it could have been previously removed by someone, and so on. There is no end to the alternative worlds that could have been actualized to testify that the cat-world we brought into existence was neither necessarily that particular "live cat" nor that particular "dead cat" but quite possibly something else.

However, what I have presented is still an oversimplified picture. Finding the cat either "alive" or "dead" at the outset appears as much a matter of chance as tossing a coin. However, quantum theory presents another picture entirely. Cat, apparatus, box, and you and I according to quantum theoretical considerations exist in what is called Hilbert space of *n* dimensions, which is quite straightforward and determinate. The state of the system is specified by a vector in Hilbert space, that vector being the combination of a vast number of vectors within subdomains. One of them may correspond to "live cat" and another to "dead cat." In other words, the state of the system can correspond to a combination of these two mutually exclusive outcomes. The only thing that can be said about the entire device before the box has been opened is that there are a vast number of possible Hilbert space vectors one of which will have been actualized when interaction occurs with some other conglomerate of Hilbert space vectors that previously has been actualized through some interaction with something else, which in turn has been actualized by something else, and so on, to the ends of the universe—a mind-boggling Chinese box state of affairs.

So at the upper left-hand and the lower right-hand corners of our transformed figure 3 we have "live" (white) and "dead" (black), the two possibilities. At the lower left-hand and upper right-hand corners we have "both-and" as possibilities not yet actualized into either one thing or the other. And at the upper right-hand corner what is actualized could always have been something other than what it is, and there always exists the possibility that it will have been something other than what it was for us. So what is finally actualized will be "neither" simply one thing "nor" the other, for there exist between the neither and the nor an indefinite number of alternate possibilities that either could have been or possibly will have been actualized. If "both-and" of our revised figure 3 quadrant aborts the principle of noncontradiction, "neither-nor" aborts the principle of the excluded middle.[3]

One might wish to conclude that the Hilbert space vectors within the box before the lid has been opened could as a very rough visual counterpart be conceived as something in the order of a given point in Klee's cosmic bundle in figure 8 depicting "dead cat" and the knot of lines in chaos depicting "live cat." In other words, "dead cat" as a possibility is inert, an object moving along its world line with all the other inert objects lying contiguous with it in its frame of reference. As such it could be represented by a single point in phase space. In contrast, "live cat" as a possibility is active: it is now mobile and roaming about its environs, now motionless for a few seconds, and so on. Its world line would consequently be represented by a random bunch of squiggles at the lower extremities of the box. Its representation, like Klee's chaos, would give the implication that any point along the line would represent a possibility of the position and momentum of the cat if and when $\sqrt{-1}$ might have been activated in favor of our quantum pussy, its jail sentence might have been ended, and it might then have seen the light of day. In this phase space picture the squiggles as possibilities could be subjected to a 50/50 decision, like a mere coin toss. Toss the coin, heads happens to turn up, and "live cat" is actualized when the box is opened.

Wrong conclusion again, however. Schrödinger's Hilbert space vectors are incompatible with phase space. In the example of phase space the squiggly lines are not real. They are only possibilities for actualization of "live cat" during successive stages of the animal's life. There was either "live cat" or "dead cat" as a possibility for actualization at each of these stages, and the series of possible actualizations makes up the squiggles. In Hilbert space, on the other hand, there is neither "live cat" nor "dead cat" until something has been actualized. But that is not all, or better put, there is nothing at all: there is simply nothing to be mapped out, nothing but no-*thing,* and a possible actualization, whenever that may be, will have been the result of a combination of all possible vectors defining "dead cat–live cat" possibilities during all the moments the box remained closed. In other words, there are no squiggles to be diagrammed, there is no diagram (a sign of Firstness) for there is not (yet) any-*thing* at all. There is no series of possibilities for actualization at successive possible points in time. There is no more than an abstract set of vectors, a combination of different outcomes one of which can be actualized if and when the box may have been opened. The decision is no toss of the coin, then, for there is not even any coin. There is simply emptiness.

This is quite a strange universe, when one stops to think about it. The Hilbert phase space world, as I understand it, is absolutely determinate. But the world of the outcome of a range of possibilities is absolutely ambiguous. Put the two together and we have both absolute determinism and absolute ambiguity or indeterminism, or neither absolute determinism nor absolute ambiguity, if you wish. Or if you prefer neither of the two options, perhaps you can dance to the shimmering rhythm of $\sqrt{-1}$ by embracing both the "both-and" and the "neither-nor." In whichever case, it would seem that all logic and all reason have been abandoned. And in a sense so they have. However, the fact remains that we are usually able to get along quite comfortably in our everyday macroscopic world of cats and other things because the actual furniture of our world and events in which they are engaged are neither excessively ambiguous nor are they completely determinate. The furniture of our world and their respectiive events are determinate enough so as to enable us to perceive and conceive some regularities in our world and thus develop a set of expectations regarding what will be actualized in the future. And they are ambiguous enough so as to occasionally surprise us, to shock us out of our otherwise lethargic states and render our world interesting.

So, What Is the Point?

In all honesty I should mention that a few quantum physicists respond to the "Copenhagen interpretation" as outlined in the preceding with the query, "Is that all there is?" and they answer their own question with a resounding "No! If quantum theory allows for ambiguity, it is incomplete, hence the Copenhagen interpretation is incapable of giving the entire account."[4] Most card carrying members of the Copenhagen fraternity, in contrast, give the rejoinder, "The universe by its very nature is ambiguous. We have learned to live quite comfortably with this uncertainty, thank you, and would suggest that you try to do the same." The debate continues as of this writing, with no end in sight. But actually, the quantum question is another issue and foreign to our present concerns except as a useful metaphor, or parable if you wish, illustrating the coming and going of Beckett, Borges, Calvino, and numerous artists and their surfaces and depths and colors and the lack thereof.

Moving on, then, consider, if I may be indulgent of a bit more of your patience, figure 24. The pair of concentric circles at A are possibly just that: a pair of concentric circles. They might also possibly be a shallow pan seen from above, a cylinder seen from one end, a short beveled rod seen from the small end, and many other things as well. In other words, A is a two-dimensional combination of visible curves as mere possibility, as if they were invisible, hardly more than emptiness; yet they can be actualized into three dimensions as a number of things. Upon a slight rotation of A, suppose this combination of curves possibly becomes actualized, at B, as either a shallow pan in three-dimensional space facing toward us or away from us, or as an ambiguous drawing in two-dimensional space, a sort of Möbius strip with two twists instead of one. At this zone, standard logic ordinarily either prevails in the mind of

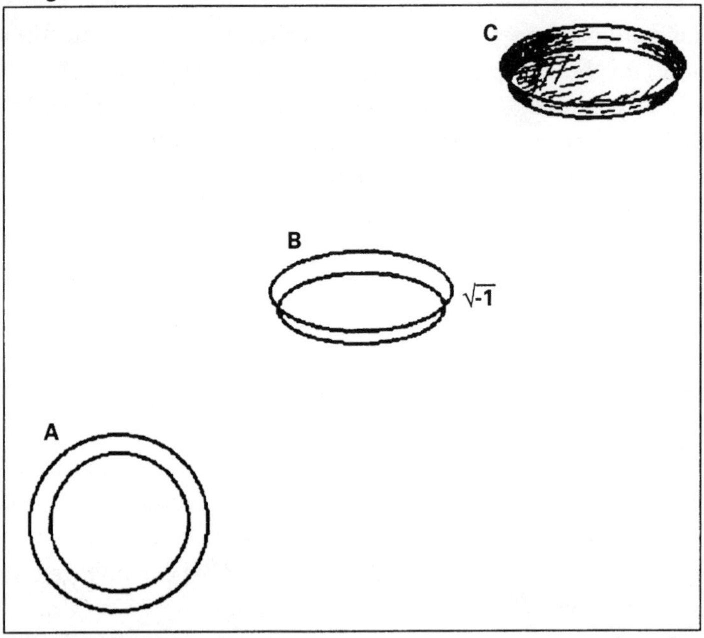

Fig. 24. From lines to an object

its user or things become undecidable. If the object is made out to be undecidable, the undecidability can more often than not be smoothed over by synthesizing the two possible pans into one pan (like the undecidability of $\sqrt{-1}$ that can be rendered more comfortably decidable as the unambiguous i [= $\sqrt{-1}$] or like the two cube possibilities that are taken as one cube, the Necker cube). In other words, if B happens to be actualized as a shallow pan, it can be seen either from above or from below, commensurate with $\sqrt{-1}$. As a shallow pan, it is one thing and one thing only, much in the order of i (= $\sqrt{-1}$), which is neither positive nor negative: it just is. Like the ambiguous Necker cube, the pan as seen from above is incommensurable with the pan as seen from below. They are two entirely different pans, and, in view of Dobbs's conclusions regarding the Necker cube, transformation of one pan into the other is logically impossible in three-dimensional space; hence it requires transformation through an infinite number of mathematically conceived points within an extra dimension of space conceived as a mathematically imaginary time dimension.

The abstract pan of B, in terms of the two possibilities of its actualization as a pan seen from above or a pan seen from below, is given greater precision in the sense that someone quite comfortable with cubist renditions of the world can, by scanning the object, see (imagine) it to be three dimensional, thus taking its one-dimensional and two-dimensional depictions into his or her perceptual and conceptual embrace. Now there is no real ambiguity, for the synthetic grasp encompasses ambiguity (in the sense of $\sqrt{-1}$ taken to the more abstract level, i).[5] In another way of putting it, the differentiated ambiguity has been de-differentiated and returned to its more fundamental perceptual and conceptual soup; then it is re-differentiated into a synthetic whole.[6] In this manner, scanning a Picasso cubist painting is like seeing the three-dimensional object from various angles simultaneously, as if one were conceiving $\sqrt{-1}$ as i, but with awareness that i contains a vacillating to and fro of values, a tension that can be released solely by viewing only one particular aspect of the painting rather than its myriad aspects. In this sense, A can be viewed as both deterministic and ambiguous—recall the Schrödinger cat example. When one scans and becomes aware of the whole of the Picasso painting as if seeing it in three-dimensional space, it is potentially rendered relatively deterministic. On the other hand, when viewed as bits and pieces in two-dimensional space, it remains ambiguous, a set of possibilities for actualization at some space-time juncture or other.

Such abstracting and synthesizing send us off in the direction of C in figure 24. The northeastern and southwestern corners of the square are, as they stand, nonclassical, while B, either one thing or the other, is quite classical though nothing is (yet) actualized—recall the relations between standard and fuzzy logic in chapter 10. However, the multiple differentiations of the object introduce chiaroscuro, and the object's three-dimensionality becomes more readily apparent at C, where some-*thing* can become what it would ordinarily become under the circumstances. In other words, along the lines of Peirce's Firstness, Secondness, and Thirdness, what might be at A is defined as possibly either the one or the other at B that most likely will have been, would have been, or should have been, such and such at C. Now, the ambiguity can be disambiguated; one of the two values of the pan can be canceled, and the other becomes simply the way things are—though the aware viewer, with memories of his or her state of mind when passing from A to B and from B to C, knows that the object could have been actualized as its enantiomorphic twin and that it is possible that in the future it will have been actualized as something else altogether. Scanning the possibilities, in the beginning, is due to the central portion, B or $\sqrt{-1}$, where the either-or of standard logic ordinarily reigns. Proceeding from this point, worlds will have been constructed. (From the illogical "both live cat and dead cat" there arises the logical possibility of "either live cat or dead cat," and whichever one of the possibilities is actualized, then at some future time and place there is no absolute guarantee that something else will not have emerged that, rather illogically, will have been "neither that particular live cat nor that particular dead cat.")[7]

Calvino (1986, 9) once wrote that "just as Zeno of Elea, by refusing to accept space as continuous, ended up by separating Achilles from the Tortoise by an infinite number of intermediate points," so "every analytic process, every division into parts, tends to provide an image of the world that is ever more complicated." This is the inverse case of Dobbs's refusing to accept the Necker cube as a pair of discontinuous manifestations by constructing a mathematical continuum of an infinite number of intermediate points between the two possible cube manifestations. Moving from the central portion of figure 24 downward—where the either-or resides and standard logic is still able to exercise its force—we have analysis as a breakdown, a collapse to two dimensions of the cubist sort into a set of lines that will have been construed as two-dimensional and even three-dimensional upon proper differentiation, dedifferentiation, and redifferentiation. At B, everything becomes surface: a juxtaposition of elements possibly differentiated by relations of contiguity rather than depth, though depth remains implicitly or tacitly acknowledged (that is, surface is highlighted, but depth cannot be forgotten). Here, that which has been differentiated is susceptible to dedifferentiation but not without leaving a trace of the object's previous differentiation into either one thing or another thing.

Moving upward in the direction of C, we have the emergence of more definite forms in terms of redifferentiation of what has been dedifferentiated. Now, what was ambiguous (differentiated) becomes, by synthetic scanning of the two possible manifestations of the object, more determinate—although things could always have been otherwise, and in the future it is possible that they will have been otherwise. At B there was analysis, differentiation, a breakup of the object into various possible manifestations. Then the object was synthesized into a whole of more definite character. In Zeno fashion, breakup of a relatively complex phenomenon into discontinuous segments at B could continue to infinity. In a finite world of helplessly limited cognizing humans, in contrast, the process comes to a tentative and temporary halt when there is either private idiosyncratic acknowledgment or community agreement that the object has been properly perceived and conceptualized at C as a synthetic whole. For example, it becomes quite clear that the pan in figure 24 has depth; that is, depth is highlighted. But it is not highlighted without the lingering acknowledgment that it could not have been so highlighted unless it had been seen as a surface at A and B; that is, as undifferentiated and differentiated breadth without there yet being any dedifferentiation and redifferentiation to produce depth at C. In other words, there was breadth before there could be depth.

Putting this situation in another framework, the actualization of an object of depth at C from an ambiguous object of breadth at B is like a combination of figures 5*a* and 6. In figure 6, the gray fuzz at A can be differentiated by the $\sqrt{-1}$ operator either into exclusively white "dots" at B or black "dots" at C or into intermittent patches of white "dots" and black "dots" at D. In the latter case we have an operation in figure 24 comparable to Vasarely's Harlequin of figure 5*b*, in which, by a distortion of the two-dimensional checkerboard of blacks and whites, a three-dimensional being emerges, depth arises from breadth. In other words, in figure 24 the two possible pans without depth, by means of the actualization of a spread of intermittent black patches and white patches, are dedifferentiated and redifferentiated such that they are given depth, with a concomitant disambiguation of the figure into a synthetic whole.

One might wish to conclude that B is largely the zone of perception and conception of our supernominalist, Funes.[8] But he sees things "out there" quite simply as he sees them: particulars unrelated to anything else. In contrast, the Tlönians, inveterate idealists of the solipsistic sort, perceive and conceive particulars, to be sure, but they are strictly "in here" at the outset, later to be projected onto the world "out there" at B, with actualization of both possible manifestations of the ambiguity.

Then, as actualizations, there is a move toward C, with awareness that the one manifestation is real along with its refutation, the other manifestation, and that all manifestations are real along with their own refutations. All metaphysical doctrines for the Tlönians must contain their own counterdoctrines; the ambiguities are embraced as undifferentiated or dedifferentiated wholes, like accepting $\sqrt{-1}$ as i (in other words everything is interrelated with everything else). In the Tlönians' world we have nomadism of the Deleuze–Guattari sort at its screaming best: deterritorialization to the extent that the world is differentiated into all possible manifestations and fused into essentially a dedifferentiated whole, with nothing prioritized over anything else. But actually, B taken as ambiguity is the wrong way to begin: it entails much more than has been implied to this point. So let us redirect our focus. If at A everything is taken as a whole, in much the sense of the Tlönians' embrace of all possible worlds in simultaneity, then we have the likes of Tzinacán and of the narrator experiencing the Aleph: there is the hoped for—and presumably actualized—condensation of the many into a drop, a singularity, a point in space (see also in this regard Calvino's "A Sign in Space," which appears in *Cosmicomics* [1968], 31–49). Here, we are in the sphere of *quality* (that is, of feeling, sentiment, emotions, hopes and fears, desires and inclinations).

How so? Well, a beginning is not easy. Perhaps Kazimir Malevich's words on Suprematism may help:

> Under Suprematism I understand the supremacy of pure feeling in creative art.
>
> To the Suprematist the visual phenomena of the objective world are, in themselves, meaningless; the significant thing is feeling, as such, quite apart from the environment in which it is called forth.
>
> The so-called "materialization" of a feeling in the conscious mind really means materialization of the *reflection* of the feeling through the medium of some realistic conception. Such a realistic conception is without value in Suprematist art. . . . And not only in Suprematist art but in art generally, because the enduring, true value of a work of art (to whatever school it may belong) resides solely in the feeling expressed. . . .
>
> An objective representation, having objectivity as its aim, is

> something which, as such, has nothing to do with art, and yet the use of objective forms in an art work does not preclude the possibility of its being of high artistic value.
>
> Hence, to the Suprematist, the appropriate means of representation is always the one which gives fullest possible expression to feeling as such and which ignores the familiar appearance of objects.
>
> Objectivity, in itself, is meaningless to him; the concepts of the conscious mind are worthless.
>
> Feeling is the determining factor . . . and thus art arrives at non-objective representation—at Suprematism.
>
> It reaches a "desert" in which nothing can be perceived but feeling. (Malevich 1959)

Feeling, or quality, pure quality, or at least as pure as one can get. Feeling before objectivity and subjectivity have their way. Here, there is no differentiation (not yet at least) of black from white or white from black or any-*thing* else. There is just feeling, as if A and the two manifestations of B were one, as if taking Malevich and Rothko and Tobey in figure 5*a* together and experiencing in any of the possible manifestations all possible manifestations: white on white, black on white, a patch barely distinguishable from its surroundings, a mere suggestion of something, some-*thing,* a desert, the Arabian king's stark labyrinth, hardly any-*thing* at all except an innumerable, undifferentiated two-dimensional sheet of grains strewn on the desert floor. This is, of course, the ultimate of simplicity, but it is simplicity within complexity of the most complex sort. Malevich again:

> The black square on the white field was the first form in which non-objective feeling came to be expressed. The square—feeling, the white field—the void beyond this feeling.
>
> Yet the general public saw in the nonobjectivity of the representation the demise of art and failed to grasp the evident fact that feeling had here assumed external form.
>
> The Suprematist square and the forms proceeding out of it can be likened to the primitive marks (symbols) of aboriginal man which represented, in their combinations, *not ornament but a feeling of rhythm.*

> Suprematism did not bring into being a new world of feeling but, rather, an altogether new and direct form of representation of the world of feeling.
>
> The square changes and creates new forms, the elements of which can be classified in one way or another depending upon the feeling which gave rise to them.
>
> When we examine an antique column, we are no longer interested in the fitness of its construction to perform its technical task in the building but recognize in it the material expression of a pure feeling. We no longer see in it a structural necessity but view it as a work of art in its own right. . . .
>
> The Suprematists have deliberately given up objective representation of their surroundings in order to reach the summit of the true "unmasked" art and from this vantage point to view life through the prism of pure artistic feeling. (Malevich 1959)

Feeling, plastic feeling, incessantly transient feeling, timelessness in the process of carrying over into space, spacelessness in the process of carrying over into time. This is comparable to Peirce's passage from pre-Firstness to Firstness, pure feeling to the possibility for sensing, then seeing some-*thing* as such and such (Secondness), and finally, to saying *that* it is so-and-so (Thirdness). What follows is gravitation upward along the diagonal of figure 24 from A toward the differentiated world of B coupled with C with which we are more familiar.

If B is more closely approximating, then and constructions the likes of Jackson Pollock's drip paintings can be in the act of becoming: a confusion of squiggles not yet clearly differentiated into many things or virtually any-*thing.* And at B, the ambiguous either-or of Escher comes into play. Moving upward from B toward C, we take a stroll through cubism, and beyond C, to art of the presumably realist or neorealist sort. Here we also have the likes of Borges's helplessly scurrying rats in the Library of Babel, the narrator who becomes obsessed with the Book of Sand, or that other narrator obsessed with the Zahir. Here, at least, there is a move toward the possibility of some order to things, of reterritorialization, differentiation into regularities, generalities. But, of course, the end of the road will never be within reach: the struggle for space, for time, for absolutely precise and decidable chiaroscuro, for the one and only perspective, will not find closure in the kingdom of this world. From Calvino's amorphous mass to crystals to aperiodic crystals and finally to mitosis and increasingly complex forms of life embodied in meiosis, everything becomes, is in the process of becoming, part of what is.

From another vantage point, Klee's formal elements of graphic art, dot, line, plane, and depth, become manifest. But it is no longer merely a matter of space. Time is now squarely in the picture:

> When a dot begins to move and becomes a line, this requires time. Likewise, when a moving line produces a plane, and when moving planes produces spaces.
>
> Does a pictorial work come into being at one stroke? No, it is constructed bit by bit, just like a house.
>
> And the beholder, is he through with the work at one glance? (Unfortunately he often is.) Does not [Anselm] Feuerbach say somewhere that in order to understand a picture one must have a chair? Why the chair? So that your tired legs won't distract your mind. Legs tire after prolonged standing. Hence, time is needed. Character, too, is movement. Only the dead point as such is timeless. In the universe, too, movement is the basic datum. (What causes movement? This is an idle question, rooted in error.) On this earth, repose is caused by an accidental obstruction in the movement of matter. It is an error to regard such a stoppage as primary. (Klee 1959)

Fire kindles the impulse to create, Klee tells us. And creation is in time. The hand leaps to the canvas and deposits a spot, then retracts. Another spot suddenly appears, as if out of the clear blue. Then a mark, perhaps even a line and a contour. And finally, a space is in the process of becoming. All this happens not merely by switches from one dimension to another but by fractal dimensions, bit by bit, curve by curve, the endotopic in incessant struggle with the exotopic, darkness in perpetual agonistics with brightness. After the work is completed and open to the proper gaze, the beholder's eye "moves about like an animal grazing, follows paths prepared for it in the picture (in music, as everyone knows, there are conduits leading to the ear; the drama has both visual and auditory trails)" (Klee 1959). This movement is comparable to the onlooker's "scanning" the work to integrate many possible (different) manifestations into one synthetic package (by redifferentiation). In the case either of the artist's activity or the viewer's activity, the work is born of time and movement, and it lives through time and movement.

There is nothing but rhythm, which brings forth the visible from the invisible.[9] Differences that hardly make a difference can be fused and confused so as to engender fuzz, exemplified, for example, by Claude Monet's *Rouen Cathedral* (1894), a static yet dancing image as if seen during different times of the day in a moment. Or there can be Vasarely's Harlequin engaging in a swirling dance as it lifts itself into another dimension. Or there can be a series of static moments portrayed as if each were part of a whole, in the order of Marcel Duchamp's *Nude Descending a Staircase* (1912). And so on.

Form, Tone, Color

The preceding section in a rough way recapitulates the historical becoming of art up to and including the present (Ehrenzweig 1975). Very schematically put, the medieval period painted from a flat bird's-eye view; during the Renaissance, form lost its fixity and change of perspective due to depth came into the picture; during the baroque period tone became more subtle with the play of chiaroscuro; and modern art up to the present brought about awareness of the subtlety of color, especially since the time of van Gogh and the Fauvists (Duve 1991).[10] In other words, we have a schematic picture of the transition from the lower portions of the diagonal in figure 24 to the upper portions. But of course the situation is more complex than that, infinitely more complex. If one considers figure 24 in terms of a space containing a superposition of all possibilities, one begins to get the picture. The picture is infinitely complex, however; hence any "minimalist" rendition of it cannot but do it some injustice. Nevertheless, I feel duty bound to make an attempt at just such a rendition. Please bear with me, if you will.

It is now a given that we rarely, if indeed ever, perceive an object without some distortion. With each perceptual grasp, the object is somewhat different than what it was. A table set with circular plates, silverware, long-stemmed wineglasses, and napkins folded roughly into equilateral triangles, if seen at various angles, will give the image of oval plates, chubby knives, forks, and spoons, short-stemmed wineglasses, triangles with bases much longer than the sides, and so on. Knowing what shapes these various objects have in our quite constant cultural world, however, we are almost immediately aware of the plates' circularity, the slenderness of the silverware, and so on. Perspectival grasps are virtually infinitely variable, but there is an acceptable constancy of form that allows us to see things in terms of certain fixity under ordinary circumstances. It is chiefly due to the Renaissance obsession with depth, the three-dimensional "binocular perspective" that vanishes to a point at the infinite stretch, that has made us hyperaware of the distinction between perception—invariably distorted to a greater or lesser degree—and the relative stability of the furniture of our world's forms.[11]

Brightness and darkness, the distortions of an object's apparent tone due to the vagaries of its illumination or lack thereof—which provide the makings for chiaroscuro—were elevated to the public's attention most prominently during the baroque period. A black object in direct sunlight might appear lighter than a gray object in a poorly lit room. Ordinarily the viewer (say, a female viewer) does not notice many of these subtly to radically distinct variations of tone but takes them according to her habitualized expectations regarding their nature. In other words, she tends to ignore the distortions of tone and see things as she expects them to be and thinks they really are—even though she may be quite wrong in her preconceptions. As far as she is concerned, her black things are black and her white things are white, regardless of how many different frequencies of light are bouncing off them and striking her retinas: she perceives (constructs) a constancy of color in spite of the fact that she actually sees variations. Likewise, she sees the constancy of the form of objects, in spite of the fact that the frame of reference of her perception has obviously distorted them. Ehrenzweig (1975, 150) writes in this regard:

> As long as the painter and his public have not overcome the constancies of form and tone there is no point in the painter reproducing scientifically the repressed distortions of form and tone. Hence the naïve painter, before the advent of perspectivic and Chiaroscuro painting, reproduced on the canvas the constant form and tone of an object which he actually saw, not what he merely knew to exist. The customary distinction between a "perceptual" type of painting which reproduces the actual perception and a "conceptual" type which reproduces the object as it ought to look according to our knowledge of its constant properties, is at best inexact because it does not take into consideration the possible dynamic changes which have taken place in our perception since the Renaissance.

A in figure 24 is the shape of the dish with which we are familiar, and it is the shape that a pre-Renaissance painter could well have presented on the canvas in spite of seeing it as ovaloid from an angle. Renaissance perspectivism would likely present the pan as something in the order of B but as a solid and unambiguous object, with the expectation that the spectator of the work would be able to fill in the lacking information, reconstruct a circular pan in his or her mind, and label it "pan" if necessary, in terms of its correct form. Supposing the pan were off-white in color, C would represent, due to insufficient lighting, the pan as it might be seen in terms of chiaroscuro effects. The product of these effects would be painted by the baroque artist, which the aware observer would then be able to rectify in his or her mind.

Over generations and generations of cultural inculcation, we have become so accustomed to preconceiving perspectival and chiaroscuro effects as the one and only manner of representation that we automatically, and without giving the matter any thought, take them to "correspond to" and "refer to" the world as the way things are, clearly and simply.[12] After the turn of the present century cubists and other painters began presenting three-dimensional objects as spread out on a two-dimensional plane, much in the order of A in figure 24. Now the spectator saw objects presumably as they "really are," that is, as they "should be" or "would be" were the spectator privy to a grasp of many perspectives from multiple frames of reference in simultaneity and without the spatial and chiaroscuro distortions. In other words, the objects were presented without distortions and hence without the effort on the spectator's part to reconstruct them, to re-present them in his or her mind if that were necessary, and to slap a linguistic label on them.[13]

Euclidean geometry, classical science, and literary and artistic endeavors, then, have taught one to brush aside what one actually perceives, reconstruct it, and conceive it as it "should be" according to the prevailing conventions. During the present century, non-Euclidean perspectives, postclassical science, and avant garde literature and art have shifted the focus. Things are not what they seem to be, and what they are in their entirety is beyond our capacities for perception from any particular perspective, though a juxtaposition of multiple perspectives would allow us to see things as "now *this,*" "now *this,*" and so on, or in the Deleuze–Guattari disjunctive synthetic mode, as "either *this,*" . . . or, . . . or, and so forth. Yet, there could be no all-encompassing grasp. Relativity and quantum theory, post-Gödel metamathematics, nonstandard logics, and much late modern and postmodern literature and art have suggested as much.

But habits are habits, hard to break once they have settled in to afford us the security of comfortable pathways of least perceptual and conceptual resistance. Over a century and a half was required for Copernicus's revolution to be considered nonparadoxical, and even more time for it to become embedded in the coming and going of everyday activities. Indeed, Erwin Panofsky (1960) argues convincingly that our culturally conditioned perceptual system continues in its effort to compel us to see Euclideanly, though however comfortable Euclidean perception may be, it is no guarantee of "objectivity" (see also Heelan 1983; Hopkins 1973; Lucas 1969). Fortunately, we still have quite some time to go with respect to our slowly emerging world perspective. So not to worry, for $\sqrt{-1}$'s Siva dance will never cease, the emergence of lines, contours, smudges, and shapes from the lower extremity of the figure 24 diagonal continues, and the push toward the upper extremity is untiring. After all, what are a few generations of insignificant antlike creatures on the wastelands of the Earth's surface within the vastness of the universe? Time will continue its march, in spite of our occasional wishes to the contrary.

Relating this to the Schrödinger cat case, certain possibilities will generally be actualized in accord with particular conventions, depending upon specific contexts. But there are many unforeseen possibilities, some of which could have been actualized but were not. Thus the possibility exists that whatever will have been actualized could have been some-*thing* other than what it is; hence in a given future moment there will possibly have been neither the neither nor the nor but some-*thing* else. The southwestern corner of the Schrödinger cat scheme is virtually a continuum of possibilities, and between any two actualities at the northeastern corner there is always the possibility that some-*thing* different will have been actualized. There is contradictory identity at one extreme and incomplete, constantly emerging identity at the other extreme.

. . . And Color

Oh yes, a final important matter has been left dangling: the emergence of color. You will recall Simplicio's protest over Sagredo's having ignored colors, after which Sagredo rightly pointed out that many issues should be aired out before taking up that problem. Well, I guess the time has come to put up or shut up, however brief the exposition must be.

Let me begin this way. The lower left-hand corner of figure 24, where Tobey's work was placed in figure 5*a*, affords a glimpse regarding the possibility of color subtleties as brought out by impressionist painters. No . . . no, that is not quite it. I must rescind. I really should begin at a more basic level, if that is possible. In an effort to bring about such a beginning, then, I will initiate my story with a few scarce words on colors, their absence in blackness, and their ubiquity in whiteness. Take a piece of obsidian. It is jet black and has a gloss; it presumably absorbs virtually all the spectrum's frequencies. But not really, for the glossy portions reflect light at certain contoured spots, presenting a glistening, at times almost a glowing, countenance. Now break the obsidian into pieces approximately the size of a pinhead and cluster the bits together. The cluster is no longer black but gray because of the intermittent patches of reflected and absorbed light. Pulverize the bits and pack the fine pieces together and a dull light gray, and from some angles almost white, tone meets your eyes. You are now seeing something in the order of Tobey's painting from a distance, except that Tobey's work contains specks of pink and other colors. But this is also true of your pulverized obsidian—as we once might have read from the pen of D. H. Lawrence, black contains a little white and vice versa, and white conceals the entire spectrum of colors.

Take a pinwheel with numerous yin/yang patches of blacks and whites, spin it, and gaze upon the emerging colors. Actually, whichever colors are seen depends upon the observer—we are all constructivists in our own way—and they vary with the speed and direction of the rotation and the pattern of the light source (Rossotti 1983, 130–42).[14] Black and white, if they are combined and the distinctions are made so fine that there are no differences that make a difference, yield gray. Likewise, two primary colors, say, red and yellow, if combined with their orthogonal, that is, a 90° rotation to the adjacent axis, $+\sqrt{-1}$ or $-\sqrt{-1}$ on the Argand plane (fig. 12), yield orange. In like fashion blue and yellow give green, and red and blue give purple. In other words, just as the oscillating, ambiguous $\sqrt{-1}$ can be expressed in a straightforward way by i, so also two primary colors can come together to engender a third tint that is neither the one nor the other but something else. On the other hand, light from the complementary colors, when combined, yields white light, that is, no color at all and all colors at once. At the same time, opaque presentations of the complementary colors cannot be combined to produce something else. They *are* that something else: mutually exclusive elements both of which cannot be present in simultaneity. Stare at a red patch, then look at a white wall, and its complementary, green, appears; do the same with a blue patch and yellow appears, and so on. Complementarities are like two 90° rotations on the surface of the Argand plane to give their opposite quality: they rest at 180° with respect to one another. Complementary colors are both one and the other but not both at the same time, and so over time they are not exclusively either one or the other; hence in a very real sense they are neither the one nor the other (see in this vein Goethe's study of colors as discussed in Matthaei 1971).

Calvino's "Without Colors," which appears in *Cosmicomics* (1968, 51–60), is a compelling study of the becoming of colors as secondary qualities. Qfwfq and his mate, Ayl, live in the lifeless gray world of primary colors of scientific objectivity. For Qfwfq this world becomes a "bit monotonous." He can "go for miles and miles at top speed," and all he can see is "gray upon gray. No sharp contrasts: the only really white white, if there was any, lay in the center of the Sun and you couldn't even begin to approach it with your eyes; and as far as really black black is concerned, there wasn't even the darkness of night, because all the stars were constantly visible" (51). Qfwfq begins searching for a more exciting world, a world other than gray. But he can find no sympathy from Ayl, for they have "different tastes, if not downright opposite ones":

> I was seeking a new world beyond the pallid patina that imprisoned everything, I examined every sign, every crack (to tell the truth something was beginning to change: in certain points the colorlessness seemed shot through with variegated flashes); instead, Ayl was a happy inhabitant of the silence that reigns where all vibration is excluded; for her anything that looked likely to break the absolute visual neutrality was a harsh discord; beauty began for her only where the grayness had extinguished even the remotest desire to be anything other than gray. (Calvino 1968, 54–55)

Eventually Qfwfq's world becomes that of a blue sea, red sun, yellow stars, green trees and grasses, red, blue, and yellow fish, and so on. He tries to communicate his newly discovered (invented, constructed) world to Ayl but to no avail. They now live within virtually incommensurable frames of reference: she in the "objective scientific" domain of primary qualities, he in the "subjectivity" of secondary qualities. For good or for bad, he realizes that Ayl's world is lost to him forever. He cannot even imagine it any more. Nothing remains of her world, "except perhaps that cold wall of gray stone" (Calvino 1968, 60).

In the sense of Goethe, it is to be expected that ordinarily the eternal struggle between black and white to produce a fusion yields a virtually uniform primary quality of grayness (as illustrated in fig. 5*a*). But Goethe also tells us that things can be otherwise, at least for the constructive faculties of human perception, which are capable of producing secondary qualities rather than their simply existing "out there" and submissively awaiting our ecstatic contemplation of them. The black-white agonistics, in addition, produces—with our participation—a fission of colors to give us the entire panorama of nature's spendor. Edwin Land's (1959, 1964, 1977, 1983) "constructivist theory" of colors is quite in line with Goethe as well as Klee and an inspiration to scholars from fields as distantly removed as biology (Varela 1984; Maturana and Varela 1980, 1987), philosophy (Lockwood 1989), intellectual history and history of science (Berman 1981; Hayward 1984, 1987), psychology and neurology (Sacks 1995; Varela, Thompson, and Rosch 1993), and the physical sciences and the humanities (Zajonc 1993; Rossotti 1983). After years of thought and experiment, Land, inventor of the Polaroid Land Camera, concluded that we judge colors by the company they keep. We tacitly compare them to one another and revise our appreciation of them according to the time of day, the light source, the general context, and our memory of comparable situations.

In carrying out one of his experiments, Land made two black and white images by using a split-beam camera to project them upon a screen from the same viewpoint and through the same lens. In order to make these images, he used two filters, a red one of longer wavelengths and a green one of shorter wavelengths (two streams of "white light" filtered by the two complementary colors produce "black light," hence the black and white images). Then he reconstructed the images, passing the first one through a red filter while the second one was sent unfiltered. We would expect this operation simply to produce a pale-pink image. But the unexpected occurred. The photograph of a young woman appeared in full color: blond hair, blue eyes, a coat of red and blue-green. Where did the colors come from? They were not in the original black and white images for any recently arrived observer who was unaware of their processing. So how did they suddenly pop up? Could it be that they were a striking example of Goethe's "color illusions"? Land's response was positive, for, he concluded, the colors we see do not depend upon wavelengths or frequencies: they are *constructed* by the brain-mind.[15]

Land's theory generally entails a "color cube" much like that presented in figure 25, based on short to medium to long wave reflectance, from darkness to lightness. *Reflectance* is a measure of the proportion of light of a particular set of wavelengths that is reflected back to meet the eyes. White surfaces reflect the majority of all visible wavelengths; black surfaces reflect very little, with the colors of the spectrum lying within the three-dimensional color state (or phase) space in the block depicted in figure 25. If it is said that a surface reflects, say, green, that is the color it elicits in the minds of the vast majority of observers under certain specified conditions. According to Land, it is simply not the case that the color most observers see in terms of the reflectance of the surface of an object depends on the frequency of the incident light waves on their retinas. Rather, if a colored surface is perceived in a certain way, according to Land's "retinex theory," the color perceived is the result of a combination and comparison, according to the tacit inferences of the ever active mind, of the surface's short, medium, and long wave reflectance. In other words, the color perceived depends not on stimulus wavelengths and intensities but on reflectance, illumination, chiaroscuro, that is, on the entire context within which the color is perceived.[16]

This reflectance depends upon what percentage of the light is reflected back when the object is illuminated by light of short, medium, or long wavelengths, in relation to the intensity of light in the surroundings. For example, an object will appear "blue" if, relative to its immediate surroundings, it has a high reflectance of short wavelengths and a low reflectance of medium and long wavelengths. This comparative reflectance is the object's lightness—not luminescence, which is a measure of the wavelengths of light bouncing off the object. If the "blue" object is backgrounded with black, then its lightness will be relatively high; if it is backgrounded with white, then its lightness will be low. In other words, the perceived color depends upon whatever the context may be. As the context goes, so also the perceived color. In this manner, the brain-mind is able to compensate for the difference between the color of the object that has been elevated to the focus of attention and the color of the various objects making up its context.[17] This allows for the perception of constancy of color, as discussed previously regarding

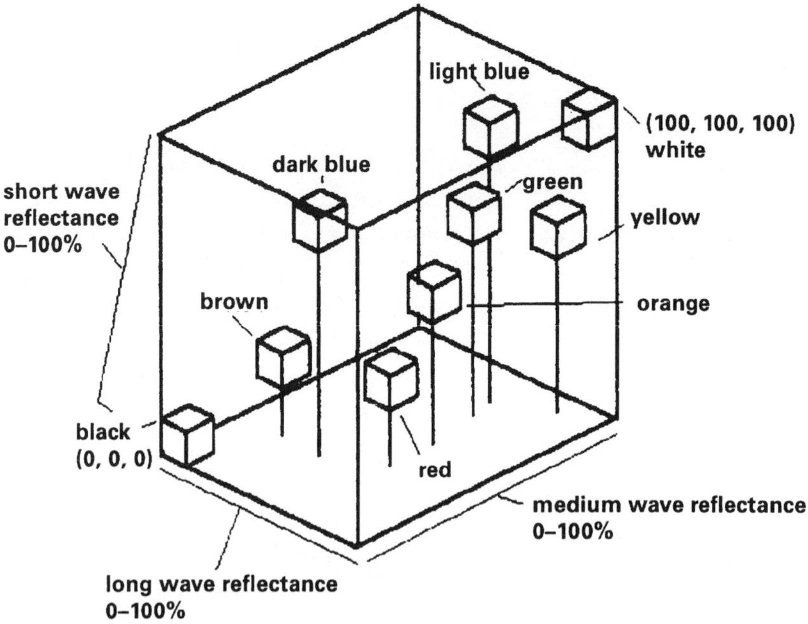

Fig. 25. The Land color cube

Ehrenzweig's studies of painting. Monet's *Haystacks* (ca. 1896) painted during different times of the day, for example, are all different. Yet in our everyday existence our brain-mind ordinarily compensates for these differences—the direct noonday sun, then a few white clouds, then gray and black rain clouds during the late afternoon, and the waning light during dusk—in order to see the haystacks as the "same haystacks" of virtually the "same color." If this were not the case, life would be unbearably complex; everything would be at each space-time juncture something else; we would suffer the fate of poor Funes the Memorious; we would never be able to make the shift toward C of figure 24.[18] But, unlike Funes, we obsessively—albeit at tacit levels—construct the world we see, *our world, our world* as a result of context-dependent seeing and conceiving. *Our world,* it appears, is tantamount to that of Goethe's struggle between black and white, which, interactively, enables us to see colors. That is, we interact with black and white in such a way that we construct the kaleidoscope of brilliance that we see as *ours* (Matthaei 1971; Zajonc 1993).

In sum, if we wish for a grand finale to the preceding pages, a culminating thought experiment, a synthesizing image that knows no rest and can be limited to no determinate boundaries, let us correlate and integrate figures 1, 3, 4, 5*a*, 6, 9, 11, 12, 13, 19, 21, 22*a*, 22*b*, 24, and 25 into figure 26. Notice the struggle between black and white, the differentiation of gray into blacks and whites and their dedifferentiation back into gray, and the possibility for redifferentiation once again, and then differentiation into the entire range of colors. Notice that if we collapse the central square of figure 26 virtually to a point, three-dimensionalize the cycles, and put the entire figure in circular motion, we will have a torus—which, by the way, has often been suggested as a topological model for the space-time continuum. Notice also that $\sqrt{-1}$, occupying the center, would be the engenderer of rhythm setting the entire form in

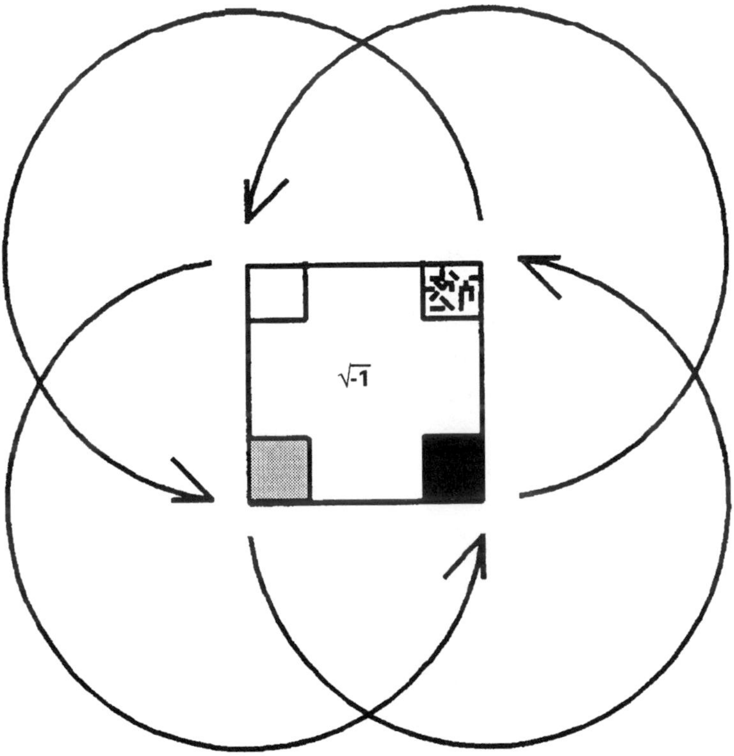

Fig. 26. The plane as process

motion: the center, *our center,* everywhere and nowhere, wherever and whenever we are in the process of becoming.

This combination into figure 26 of the key figures in the present disquisition should, I would hope, give us a sense of the space and time within which the entire drama of the works of Beckett, Borges, Calvino, Eco, and Pynchon, and those of the scientists, philosophers, and painters herein discussed, is played out. We should have a sense of the space and time within which the somewhat mystic quality of suprematist art spills out into multiple dots, smudges and smears, lines, contours, shapes and volumes, tones and hues and intensities to illustrate the flux and flow, the woofs and warps, of their quivering, scintillating, undulating dance. We should, as a consequence, have a feeling for the complementary interrelationships between *inside* and *outside, here* and *there, matter* and *form, continuity* and *discontinuity, timelessness* and *temporality, fixity* and *change, order* and *chaos,* and above all, *simplicity* and *complexity.* (I highlight complementary interrelationships in order to emphasize the yin/yang quality of all pairs of terms actualized throughout this study. Oppositions? Yes. In a manner of speaking. But not really. For they are complementarities: the one term necessarily contains something of the other term and vice versa, and both are always already found in a tight embrace, as they did their dance, as they do their dance, as they are in the process of dancing, dancing, as they will always already have been dancing, dancing, dancing. . . .)

Of course, the circle of creators and their creations discussed throughout this book might have been expanded indefinitely to include all writers and scientists and philosophers and artists, all theories and their exemplification, all books and their interpretations, and, in fact, the whole of everyday living. All these might have beens would have been an interminable story, however, a story defying compression into the pages of any single and hopelessly inadequate book, especially this one. For, the story goes on, though it can't go on, yet it can't help but go on, so it goes on . . . somehow . . . somewhere . . . somewhen . . .[19]

Ends and beginnings—there are
no such things
There are only middles
—Robert Frost

Notes

Introduction

1. For a general survey of the relations between science and painting in this regard, see Henderson 1983, Richardson 1971, Shlain 1991, Szamosi 1986, Waddington 1970, and White 1967; specifically on the turn-of-the-century period, see Butler 1994 and Kern 1983. Regarding "minimalism" in its relation to the complexity of which Sagredo speaks, see Slobodkin 1992 (118–23).

2. The terms *possible* and *actual* are used in the sense of American philosopher Charles S. Peirce's categories, Firstness, Secondness, and Thirdness, with respect to which more will be written in chapter 11.

3. I will provide a brief discussion of colors in the final chapter of this volume.

4. I will briefly expound on this phenomenon with respect to the work of Ludwig von Boltzmann in chapter 1.

5. Since I will often allude to Vasarely's work in this disquisition, I include a black and white copy of it here. I should mention that the Harlequin was originally in red and yellow-orange patches, but for my purposes in this disquisition I am using a black and white rendition of it, courtesy of the Denise René Gallery in Paris.

6. I sensed the appropriateness of including the material in quotes directly from Benardete (1964, 252), whose treatise on the paradoxes of infinity is pretty much in the spirit of this dialogue.

7. Actually, the "last encounter" between Simplicio, Sagredo, and Salviati is found in J. M. Jauch's *Are Quanta Real?* (1973), consisting of a meditation on the nature and implications of the quantum universe. While I cannot be so presumptuous as to suggest that my modest contribution is worthy of being placed beside a composition by the renowned physicist, perhaps the two dialogues could be looked upon as complementing one another in a rough sort of way.

Chapter 1

1. It should be mentioned in passing that in addition to the entropic "arrow of time" toward increasing disorder that threw the timelessness of classical mechanics into disarray, Darwin's theory of evolution presented an "arrow of time" pointing toward increasingly complex and relatively ordered biological systems that apparently contradict the second law of thermodynamics. While the Newtonian universe was a perfect machine, the universe of thermodynamics was running down. On the other hand, Darwin's biological universe was becoming more, not less, organized through time, as

simple creatures evolved into more complex ones (Coveney and Highfield 1990). Discussion in this inquiry of life processes will be held in abeyance for now, however, to reemerge in chapters 8 and 9 with a discussion of Ilya Prigogine's physics of complexity and in chapter 11 with a discussion of a very important aspect of Calvino's work.

2. See Denbigh 1975 (79–82), Denbigh and Denbigh 1985, Horwich 1987 (60–67), and Layzer 1990 (71–76).

3. Figure 6 does not take velocity into account, and for good reason. Velocity is a function of the change of position of an entity in time, and the Cartesian plane, where everything is already there in a static instant, simply cannot account for full-blown time. That is to say, the notion of time is available only from some perspective outside the plane. As we shall note in the chapters that follow, the Cartesian depiction is in this sense contrary to the notion of process as the qualifier for nature, and it is contrary to Prigogine's physics of complexity according to which time occupies a central role in the affairs of the universe.

4. The reader may note a certain relationship between figure 6 and the following: automata theory (for example, Atlan 1974, 1981a, 1981b, 1984), also John Horton Conway's "Game of Life" developed in the 1970s (see Gardner 1983; Poundstone 1985), and especially recent work in "artificial life" (for a popular account, Levy 1992; for more technical expositions, Langton 1989; Langton et al. 1992). While my "minimalist" depiction might bring these parallels to mind, the general thrust of my inquiry is something other: toward concrete life processes and everyday living.

5. It might be added at this juncture that for this reason, among others, Boltzmann was generally found unacceptable by the scientific community, which undoubtedly contributed to his bouts of depression and ultimately to his suicide.

6. The case of Musil's man without qualities is found in much literature up to and including that of the beat generation in terms of a loss of the self in the transition from the rugged Faustian individual to the individual characterized by comedy and irony (Frye 1957). This trend is especially evident in Samuel Beckett and the French "new novel," as crafted by Natalie Sarraute and Alain Robbe-Grillet (see Morrissette 1962; Sypher 1962). It is also prevalent in the other arts, notably music (Cage 1961) and painting (Dubuffet 1968; see Meyer 1967). More recently, the decentered subject in poststructuralist, deconstructionist, and postmodernist debates and diatribes is taken to the extreme with the work of Baudrillard (1981, 1983), among others (see Harvey 1989; Kellner 1989).

7. Somewhat in the order of Simplicio's words, everything that is consists of signs, and the trick to successful coping consists in exacting an effective split between what is taken as signs and that to which they relate, that which is construed as real or nonsigns. Oedipa, however, is incapable of establish-

ing the necessary relations between signs and her reality hence everything remains for her as if it consisted of nonreal nonsigns, everything is as if an uninterpreted dream—the dream as an autonomous realm outside any and all relations to the real.

8. Here, especially, Cohen and Stewart's (1994) pun on the confused relationship between "complexity" and "simplicity" as "complicity" and "simplexity" surfaces. Throughout their book, these authors attempt to come to grips with issues that will remain in the spotlight of much of the present inquiry, though their exposition rests almost exclusively on mathematics and science.

9. Here, once more, we have the essence of Szilard's (1929) argument and, more explicitly, that of Brillouin (1951), as we shall note subsequently.

10. The notion of "almost perfect accuracy" carries with it an allusion to the work of French mathematician Henri Poincaré, who has posthumously been endowed with the title of "father of chaos theory." Poincaré's study of celestial mechanics in 1892 culminated in the "many body problem," which revealed serious flaws in Newtonian mechanics. The many body problem provided rigorous demonstration that there was no foolproof guarantee that a system as apparently simple as that of Sun-Earth-Moon would forever remain in its current configuration but rather, that it was possible that some perturbation in the initial conditions or thereafter would, according to nonlinear dynamics, mushroom into a catastrophic situation such that chaos could well ensue (see, in general, Poincaré 1913, 1914).

11. "Cybernetics," coined by Norbert Wiener (1948), is a derivation of the Greek word *kybernos* ("governor" or "pilot" or "steersperson"). It entails a servomechanism or controlling device that, by feedback loops, mediates between and regulates information between sender and receiver. Cybernetics is exemplified in systems such as a thermostat (mediating between room temperature and furnace), cruise control on a car (mediating between the accelerator pedal and engine rpms), a guided missile (mediating between direction and target), a text (mediating between the reader and his or her background knowledge), and signs in general (for a brief historical account of the early years of the cybernetics movement, see Pierce 1961 and Singh 1966).

12. Perhaps the term *entropy,* it was felt, could give scientific validity to the concept of information. Perhaps it could even give it respectability. Actually, the concept soon became a best-seller, and the age of information was on its way. The problem is that the idea of "information" of the information theory sort took on undue importance—exemplified with the onslaught of the budding new medium of TV and new techniques of propaganda. Some years later, the overriding emphasis on information of the cybernetic type was dramatically forced onto the scene by way of the Nixon tapes, which seemed to take on more importance than the human beings

who created and presumably manipulated them. Today, faith in the bits and nothing but the bits is a sort of perverse revival of the ancient dream of the universe as a book, except that the book has become a vast cybernetic hypertextual network the complexity of which is beyond the wildest dreams of classical science.

13. For a discussion of the changes in the use of the term *entropy*, see Denbigh and Denbigh 1985.

14. Surprise must be highlighted in this context, for it is crucial in any consideration of a message's meaning in terms of what was expected of a message as a set of possibilities and what actually happened to pop up as the actualization of one or more of those possibilities, whether expected or bringing about a slight to jolting shock of surprise (for further along these lines, in philosophy of science see Popper 1972; in Peirce studies, Merrell 1991b, 1995a; in cognitive psychology, Neisser 1967, 1994; Bruner 1957, 1986, 1987, 1990).

15. In addition to Howard Gardner's (1987) monumental synthetic volume, see early warning signs regarding the success of AI in Weizenbaum 1976 and, more recently, Campbell 1989; for the philosophical problems surrounding the AI project, Dreyfus 1994, Dreyfus and Dreyfus 1986, Pagels 1988, and Robinson 1992; and for an example of the eternally optimistic view regarding the capacity of machines to become comparable to minds, Collins 1990 and especially Nadeau 1991.

16. In this regard see especially Putnam 1988, an early proponent of the computer-brain analogy who later gallantly made an abrupt shift, confessing to the shortcomings of his previous ways—an admirable, and rare, move in academia.

17. At the heart of the problem regarding the term *information* of information theory is the distinction between "information" as nothing more than a consideration of statistical probabilities—Shannon's use of the term—and "information" giving rise to contextualized, embodied meaning—which rests outside the domain of information theory, properly speaking. Attention to this problem, however, would lie outside the scope of the present disquisition (for further, however, see Dretske 1981).

18. Unfortunately, in the likes of David Porush (1989), among many, we find an ongoing infatuation with cybernetics. He tells us that "without exaggeration we can say that cybernetics is the paradigmatic postmodern science. It views the universe as a set of interconnected systems of energy, matter, space, and time all of which can be described in terms of (or reduced to) how much information those systems transmit or contain" (379). So it would appear that at least in some circles the problem still stands.

19. This evokes recollections of Heidegger's critique, as well as that of more recent phenomenologists, of technology and cybernetics (see Dreyfus 1994). But in spite of this scathing criticism, cybernetics (or AI and even so-

called artificial life) will be satisfied with nothing short of a well-defined universe, with virtually no inconsistency, no incompleteness.

Chapter 2

1. Here I follow Peirce's (1931–35, 5:448 n) notion that the universe is a "perfusion of signs," if it does not consist exclusively of signs. And when Peirce writes "signs," he means to include those linguistic as well as those nonlinguistic or extralinguistic.

2. I use the term *linguicentrism* in Merrell 1997 as a wedge with which to get a critical hold on logocentric and many presumably nonlogocentric practices insofar as they share an overriding prioritization of, and occasionally an obsession with, language as the chief—and it even appears at times the exclusive—source of all that makes for genuine understanding.

3. "Icons" and "indices" make up two of the legs of Peirce's basic sign tripod, including also "symbols," whose most effective medium consists of linguistic signs and the signs of artificial languages.

4. In fact, in *Simulations* (1993), Baudrillard comes quite close to Eco's views on "fakes and forgeries."

5. Perhaps Borges himself is a realist temperament disguised as a nominalist, then. In this light, judging from his frequent intellectual excursions into mysticism, it should come as no surprise that he often longed for a view sub specie aeternitatis, such as that afforded by the Aleph, for example. On the other hand, Borges once remarked that all perspectives, all classifications of the world, are nothing more than convenient intellections (Borges 1953, 18–19).

6. Along these lines, Calvino presents an interesting comment in "All at One Point," which appears in *Cosmicomics* (1968, 43–47), regarding astrophysicist Edwin P. Hubble's theory that the entire universe was originally concentrated in a single point. But unlike the Aleph, which takes up a golf ball–size piece of space, within the point there is no space at all. On the other hand, the sign of "A Sign in Space," also in *Cosmicomics* (Calvino 1968, 31–39) is comparable to the Zahir in that each sign is a reiteration of some original sign, a sign of something, of itself or something else, anything else. Qfwfq, the hero of the story, leaves a mark, a sign, at his starting point as he embarks on his travels through space. The sign stays where Qfwfq left it to mark that spot, and at the same time it marks Qfwfq: "I carried it with me, it inhabited me, possessed me entirely, came between me and everything with which I might have attempted to establish a relationship" (33). As signs proliferate in Qfwfq's universe, it becomes

> a general thickness of signs superimposed and coagulated, occupying the whole volume of space; it was constantly being dotted, minutely, a

> network of lines and scratches and reliefs and engravings; the universe was scrawled over on all sides, along all its dimensions. There was no longer any way to establish a point of reference: the Galaxy went on turning but I could no longer count the revolutions, any point could be the point of departure, any sign heaped up with the others could be mine, but discovering it would have served no purpose, because it was clear that, independent of signs, space didn't exist and perhaps had never existed. (Calvino 1968, 39)

The circle is thus completed. All space, without beginning or end, whose center is everywhere and whose circumference is nowhere, is tantamount to the compacted Aleph, or better, to Calvino–Hubble's original universe point. The end becomes the beginning and the beginning the end. The Aleph or point contained within the whole is itself the whole. In contrast, the Zahir is a linear series of all the particulars, the sum of which makes up that whole, which is coterminous with the Aleph.

7. This allusion to signs that might follow, signs that actually follow, and signs that should follow corresponds to Peirce's three categories of sign processing, Firstness, Secondness, and Thirdness, about which, as has been mentioned, more will be written subsequently (see Almeder 1980; Hookway 1985; and Savan 1987–88 for a discussion of the categories).

8. The topics mentioned here will be discussed as the pages of this inquiry unfold, bringing forth the notion that all series, originally of linear nature, will become multiply complex along interconnected nonlinear and fractally dimensional pathways.

9. In light of the function of the membrane as a sort of biological operator separating inside from outside, *this* from *that,* see Varela 1984 and an exposition of Varela's concept from a semiotic perspective in Merrell 1996, also briefly to be presented in chapter 11 of this volume.

10. From the launching pad of "Blood, Sea," Calvino moves on to the suggestion of symmetry breaking in "Crystals," then to "Mitosis," with its unicellular ecstasy and mystical experience of subdivision of time, and "meiosis," which includes the self-reproduction of many cell organisms. But that is another story to be told at a later time in this inquiry when Prigogine begins to take center stage.

11. This notion of the interconnectedness of all things has been for some time the gradually emerging view of the universe from quantum physics to biology (for readable accounts see Bohm 1980; Davies 1988; Herbert 1985; Pagels 1982; also, in relation to semiotic theory, see my presumptuously entitled "semiotic trilogy" in Merrell 1991b, 1995a, 1996).

12. This form of complexity is germane to Deleuze's (1993) concept of the *fold.* The fold is nonlinear and nonexistential, since it implies the concept of multiple surfaces. As radical nonlinearity, it is nondialectical, for there are

no legitimate singularities against which antisingularities can be engendered, such engenderment giving rise to synthetic or composite singularities: the fold gives freedom to actualization in many directions. And the fold is non-Cartesian, for, due to its multiple overlappings, intertwinings, and mutually interpenetrating mergings, dimensions do not come in simple ones, twos, and threes, but rather, as we shall observe, they are fractal (Badiou 1994). Dimensional events along a given plane are drawn up by multiply directional vectors, tensions, and tendencies, and Deleuzean "lines of flight" within the whole rhizomic mess. That is to say, there is a multiplicity of dimensional trajectories following deterritorializing vectors within the whole space-time field, driving any and all semiosic processes to the abyss of chaos: every-*thing* is always already becoming some-*thing* else, possibly every-*thing* else. Each line of flight is part of the rhizome, which, as multidimensionality, brings about perpetual transformation (translation) of all lines of flight as they converge, diverge, involute, convolute, and become increasingly intertwined (this topic will be taken up in chap. 8).

13. The terms *grue* and *bleen* come from Nelson Goodman's notorious "New Riddle of Induction" (1979)—to be discussed briefly in chapter 6—which plays havoc with the idea of clear and distinct percepts, concepts, and classificatory schemes and conceptual frameworks and puts us much in line with the notion that process, not product, is the universe's way, in spite of our wishes to the contrary (for further with regard to Goodman, see also Merrell 1985, 1997).

Chapter 3

1. I might add that Eco's library from the outside, objective view is very much unlike Borges's library of the "Library of Babel," which appears in *Labyrinths* (1962, 51–58), whose fastidious repetition of hexagons produces a massive beehive cluster of spatial symmetries, though there are asymmetries of another sort, to be discussed in chapter 6.

2. For further on this and other fascinating aspects of Escher's work, see Bool et al. 1981, and Coxeter et al. 1986.

3. As we shall note in the next section, a four-dimensional being would see Scharlach's trajectory—or our own for that matter—in an instant, much as Schlarlach saw that of Lönnrot in an instant.

4. Recall also the distinction between the prisoners of the Prisoners' Paradox, who existed in a time of past/now/future, and their lawyer's timeless time of before/after.

5. This allusion to "time in an instant," an apparent contradiction, is "time spatialized" in the sense of Einstein's space-time continuum making up his conception of what Hermann Minkowski dubbed the "block universe." The timeless "time" of the block universe within which any notion

of time is no more than a figment of the imagination is that against which Ilya Prigogine reacted in his argument for irreversibility, in a universe of self-organization—to be discussed in chapter 8.

6. That a "single repetition" suffices to prove the nonexistence of time is precisely Borges's method for refuting time by means of a ("minimalist") collapse of the combined thought of Hume and Berkeley in "A New Refutation of Time," which appears in *Labyrinths* (Borges 1962, 217–34).

7. The preceding reference to C. Howard Hinton is timely, since it bears on this concept of a fourth dimension. During the revival of mysticism around the turn of the present century, some observers thought the fourth dimension the ideal candidate for the source of mystical experiences. Chief among the books written in that period was Hinton's *The New Era of Thought* (1888), where he claimed that it is possible to acquire a mental grasp of a dimension of space higher than that in which we live and breathe. He provided in his book a series of mental exercises along with twenty-seven three-dimensional colored cubes that fit together into a single cube, which would supposedly be four-dimensional, a "hypercube." If one could learn the relations between all the cubes, Hinton promised, then one could know the fourth dimension.

Chapter 4

1. It should be pointed out at this juncture that the very idea of a monolithic theory to end all theories goes against the grain of the current "postmodern turn." Postmodernists challenge global theories, ideologies, and worldviews of religious, philosophical, political, or social origin. Many dismiss marxism, Christianity, fascism, capitalism, liberal democracy, secular humanism, and even feminism and minority movements as logocentric and totalizing, metanarratives that rest on assumptions and beliefs that are not necessarily any less uncertain and potentially any less damaging than astrology, witchcraft, and cults of all sorts. Although one would not expect that physics could be placed in the same category with these practices, nonetheless, the search for a grand unified theory in the sciences is part and parcel of modernity's dreams and not in keeping with the postmodern turn (for a conception of what might be the characteristics of a genuine postmodern science see Toulmin 1982, 1990).

2. One should add that the Watson–Crick reductionistic-mechanistic view has culminated in the Project Genome dogma (Alberts 1988; Davis 1991; Kevles and Hood 1992), which has recently been placed on the chopping block (Lewontin 1992; Cook-Deegan 1994), and if not discredited, at least it is the focus of serious warning signs (Pollack 1994).

3. See Mandelbrot 1985 and Peitgen and Richter 1986.

4. This intuitive idea that the part contains the whole is notorious, from

mythology to religion to the arts, the sciences, and mathematics. One pearl of the Buddhist string of pearls mirrors the whole; one wave of the ocean contains the entire ocean; in William Blake's grain of sand—or the Aleph or the jaguar's spots—the entire universe can be seen; a particular type of space-time singularity can contain the entire universe; infinities can contain themselves . . . and so on. The problem is that this conception can lend itself—albeit erroneously—to that human obstinacy for simplifying matters. For Darwin life is a dog-eat-dog struggle; for Marx human society is class struggle; for B. F. Skinner and rat psychologists behavior is stimulus-response; for Freud every protuberance is phallic in the zone of the libido and every male child wants to kill Daddy and jump into bed with Mommy. Generalities can be, of course, simplifications of virtually unintelligibly complex matters, but often they are necessary simplifications for effective communication in our everyday affairs. The only reasonable counsel might be to resort to simplicities whenever necessary, while maintaining suspicion regarding them.

5. See Wigner 1965 and Lee 1966; for a layperson account, Zee 1986.

6. It bears mentioning that John D. Bernal (1967) also believes crystallography holds the key to molecular biology, as did Louis Pasteur. For a general survey of current hypotheses regarding the origins of life, see Casti's (1989) colorful method of pitting scientific hypotheses against one another as if they were making an appeal to a jury in the courthouse, with Casti's own verdict at the end of the procedure.

7. It is worthy of note that the very orthographic depiction of Calvino's hero resists symmetry breaking of the aperiodic sort. If one bisects qfwfq with a virgule at the peak of the *w*, one has a mirror-image symmetry.

Chapter 5

1. Reference to "radical empiricism" is directly from William James (1967), as is reference to "pluralism," but the general notion of "pluralism" as I use the term here is found in a diverse mix of contemporary scholars, including Nelson Goodman (1978), Hilary Putnam (1981, 1983, 1988, 1990), and Richard Rorty (1979, 1982, 1989).

2. That is to say, chaos theory, involving imaginative mathematics and heavy duty computer power, presents a universe that is deterministic—Simplicio said as much. But this determinism, though yielding to fundamental physical laws, manifests a predisposition toward complexity and, above all, unpredictability. In this sense chaos theory is actually no radical departure from classical physics. The algorithms for generating fractals can culminate in systems that are "chaotic," to be sure, and as such they are unpredictable, yet in principle—if not in practice—they are deterministic in the good classical fashion (for a survey of chaos theory, see especially Crutchfield et al.

1986; Dyke 1990; Hall 1992; Hunt 1987; and for popular and highly accessible accounts, Briggs and Peat 1989; and Gleick 1987).

3. More technically speaking, for a given object six variables in six-dimensional space are not enough; two more variables, the right- or left-hand spin of an object as it careens through space, are necessary—these variables will be discussed briefly in chapter 6.

4. In spite of Simplicio's wishes to the contrary, time and logic—a nonstandard form of logic—do mix, and as we shall observe subsequently, so do space and time, matter and time, and mind and time. They mix in phase space.

5. The allusion here is quite obviously to Brownian movement, which is random or near-random, typical of the archetypal example of molecules of a gas in an enclosed chamber.

6. As a background to the following, see Berman 1994, Kandinsky 1979, and Roskill 1992.

7. This process is also much in the order of Goethe's theory of colors—to be briefly discussed in the final stages of this inquiry—according to which color comes from the agonistics between black and white, light and shade, which enable us to "construct the visible world, and thus, at the same time, make painting possible" (Matthaei 1971:73).

8. The implication is that, since $\sqrt{-1}$ is "ambiguous," neither one thing nor the other and yet both one thing and the other, an alternative symbol, *i*, can be used in the "unambiguous" sense that it can be taken at face value as what is as it is: in this sense I would like to compare *i* to the yin/yang symbol consisting of two complementary elements making up a unified whole.

9. For the interesting development of a quantum theoretical model of consciousness and consciousness of colors, from which in part the present discussion was derived, see Lockwood 1989.

10. In this vein, Stove (1982) argues that the philosophy of Karl Popper, just as much as that of philosophers and historians the likes of Thomas Kuhn, Imre Lakatos, and Paul Feyerabend, when taken at face value, is irrational through and through.

11. For a comparable notion regarding the emergence of the poetic line from chaos, see Ammons 1982, García Lorca 1980, and Hass 1985.

12. I by no means wish to imply here or elsewhere that imaginary numbers are purely imaginary. They are every bit as real as the so-called real numbers, the chief difference being that they are not of our concrete world in which we can count oranges and apples and sheep and goats. However, to repeat, imaginary numbers are used in relativity and quantum theory, as well as in electrical engineering, to give mathematical and geometrical maps of real world physical phenomena.

13. Of course, in a manner of speaking, we have seen all this before, in figures 5*a*, 9, and 10, this struggle between opposites to bring about the emergence of all that is.

14. A rich list of works of science, philosophy, literature, and painting could at this juncture be evoked. Unfortunately, limited time and space do not permit further discussion of the issues at hand. With respect to painting, on the work of Goethe and its import to contemporary topics, I would suggest Zajonc 1993; for painters such as Monet, Manet, Cézanne, and Picasso as important precursors to Klee, see note 1 of the introduction.

Chapter 6

1. According to bifurcation theory, any and all open "far-from-equilibrium" systems (i.e., systems of disorder) reach points where there are two choices. Beyond this critical point the properties of the system can change abruptly, unpredictably, and along nonlinear paths. Such systems are most effectively accounted for, according to many investigators, by Ilya Prigogine's theory of "dissipative structures" (for Prigogine admirers from various angles, see Casti and Karlqvist 1986; Coveney and Highfield 1990; Jantsch 1980; Kellert 1993; Wicken 1987b; for a sample of Prigogine's own work, see Nicolis and Prigogine 1989; Prigogine 1980; Prigogine and Stengers 1984).

2. This is also the case of the narrator of Calvino's *If on a Winter's Night a Traveler,* who speculates that there are two paths open to him upon embarking on the project of writing a book:

> Either write a book that could be the unique book, that exhausts the whole in its pages; or write all books, to pursue the whole through its partial images. The unique book, which contains the whole, could only be the sacred text, the total world revealed. But I do not believe totality can be contained in language; my problem is what remains outside, the unwritten, the unwritable. The only way left me is that of writing all books, writing the books of all possible authors.
>
> If I think I must write one book, all the problems of how this book should be and how it should not be block me and keep me from going forward. If, on the contrary, I think that I am writing a whole library, I feel suddenly lightened: I know that whatever I write will be integrated, contradicted, balanced, amplified, buried by the hundreds of volumes that remain for me to write. (Calvino 1981, 181)

3. See, in this regard, Borges's brief essay on Pascal (1964, 189–92).

4. Whether this movement, which implies the passage of time, is real—pace Prigogine—or merely a figment of the narrator's imagination—pace Einstein—is a much debated topic, and it merits more than passing mention. Unfortunately, its adequate treatment lies beyond the scope of this inquiry (but see, from a phenomenological view, Wood 1989; and from a scientific-metaphysical view, Fraser 1982; Lockwood 1991; Whitrow 1980; and Ray 1991).

5. Taking an additional step, in Landow 1992, 1994 we find the idea of the text, that is, the hypertext, as labyrinth.

6. The fabled "view from nowhere" (Nagel 1986)—which has been the goal of Western science—is the view desired by Palomar. It is a window to the world, a "split into a looking world and a world looked at" (Calvino 1985, 114). But in such case where is the "I"? Is it not part of the world? If so, is it not true that it cannot look at itself looking at the world? And even if it could do so, would it not need another "I" looking at that "I," and so on? Palomar tries to look at the outside from his outside vantage, but nothing but the "usual quotidian greyness surrounds him" (115). He then decides to wait for one of those rare and lucky coincidences when the world wants to look and be looked at in the same instant, which he believes he can take advantage of and see the whole of things. But to no avail. He remains imprisoned within his finitude.

7. Allusion could be made here to color schemes the likes of Land's (1977) color field, Munsell's (1946) color tree, or Ostwald's (1969) double cone color solid, and especially to Berlin and Kay's (1969) classical study of linguistic categories for colors (see also Teevan and Birney 1961). Color as a concept as and a continuum that can be cut up by language in indefinite ways has engaged the attention of thinkers and scientists as distinct as Aristotle, Pliny, da Vinci, Descartes, Newton, Goethe, and, in our own century, Wittgenstein and Goodman, to name only a few. Recent studies such as those by Hardin (1984, 1988), Ramachandran (1987), and Ramachandran and Gregory (1978) and the synoptic work of Rossotti (1983) testify to the fact that the issue is far from resolved. In this modest essay I cannot hope to offer a disquisition on color. Rather, I attempt to bring commonalties of thought and speculation from various directions to bear on the central topic at hand (however, for more with respect to color in studies of art see Gombrich 1985; Goodman 1976; Kandinsky 1947; Merleau-Ponty 1964; Mitchell 1994; in psychology, Jameson and Horvish 1989; Pirenne 1970; Thompson 1995; Zeki 1983, 1985, 1993; in philosophy, Hilbert 1987; Landesman 1989; Lockwood 1991).

8. See, regarding what some observers consider the threat of solipsism and nihilism in postructuralist-postmodernist thought, Neville 1992; O'Neill 1992, 1995; and Rosenau 1992. It is by now quite obvious, I would expect, that I see no such threat.

9. Paracelsus, the eccentric astrologist-healer, followed the dictum that the patients are the healer's textbook and the sickbed is his or her study. He is reported to have once thrown the medieval book of medicine into a fire and replaced it with the "book of nature," the "doctrine of signatures."

10. Galileo's "gaze," admittedly, is the product of Western thought that is currently under criticism, or deconstruction. According to poststructuralism as well as postempiricist philosophy of science, the subject's presumed neu-

tral observation of the object is a false doctrine, patterning the erroneous notion that mind is the mirror of nature. Nature, we are now told, is not immediately accessible to mind, but only mediately so, and there is no presence, in all its plenitude, nor is the object identical with itself. There can be no more than differences, for one cannot read the same book (universe) twice. This, however, should not eliminate the idea of a book-world metaphor, which, more appropriately in poststructuralist jargon, should be a text-world (Derrida 1978, 294–300).

11. Borges, it bears mentioning, remained fascinated with the book of nature image. Examples are legion. On various occasions, Borges refers to Bacon, Bloy, Carlyle, Novalis, and de Quincey, all of whom registered more than a passive interest in the book metaphor. He notes the Christian idea that the divinity wrote a book or, rather, two books, one of which was the universe (Borges 1964, 119). In stories that appear in *Labyrinths* (1962), "The Secret Miracle" (88–94) and especially "The God's Script" (169–73), God is a single letter, and the world is that letter generated into a single book, which contains God as an immanent being in the Spinozistic sense. The protagonist of "The Garden of Forking Paths," which also appears in *Labyrinths* (1962, 19–29), speculates on the possibility of an infinite book and an infinite maze, both of which are one, and one with the universe. In "Tlön, Uqbar, Orbius Tertius," also in *Labyrinths* (1962, 3–18), Uqbar is described in only one encyclopedia, in which there is bookish reference to the imaginary Tlön, which later takes on a reality eventually to become the world, *our world:* a text. And in *The Book of Sand* (1978) "Undr" (81–87) and "The Mirror and the Mask" (75–79) entail a search for the ultimate expression of all things with the power of a solitary word. Suffice it to recapitulate the views of two Borgesian critics: the predominant theme of Borges's fictions is that literature is as much being as the world (Lefebve 1964), and, if the world is a book, as Borges tells us, then all existent books as a collection compose the world (Blanchot 1959).

12. Significantly, in this regard, Calvino's narrator in "The Night Driver," in *Labyrinths* (1969, 128–36), transforms himself into a message of himself, thereby becoming indistinguishable from any and all other signs.

13. Contrast, for example, the rigid architectonics of Borges's library with the pentagonal, heptagonal, octagonal, that is, aperiodic, structure of Eco's aedificium in *The Name of the Rose* (1980).

Chapter 7

1. Indeed, we read that "the cell, the aperture, the corridors along which the jailer comes twice a day with the soup and the bread could be simply tiny pores in a rock of spongy consistency" (Calvino 1969, 138).

2. We have read this story before, of course, from the narrator of Borges's

"Library of Babel," who is in the library and in the act of writing the short story about the library that the reader is in the act of reading, also presumably from within that selfsame library.

3. I must mention, once again, that tomes have been written on this topic and I stand nary a chance of resolving the problem here. My goal does not include solving the problem of time, but rather, I wish to evoke at least a feel for the problem as I see it related to and portrayed in the theories, ideas, fictions, and intuitions I discuss.

4. The reader has surely noted the anomaly in the remark that the book is "no more or less than infinite" and in language's incapacity to account for the imageless image of any-*thing* containing an infinity of *things*.

5. Infinity, I must clarify at this juncture and with respect to the preceding discussion of Cantor, comes basically in two shapes: *actual infinity* and *potential infinity*. We might wish to experience a chalk mark on a blackboard as continuous (an actual infinity), but it is not, for it merely fails to reveal noticeable gaps to the naked eye. It is no more than the visible expression of a potential. Zeno's paradoxes are predicated on the concept of a potential infinity (a never-ending succession of steps). However, since the implication of actual infinity underlies this concept, Zeno is not a finitist, and finitistic arguments, which merely banish the idea of the actual infinite altogether, cannot legitimately dispel him (Benardete 1964, 13–20).

6. I am assuming, of course, that we are taking infinity to be of the potential rather than the actual variety.

7. Or in another manner of putting it, it is comparable to Borges's mystical, quasi-mystical, or perhaps simply imaginary "experience in death" (see Merrell 1991a).

Chapter 8

1. Prigogine's "physics of complexity" is not to be confused with "chaos physics" (for a layperson account of chaos physics, see Gleick 1987 and for the physics of complexity, Prigogine and Stengers 1984, as well as the list in note 1 of chapter 6). Paul A. Harris (1993, 638) writes that "Prigigone's work has proved appealing to those in the humanities because he has framed his research in chemistry, physics, and cosmology as a reaction against the 'timeless' world of classical physics." Chaos physics, chiefly I would suspect due to its attractive computer-generated visual images, has also been of considerable interest among humanists, in spite of the fact that since it is in principle deterministic—though chaotic events are unpredictable—it still remains with one foot in the doorway of the classical paradigm.

2. Throughout the entire Latin American area, in a diversity of ways, that dreamy dichotomy between the so-called third world and the "developed world" has remained inordinately cloudy. Actually, languages, cultures, and

economic development have presented a confusing picture. Few cities are more cosmopolitan and European than Buenos Aires; few cities are further from Western Enlightenment ideals and suffer more from human degradation than Haiti's capital, Port-au-Prince. Latin America obstinately resists any and all neat classificatory schemes. It is complexity taunting the crystals of taxonomic simples, virtual chaos somehow maintaining itself in a dynamic balance in spite of the overwhelming odds against its so doing (see Zea 1963, 1969, 1992).

3. While the term *pluralism* is most common to postmodern parlance, many Latin Americanists are more comfortable with *hybridization,* brought to prominence especially by Néstor García Canclini (1993, 1995), and *transculturation,* coined by Fernando Ortiz (1947) in his seminal study of the Cuban sugar economy and in recent years put to use by Ángel Rama (1982).

4. Once again, I fear that with neither time nor space to develop the theme here, I can hardly do more than add in passing that Beckett's words suggest interaction between knowing *how* (which entails tacit, habitualized actions and responses) and knowing *that* (which pertains chiefly to conscious mental acts that can be by and large articulated) and between perceiving something *as* something in particular and perceiving *that* such and such is the case regarding this something, concepts that I owe chiefly to Norwood R. Hanson (1958, 1969) and Michael Polanyi (1958) and have developed further in Merrell 1991b, 1995a, 1996, 1997.

5. The Unnamable's interior monologue may go on to infinity, for all we know. If it were to do so, we might describe this novel as "a curve having one of its axes as an asymptote. In other words, as *y* (the content of the novel) approached infinity, *x* (the length of the novel) would approach nearer and nearer to zero. Content zero, length infinity—these are the mathematical limits of the novel" (Mercier 1959, 145).

6. For more on Beckett in regard to the notion of indifference, see Merrell 1985.

7. What is suggested in the combination of Beckett's characters and Benítez-Rojo's repeating island is a frame of mind that is quite contrary to the classical view of things. Isabelle Stengers and Ilya Prigogine observe the remarkable moment of this profound change in our scientific—as well as nonscientific, I might say—concept of nature and of human societies and their relation to nature. They issue a call for a revised conception of human-nature and human-human relations, with an eye toward abolishing the age-old distinctions between aesthetic, ethical, and scientific values. They observe that societies

> are immensely complex systems involving a potentially enormous number of bifurcations exemplified by the variety of cultures that have evolved in the relatively short span of human history. We know that

> such systems are highly sensitive to fluctuations. This leads both to hope and a threat: hope, since even small fluctuations may grow and change the overall structure. As a result, individual activity is not doomed to insignificance. On the other hand, this is also a threat, since in our universe the security of stable, permanent rules seems gone forever. (Prigogine and Stengers 1984, 212–13)

8. Arborescence/rhizome, to be sure, is a binary opposition. However, it is a new form of binary: a complex contradictory-complementary binary, no simple binary. Deleuze and Guattari quite rightly contend that we must dissolve all the customary binaries we can, but, since we cannot do without binaries and expect to speak and write, we should at least strive for new forms of binarism.

9. It bears mentioning that Deleuze and Guattari are quite ready to concede that "trees have rhizome lines and the rhizome points of arborescence." This is the point made previously that with an infinity of lines, chaos can ensue. A tree structure consists of binary forking lines. That much is clear enough. However, when the bifurcations become infinite in number, and nonlinear to boot, rhizomicity and chaos can result (Deleuze and Guattari 1987, 34).

10. Comparably: "Ask any molecule," writes John A. Wheeler (1980, 352), "what it thinks of the second law of thermodynamics and it will laugh at the question." Or, ask a molecule if it is moving at random, and it will say that it is not, that it has free will (Park 1980, 60). If we observe this molecule's behavior among a collection of like molecules, it will appear to us chaotic, but if we place 6×10^{23} of them in an enclosed chamber, as an aggregate they become relatively predictable in terms of temperature, pressure, and volume, following Boyle's laws of gases.

11. Deleuze and Guattari (1987, 9) offer the example of a colony of ants as a rhizome. The movements of each individual ant appear for all intent and purpose random, yet as a whole, the colony appears comparable to a sentient organism (see also Hofstadter 1979).

12. It will be noticed that I used Deleuze and Guattari's terminology somewhat differently in Merrell 1991a. That, however, was another time and another context, which called for slightly variant semiosis.

13. In this section I outline a slight variation of Deleuze and Guattari's three syntheses in accord with the assumptions I have heretofore developed in my inquiry, I hope without unduly violating the subtleties of the Deleuze–Guattari scheme.

14. Deleuze and Guattari's "body without organs," taken from playwright and poet Antonin Artaud, is the body words enter without substance or meaning but as pure signs (signifiers), the body that in turn spews forth a salad of words in entire sonic blocks, without any apparent sense or reason,

though there may be plenty of rhyme. The nature of the body without organs bears a close relationship to figure 4, I would suggest, especially in light of the fact that Deleuze and Guattari (1987, 149) relate it closely to the "Dogon Egg" of Asian thought as a relatively undifferentiated "distribution of intensities."

15. Perhaps the most notable literary examples of these Caribbean characteristics are found in Guillermo Cabrera Infante's *Three Trapped Tigers* (1971) and Luis Rafael Sánchez's *Macho Camacho's Beat* (1981).

Chapter 9

1. Interestingly enough, this is physicist David Bohm's (1980) conception of the becoming of discrete quantum events to make up what was referred to in chapter 6 as the "explicate order" from the continuum of possibilities he calls the "implicate order."

2. Deleuze and Guattari readily and quite wisely concede that there are intermediate stages between what might otherwise be taken as a simple binary, felt/fabric, such as knitting, crochet, patchwork, quilting, and so on, and they offer plenty of examples of these intermediate stages in music, mathematics, and many everyday activities.

3. Significantly, Deleuze and Guattari have little use for any pretentious universal semiology or semiotics that seeks to explain everything in terms of linguistic signs. They admire C. S. Peirce (see Guattari's *Molecular Revolution: Psychiatry and Politics* [1984] and Deleuze's *Cinema 1: The Movement-Image* [1986]), but do not cater to his extending semiotics beyond the strictly human use of signs—in this respect, I would suggest, they are still to a certain degree prisoners of linguicentricity.

4. I refer here to rotational group symmetry according to group theory, which is a convenient way mathematically and geometrically to account for crystal lattice structures (see Weyl 1952).

5. See Popper 1983, Hacking 1965, and Mellor 1981, on "propensity" theory.

6. I suppose it could be said that "perpetual consummation" is appropriate in light of Derrida's (1981) use of the term *hymen*. The "hymen" affords the image of "neither desire nor pleasure but between the two. Neither future nor present, but between the two" (Derrida 1981, 212–13). Once again, it would appear, we have an alternative to the hairline of figure 4.

7. The purely aleatory view of the emergence of life is Jacques Monod's (1971) thesis. For Monod and other like-minded biologists of neo-Darwinian ilk, the probability of life as we know it emerging from lifeless matter is computable to virtually zero; nevertheless, life exists, which is nothing more nor less than a sheer miracle.

8. Roger Penrose (1989), known for his work on black holes and space-

time singularities, demonstrates how it is possible, with his "Penrose tiles," to tile a plane with fivefold symmetry using two shapes: a fat rhombus and a thin rhombus. The remarkable pattern produced is orderly as a whole, though there is no periodicity when considering its parts. Penrose's tiling pattern visually accounts for "quasi-crystals," which he proposes as a model for his quantum world. Quasi-crystals, it would appear, are in this manner a sort of quantum theoretical counterpart to Schrödinger's biological aperiodic crystals.

9. On the "myth of simplicity," see the work of Mario Bunge (1963), scientist, philosopher, and (as was Borges) Argentine.

10. For the complexity of this problem, see, from a philosophical view, Nagel 1974.

Chapter 10

1. It hardly needs saying that in a few short pages I cannot hope to do justice to the concept of fuzzy logic. What is of interest here, however, is the idea of fuzziness, not a thorough exposition of fuzzy logic, in regard to Sagredo's response to Simplicio's argument as embodied in figures 1, 3, 4, and 5*a*, and to the problem of complexity and simplicity in particular aspects of science, literature, and painting insofar as they are presented in this inquiry.

2. It is worthy of note that in McNeill and Freiberger's conception (1993, 28–34), the very idea of fuzziness is to a large degree a matter of cultural conventions, and in part for that reason it is more developed in Japan, where ambiguously complementary yin/yang thinking is the tradition, than in the United States.

3. Simplicio's moment of elucidation is a relevant topic that cannot be included in this inquiry, given limitations of time and space. In passing, I might suggest that the hairline in figure 4 illustrates the emptiness that gives rise to such breakthroughs. As a result of that which can emerge from the hairline, it is possible to construct new ideas and new texts indefinitely. The hairline, for obvious reasons, has a bearing on the basic principles of logic. It is comparable to what is termed Sheffer's (1913) *stroke function,* and to Peirce's *ampheks,* which he developed as a step toward a "logic of relatives" (see Berry 1952 and Merrell 1991b). Although the stroke, like the hairline, implies incompatibility or inconsistency in the most fundamental sense, Whitehead (1938, 52) tells us that it provides for "the whole movement of logic," and Ernest H. Hutten (1962, 178) writes that "[i]t is the very essence of rationality to abolish contradictions; but logic—being the most rational thing in the world—is generated by contradiction."

4. The Cartesian coordinates, one should bear in mind, do not make allowance for the Argand plane and its incorporation of imaginary and com-

plex numbers, as described in the preceding. This does not, however, prohibit consideration of Kosko's interpretation of fuzzy logic, but rather, fuzzy logic, I would suggest, is in a manner of speaking contained within the larger Argand-bearing spheres of vagueness and generality, like C. S. Peirce's category of Secondness sandwiched between Firstness and Thirdness, a topic I have fleshed out to some extent in Merrell 1991b, 1995a, 1996.

5. For the so-called semiotic square, see Greimas and Courtés 1982; for my serious reservations regarding the "square" as a viable model, see Merrell 1995a, 1995b.

6. This formulation, I should mention in passing, holds relevance to the work of George Lakoff (1987) and Mark Johnson (1987), following "prototype theory" in cognitive psychology (Rosch 1973, 1974, 1975a, 1975b, 1977); it also finds certain echoes in philosophy (Putnam 1981, 1988) and biology (Maturana and Varela 1980, 1987), as developed further in Merrell (1997).

7. Notice the similarity between figure 20 and fiigure 15, where following Klee, I used a one-dimensional loop in an attempt to depict a second and a third dimension as a result of the line passing over and under itself.

8. This two- and three-dimensional nature of the Möbius strip holds relevance to the concept of strange attractors, as discussed in chapter 5. I should also mention at this juncture that although allusion to Lacan might ring strange, he is relevant, I would submit, due to his obsession with topological forms with their implication of fuzzy or fractal dimensions. Lacan's "rail" or "line" of "metonymy" is not merely a line, but, like a line drawn around the Möbius strip to describe a circle, it is a one-dimensional line that merges into a second dimension, and the circle does not merely lie on a flat plane, but the plane merges into a third dimension. Lacan's schemes in this regard project out and beyond Euclidean geometry, classical logic, and our accustomed modes of perception and conception.

9. Compare this "extra dimension" to that of Scharlach's modus operandi in contrast to Lönnrot's relatively limited sphere of existence regarding his sleuthing activities. It is as if Scharlach were to take one of the triangles of Figure 22*b*, flip it over in a third dimension to construct the left triangle's enantiomorphic twin on the right-hand side, then flip that second triangle up and connect it to the first triangle to form a rhombus, and thereby pinpoint the location of the next crime. Lönnrot could perform the same operation only with a compass on the surface of a two-dimensional map, given his relatively limited faculties.

10. The implication is that, following from note 4 of this chapter, just as fuzzy logic is capable of incorporating binary thinking, so also vagueness and generality (or, so to speak, Peirce's Firstness and Thirdness) are capable of incorporating fuzzy logic (with neither time nor space to develop this concept here, a perusal of the three Merrell references in note 4 might do the trick).

11. On the moiré effect in op art much in regard to the use of the terms here, see Ulmer 1985.

12. Blacks, whites, grays, and drab images pervade much of Beckett's work. For example, in *Happy Days* (1961) Winnie is found up to her neck in a mound of sand, unable to move, yet she speaks her lines with emotionless resonance. In *Endgame* (1958), the stage is hardly more than blacks, whites, and grays. There are two characters trapped in trash cans, uttering vague references that create an image of tragedy, though it also seems to oscillate between no more than terror and farce.

13. From a diversity of views regarding the inconsistent nature of the mind and of the world, see da Costa 1974, Lupasco 1947, Melhuish 1967. Priest 1987, Priest et al. 1989, Rescher and Brandom 1979, and Routley 1979.

14. Along these lines, and especially regarding the thesis underlying the present inquiry, I should make particular mention of the interdisciplinary work of Varela, Thompson, and Rosch (1993).

15. Moreover, as pointed out in the preceding, fuzzy logic contains standard logic. I might also be inclined to suggest that the spheres of vagueness and generality in turn embrace both fuzzy logic and standard logic, though I do not dare make this assertion here without the supporting arguments in Merrell 1991b, 1995a, 1996—assuming those arguments are valid, of course.

Chapter 11

1. For further in regard to the concept of sign-events and thought-signs, and for much of what follows on the semiotics of C. S. Peirce, see Merrell 1983, 1991a, 1995a, 1996, 1997.

2. For a discussion of the Peircean categories, see Almeder 1980 and Hookway 1985.

3. It is for this reason that I prefer to eschew such commonplace terms as *reference, correspondence,* and *representation,* all of which are part and parcel of the utopian, linguicentric, ocularcentric idea of thought as a mirror of nature (in certain respects Peirce himself might also be charged as guilty of the mirror of nature crime; however see Merrell 1997).

4. *Imaginary* as a fourth term in this sense follows in the footsteps of Spencer-Brown's (1979) calculus.

5. This condition is, one would speculate, comparable to the status of Beckett's Worm as described in the previous chapter, except that Worm is apparently devoid of all consciousness, at least of the human sort. In this sense Worm's status would be more comparable to the bare suggestion of some-*thing* distinguished from every-*thing* else.

6. In what follows I generally draw from, among others, the work of biologist and systems theorist, Erich Jantsch (1980), who effectively brings Prigogine's physics of complexity into his narrative regarding living organisms.

7. I should point out that elsewhere (Merrell 1996) I have been critical of certain aspects of Maturana and Varela's overbearing focus on autonomy and their avoidance of the openness of organisms to their environment.

8. For intriguing accounts of the infinite regress model of consciousness, see Dunne 1934 and Matte-Blanco 1975, 1988.

9. For further, see Stewart and Golubitsky's (1992) excellent study of symmetry breaking in general and, in particular, with respect to animal gaits.

10. In *The Confessions of Zeno* Svevo describes the human counterpart to the centipede. After Zeno learned that when one walks rapidly, in a mere one-half second no fewer than fifty-four muscles are set in motion:

> I at once directed my attention to my legs and tried to discover the infernal machine. I thought I had succeeded in finding it. I could not of course distinguish all its fifty-four parts, but I discovered something terrifically complicated which seemed to get out of order directly I began thinking about it. (Svevo 1958, 397)

Of course, the organism's attention to such asymmetrical movements is unnecessary; what is done is simply done because that is how it is done, and without much ado. In other words, it is the same story: the organism is caught up in chance and is therefore not free, but it is free to do what it does naturally, because it is programmed to do so, but if it is so programmed, it is not free, yet it is free, because there is no telling when and where it will do what it does naturally.

11. Here I recapitulate Deleuze and Guattari's influence from Hjelmslev regarding the expression/content distinction.

Chapter 12

1. I really should apologize for bringing up Schrödinger's cat one more time: I have used the notorious thought experiment in various previous inquiries. Nevertheless, the example seemed so appropriate to what I was trying my best to say, I simply could not resist the temptation.

2. That is to say, the cat is "neither alive nor dead" as an absolutely determinate actualized event, but rather, as we shall note subsequently, and in keeping with the preceding narrative on figure 3 and its ramifications, alternatives between the "neither" and the "nor" can at some time and in some place stand a chance of popping up.

3. I should also apologize for the clumsiness of my attempt at a natural language rendition of the Schrödinger's-cat phenomenon. Part of the problem, as I understand it, is due to the fact that the thought experiment mixes the microscopic quantum world with our empirical macroscopic world, which according to many physicists simply cannot ride, for the macroscopic

domain, *our world,* operates under entirely distinct principles. One world might be quantum, operating with its own logic, but the other world remains quite comfortable with classical principles. Natural language, basically in tune with classical principles, goes into spasms when placed in the context of the quantum world; hence the apparently cross-eyed view of the Schrödinger phenomenon when it is related to the macroscopic domain in which we live and breathe.

4. This rebuttal entails what is by and large known as the "hidden variables" view according to which the Copenhagen interpretation is incomplete, since it cannot unambiguously account for the most fundamental events of the universe; consequently some set of hidden variables must underlie this interpretation, which should actually be the focus of the physicist's investigation (see Bohm 1980).

5. See Ihde 1973. Ihde's phenomenological experiments illustrate that many—and especially spatially ambiguous—perceptual forms can be seen as either two dimensional or three dimensional, depending on the forms' construction by means of what is foregrounded and what is backgrounded.

6. Regarding use of the terms *differentiation, dedifferentiation,* and *redifferentiation* in this particular context, see Ehrenzweig 1975—recall also the dialogue between Simplicio and Sagredo with respect to figure 1.

7. It might appear that I have exercised a switch in figure 24 from figure 5*a* and its variations, where the upper right-hand corner represented some-*thing* definitely actualized. With neither time nor space to divulge the necessary details here (but see Merrell 1997), I state that figure 24 gives a peek into the realization of myriad alternatives to what happened to have been actualized in the past, according to good logic, good reason, and customs and conventions. In other words, the neither and the nor leave the door open—since the excluded-middle principle has been abrogated—to other possibilities that might have come to pass but did not or that may yet come to pass in the future, thus testifying to the element of underdetermination in any and all perceptions, conceptions, theories, narratives, and artworks and their interpretations.

8. B also recalls what Rudolf Arnheim (1954) dubs a "Newtonian oasis," a sort of mesocosm between the infinitesimally small and the infinitely large, or in Peircean terms, between vagueness (the sphere of both-and) and generality (the perpetually incomplete domain, where something is neither absolutely one thing nor another thing but could always have been or will have been something else). As we shall note subsequently, it is Newtonian oases of Euclidean perception and conception that lift themselves from the sea of non-Euclidean chaos at the extremes in whichever direction.

9. This conception of things is also relevant to the work of Merleau-Ponty (1962, 1964, 1968), but that, of course, would be another story that merits a possible telling at some other place and some other time.

10. In Merrell 1995b, from a semiotic perspective I attach the terms *qualisignification, sinsignification,* and *legisignification* to the various periods regarding narrative texts, science, and painting.

11. I should point out at this juncture that "binocular vision" is invariably regarded in Euclidean terms (see Uttal 1978). On the other hand, Heelan (1983) argues at length, and quite effectively, that an alternative, and perhaps more natural, form of space perception is "hyperbolic," which is also more in line with current scientific conceptions of time and space.

12. Regarding such habits of perception and conception from a Kantian view, see Walker 1978, from a phenomenalist view, Mach 1959, and from the phenomenological view, Merleau-Ponty 1962.

13. Snyder (1980), for example, argues that the task of Renaissance painting is to construct scientifically an image on a two-dimensional plane that is so crafted that it gives the viewer a "quasi-hallucinatory" perspective of the three-dimensional object, a practice that was supposedly confirmed by the medieval camera obscura.

14. Paul C. Vitz and Arnold B. Glimcher's (1984) discussion of computer-generated "art" by François Morellet and Béla Julesz is worthy of note here. This "art" consists of random dots of black and white that at a distance appear gray, with which Vitz and Glimcher lose interest in light of a suggestion by Rudolf Arnheim that random happenings add up to very little of import. However robotic and mechanical such random distributions may be, however, they are at least of interest insofar as they illustrate the coming together of opposites to produce something that was originally in neither of the two poles of the duality. This coming together of opposites to produce an entire spectrum of differentiated qualities, quite significantly, was the focus of studies by Goethe and Helmholtz.

15. In terms of Simplicio's implications, a white object is "objectively" every color, yet it is no color, for it reflects all frequencies; a black object is "objectively" no color, yet it is every color, for it contains all frequencies since they are absorbed; colors, then, are so to speak "subjective" and "objective"; they are constructed by us, within our participatory universe (recall, in this light, Simplicio's observation of the black and white quadrants of fig. 3 during his dialogue with Sagredo).

16. I hardly need mention that Land's hypothesis goes entirely against the current of Renaissance Euclidean perception as described in the preceding.

17. Land, as discussed previously, demonstrated how red and white light could be mixed in such a manner that they could produce sensations of a broad range of colors. Rossotti (1983, 141) writes, regarding Land's results, that

> We do not yet know if the mechanism by which we adapt to the colour of the illumination takes place in the eye or in the brain. But there

> seems to be a series of brain cells, each of which picks up processed signals from a very narrow band of wavelengths. Between them, they span the visible spectrum; and it seems that each discernible colour (including the non-spectral purple) is associated with its own set of brain cells. On these cells, a number will respond only in the presence of the background illumination. Thus some of the cells which respond to changes in the intensity of red light (but are insensitive to changes in the intensity of other colours) will respond to red *only* if it is mixed with light of other colours; pure red light produces no activity in such cells. It may well be that observation of this type will eventually help us to understand why the perceived colours of objects depend so little on the light which falls on them.

The constancy we impute in an object as a result of the light of a certain set of frequencies depends in large part on who is looking, from what angle, and the nature of the context. The color of an actual elephant in the wild changes less with varying illumination than the color of a painting of an elephant. The colors of a realistic painting of an elephant change less than the colors of an abstract painting of an elephant with poorly defined regions of color. As the context and the mode of depiction of an object change, so also the perception and conception of that object. To a great extent we see what we are in a state of readiness to see (see Rossotti 1983, 42).

18. In another way of putting it, armed with Land's theory, we are able to reconstruct our percepts in order that they conform with Euclidean imperatives according to our culturally inculcated mode of perception and conception, and if the inculcated mode had been something different, in all probability our reconstructive faculties would have become accommodated to that mode.

19. A confession is in order, however. Scholars of colors the likes of Berlin and Kay (1969) focus exclusively on the visual, while ignoring auditory, tactile, olfactory, and gustatory sensations. It might appear—and with reason—that in this inquiry I commit the same ocularcentric fallacy, even though I attempt to evoke a rather kinesthetic feel for *complexity, simplicity, chaos,* and kindred terms. In my own defense I might write that I am presently at work on a study of the entire range of sensory modes and how they affect the signs we take in and the signs we make. It is my hope that I may be able to render this complex question a slight bit more intelligible.

References

Alberts, B., ed. 1988. *Mapping and Sequencing the Human Genome.* Washington, DC: National Research Council, National Academy of Sciences.

Almeder, Robert. 1980. *The Philosophy of Charles S. Peirce: A Critical Introduction.* Totowa: Rowman and Littlefield.

Ammons, A. R. 1982. "A Poem Is a Walk." In *Claims for Poetry,* ed. D. Hall. Ann Arbor: University of Michigan Press.

Andrade, Mario de. [1928] 1984. *Macunaíma.* Trans. E. A. Goodland. Reprint, New York: Random House.

Arnheim, Rudolf. 1954. *Art and Visual Perception.* Berkeley: University of California Press.

Ashby, W. Ross. 1952. *Design for a Brain.* New York: John Wiley.

Atlan, Henri. 1974. "On a Formal Definition of Organization." *Journal of Theoretical Biology* 45:295–304.

———. 1981a. *Entre le cristal et la fumée: Essai sur l'organisation du vivant.* Paris: Seuil.

———. 1981b. "Hierarchical Self-Organization in Living Systems—Noise and Meaning." In *Autopoeisis: A Theory of the Living Organization,* ed. M. Zeleny, 177–84. New York: North-Holland.

———. 1984. "Disorder, Complexity and Meaning." In *Disorder and Order,* ed. P. Livingston, 109–28. Saratoga, CA: Anma Libri.

Badiou, Alain. 1994. "Gilles Deleuze, the Fold: Leibniz and the Baroque." In *Gilles Deleuze, and the Theater of Philosophy,* ed. C. V. Boundas and D. Olkowski, 51–69. New York: Routledge.

Balakian, Anna E. 1959. *Road to the Absolute.* New York: Noonday Press.

Barrow, John D. 1991. *Theories of Everything: The Quest for Ultimate Explanation.* New York: Fawcett Columbine.

Barth, John. 1966. *Giles Goat-Boy.* Garden City: Doubleday.

———. 1967. *End of the Road.* New York: Grosset and Dunlap.

Barthes, Roland. 1968. *Elements of Semiology.* Trans. A. Lavers and C. Smith. Boston: Beacon Press.

———. 1975. *S/Z.* London: Cape.

Bateson, Gregory. 1972. *Steps to an Ecology of Mind.* New York: Chandler.

Baudrillard, Jean. 1980. "The Implosion of Meaning in the Media and the Implosion of the Social in the Masses." In *The Myths of Information: Technology and Postindustrial Culture,* ed. K. Woodward, 137–48. Madison: Coda Press.

———. 1981. *For a Critique of the Political Economy of the Sign.* Trans. C. Levin. St. Louis: Telos.

———. 1983. *Simulations.* New York: Semiotext(e).

———. 1993. *Symbolic Exchange and Death.* London: Sage.

Beckett, Samuel. 1953. *Watt.* New York: Grove.

———. 1955. *Molloy, Malone Dies, the Unnamable.* New York: Grove.

———. 1958. *Endgame.* New York: Grove.

———. 1961. *Happy Days.* New York: Grove.

———. 1964. *How It Is.* New York: Grove.

———. 1965. *Proust and Three Dialogues.* London: John Calder.

———. 1967. *Stories and Texts for Nothing.* New York: Grove.

———. 1972. *The Lost Ones.* New York: Grove.

———. 1973. *Not I.* London: Faber.

Benardete, José A. 1964. *Infinity: An Essay in Metaphysics.* Oxford: Clarendon.

Benítez-Rojo, Antonio. 1994. *The Repeating Island: The Caribbean and the Postmodern Perspective.* Durham: Duke University Press.

Bennett, Charles H. 1987. "Demons, Engines, and the Second Law." *Scientific American* 258 (November): 108–16.

Bergson, Henri. [1911] 1964. *Creative Evolution.* Reprint, London: Macmillan.

Berlin, Brent, and Paul Kay. 1969. *Basic Color Terms: Their Universality and Evolution.* Berkeley: University of California Press.

Berman, Art. 1994. *Preface to Modernism.* Urbana: University of Illinois Press.

Berman, Morris. 1981. *The Reenchantment of the World.* Ithaca: Cornell University Press.

Bernal, John Desmond. 1967. *The Origins of Life.* Cleveland: World.

Bernstein, Richard J. 1983. *Beyond Objectivity and Relativism: Science, Hermeneutics, and Praxis.* Philadelphia: University of Pennsylvania Press.

Berry, George D. 1952. "Peirce's Contributions to the Logic of Statements and Quantifiers." In *Studies in the Philosophy of Charles Sanders Peirce,* ed. P. P. Wiener and F. H. Young, 153–65. Cambridge: Harvard University Press.

Black, Max. 1937. "Vagueness, an Exercise in Logical Analysis." *Philosophy of Science* 6:427–55.

Blanchot, Maurice. 1959. "L'infini littéraire: L'Aleph." In *Le Livre à venir.* 116–19. Paris: Gallimard.

Bohm, David. 1980. *Wholeness and the Implicate Order.* London: Routledge and Kegan Paul.

Bool, F. H., et al. 1981. *M. C. Escher: His Life and Graphic Works.* New York: Abrams.

Borges, Jorge Luis. 1953. *Historia de la eternidad.* Buenos Aires: Emecé.

———. 1962. *Labyrinths, Selected Stories and Other Writings.* Ed. D. A. Yates and J. E. Irby. New York: New Directions.

———. 1964. *Other Inquisitions, 1937–1952.* Trans. R. L. C. Simms. Austin: University of Texas Press.

———. 1970. *The Aleph and Other Stories, 1933–1969.* Trans. N. T. di Giovanni. New York: E. P. Dutton.

———. 1974. "Los dos reyes y los dos laberintos." In *Obras Completas, 1923–1972,* 607–8. Buenos Aires: Emecé.

———. 1978. *The Book of Sand.* Trans. N. T. di Giovanni. New York: E. P. Dutton.

Bridgman, Percy W. 1959. *The Way Things Are.* Cambridge: Harvard University Press.

Briggs, John, and F. David Peat. 1989. *Turbulent Mirror.* New York: Harper and Row.

Brillouin, Leon. 1951. "Life, Thermodynamics and Cybernetics." *American Scientist* 37:554–68.

Brock, Jarrett E. et al. 1979. "Principal Themes in Peirce's Logic of Vagueness." In *Peirce Studies,* vol. 1, 41–50. Lubbock: Institute for Studies in Pragmaticism.

Bruner, Jerome S. 1957. "Going beyond the Information Given." In *Contemporary Approaches to Cognition: A Symposium Held at the University of Colorado,* 41–69. Cambridge: Harvard University Press.

———. 1986. *Actual Minds, Possible Worlds.* Cambridge: Harvard University Press.

———. 1987. *Making Sense: The Child's Construction of the World.* London: Methuen.

———. 1990. *Acts of Meaning.* Cambridge: Harvard University Press.

Bunge, Mario. 1963. *The Myth of Simplicity: Problems of Scientific Philosophy.* Englewood Cliffs: Prentice-Hall.

Burroughs, William. 1966. *The Soft Machine.* New York: Grove.

Butler, Christopher. 1994. *Early Modernism: Literature, Music and Painting in Europe, 1900–1916.* Oxford: Clarendon.

Cabrera Infante, Gabriel. 1971. *Three Trapped Tigers.* New York: Harper and Row.

Cage, John. 1961. *Silence.* Middletown: Wesleyan University Press.

Cairns-Smith, A. G. 1985. *Seven Clues to the Origin of Life.* Cambridge: Cambridge University Press.

Calvino, Italo. 1968. *Cosmicomics.* Trans. W. Weaver. New York: Harcourt Brace Jovanovich.

———. 1969. *T Zero.* Trans. W. Weaver. New York: Harcourt Brace Jovanovich.

———. 1974. *Invisible Cities.* Trans. W. Weaver. New York: Harcourt Brace Jovanovich.

———. 1981. *If on a Winter's Night a Traveler.* New York: Harcourt Brace Jovanovich.

———. 1985. *Mr. Palomar.* New York: Harcourt Brace Jovanovich.

———. 1986. *The Uses of Literature.* Trans. P. Creagh. New York: Harcourt Brace Jovanovich.

Campbell, Jeremy. 1989. *The Improbable Machine.* New York: Simon and Schuster.

Capra, Fritoj. 1975. *The Tao of Physics.* Berkeley: Shambhala.

Casti, John L. 1989. *Paradigms Lost: Images of Man in the Mirror of Science.* New York: William Morrow.

Casti, John L., and A. Karlqvist, eds. 1986. *Complexity, Language, and Life: Mathematical Approaches.* Berlin: Springer-Verlag.

Chesterton, Gilbert K. 1908. *Orthodoxy.* London: The Budley Head.

Christ, Ronald. 1967. "Jorge Luis Borges, an Interview." *Paris Review* 40:116–64.

Clément, Catherine. 1983. *The Lives and Legends of Jacques Lacan.* Trans. A. Goldhammer. New York: Columbia University Press.

———. 1994. *Syncope.* Minneapolis: University of Minnesota Press.

Cohen, Jack, and Ian Stewart. 1994. *The Collapse of Chaos: Discovering Simplicity in a Complex World.* New York: Penguin.

Coletti, Theresa. 1988. *Naming the Rose: Eco, Medieval Signs and Modern Theory.* Ithaca: Cornell University Press.

Collins, H. M. 1990. *Artificial Experts: Social Knowledge and Intelligent Machines.* Cambridge: MIT.

Cook-Deegan, Robert. 1994. *The Gene Wars: Science, Politics, and the Human Genome.* New York: W. W. Norton.

Costa, Newton C. A. da. 1974. "On the Theory of Inconsistent Formal Systems." *Notre Dame Journal of Formal Logic* 15:497–510.

Coveney, Peter, and Roger Highfield. 1990. *The Arrow of Time.* New York: Random House.

Coxeter, H. S. M., M. Emmer, R. Penrose, and M. L. Teuber, eds. 1986. *M. C. Escher: Art and Science.* Amsterdam: North-Holland.

Crutchfield, James P., J. Doyne Farmer, Norman H. Packard, and Robert S. Shaw. 1986. "Chaos." *Scientific American* 255 (December): 46–57.

Dantzig, Tobias. 1930. *Number: The Language of Science.* 4th ed. New York: The Free Press.

Davies, Paul. 1988. *The Cosmic Blueprint.* New York: Simon and Schuster.

Davis, Joel. 1991. *Mapping the Code: The Human Genome Project and the Choices of Modern Science.* New York: Wiley.

Deleuze, Gilles. 1986. *Cinema 1: The Movement-Image.* Trans. H. Tomlinson and B. Habberjam. Minneapolis: Minnesota University Press.

———. 1990a. *The Logic of Sense.* New York: Columbia University Press.

———. 1990b. *Expressionism in Philosophy: Spinoza.* New York: Zone Books.

———. 1993. *The Fold: Leibniz and the Baroque.* Trans. T. Conley. Minneapolis: Minnesota University Press.

———. 1994. *Difference and Repetition.* New York: Columbia University Press.

Deleuze, Gilles, and Félix Guattari. 1983. *Capitalism and Schizophrenia.* Vol. 1, *Anti-Oedipus.* Minneapolis: Minnesota University Press.

———. 1986. *Kafka: Toward a Minor Literature.* Minneapolis: Minnesota University Press.

———. 1987. *Capitalism and Schizophrenia.* Vol. 2, *A Thousand Plateaus.* Trans. B. Massumi. Minneapolis: Minnesota University Press.

DeLong, Howard. 1970. *A Profile of Mathematical Logic.* New York: Addison-Wesley.

Denbigh, Kenneth George. 1975. *An Inventive Universe.* New York: G. Braziller.

Denbigh, Kenneth George, and J. S. Denbigh. 1985. *Entropy in Relation to Incomplete Knowledge.* Cambridge: Cambridge University Press.

Derrida, Jacques. 1978. *Writing and Difference.* Trans. A. Bass. Chicago: University of Chicago Press.

———. 1981. *Dissemination.* Trans. B. Johnson. Chicago: University of Chicago Press.

Dick, Philip. 1966. *The Unteleported Man.* New York: Berkeley.

———. 1968. *Do Androids Dream of Electric Sheep?* New York: Doubleday.

Dirac, Paul A. C. 1963. "The Physicist's Picture of Nature." *Scientific American* 208 (May): 45–53.

Dobbs, H. A. C. 1972. "The Dimensions of the Sensible Present." In *The Study of Time,* ed. J. T. Fraser, F. C. Haber, and G. H. Miller, 274–92. New York: Springer-Verlag.

Doob, Penelope Reed. 1990. *The Idea of the Labyrinth.* Ithaca: Cornell University Press.

Dretske, Fred I. 1981. *Knowledge and the Flow of Information.* Cambridge: MIT.

Dreyfus, Hubert L. 1994. *What Computers Still Can't Do: A Critique of Artificial Intelligence.* Cambridge: MIT.

Dreyfus, Hubert L., and Stuart S. Dreyfus. 1986. *Mind over Matter.* New York: The Free Press.

Dubuffet, Jean. 1968. *Drawings.* New York: Museum of Modern Art.

Dunne, J. W. 1934. *The Serial Universe.* London: Faber and Faber.

Duve, Thierry de. 1991. *Pictorial Nominalism: On Marcel Duchamp's Passage from Painting to the Readymade.* Trans. D. Polan. Minneapolis: University of Minnesota Press.

Dyke, C. 1990. "Strange Attraction, Curious Liaison: Clio Meets Chaos." *Philosophical Forum* 21:369–92.

Dyson, Freeman. 1979. *Disturbing the Universe.* New York: Harper and Row.

———. 1988. *Infinite in All Directions.* New York: Harper and Row.

Eco, Umberto. 1976. *A Theory of Semiotics.* Bloomington: Indiana University Press.

———. 1980. *The Name of the Rose.* Trans. W. Weaver. New York: Harcourt Brace Jovanovich.

———. 1990. *The Limits of Interpretation.* Bloomington: Indiana University Press.

Eddington, Arthur S. 1958. *The Nature of the Physical World.* Ann Arbor: University of Michigan Press.

Edgerton, Samuel, Jr. 1976. *The Renaissance Rediscovery of Linear Perspective.* New York: Harper and Row.

Effinger, George Alec. 1973. *What Entropy Means to Me.* New York: Signet.

Ehrenberg, W. 1967. "Maxwell's Demon." *Scientific American* 217 (May): 103–10.

Ehrenzweig, Anton. 1975. *The Psychoanalysis of Artistic Vision and Hearing: An Introduction to a Theory of Unconscious Perception.* London: Sheldon.

Einstein, Albert, and L. Infeld. 1952. *The Evolution of Physics: The Growth of Ideas from Early Concepts to Relativity and Quanta.* New York: Simon and Schuster.

Engel-Tiercelin, Claudine. 1992. "Vagueness and the Unity of C. S. Peirce's Realism." *Transactions of the C. S. Peirce Society* 28 (1):51–82.

Federman, Raymond. 1965. *Journey to Chaos: Samuel Beckett's Early Fiction.* Berkeley: University of California Press.

Foerster, Heinz von. 1980. "Epistemology of Communication." In *The Myth of Information: Technology and Postindustrial Culture,* ed. K. Woodward, 18–27. Madison: Coda Press.

Ford, Joseph. 1983. "How Random Is a Coin Toss?" *Physics Today* 36 (April): 40–47.

Fraser, J. T. 1982. *The Genesis and Evolution of Time: A Critique of Interpretation in Physics.* Amherst: University of Massachusetts Press.

Frye, Northrop. 1957. *Anatomy of Criticism: Four Essays.* Princeton: Princeton University Press.

Fuentes, Carlos. 1969. *La nueva novela hispanoamericana.* Mexico: Joaquín Mortiz.

Galilei, Galileo. 1957. *Discoveries and Opinions of Galileo.* Trans. and with an Introduction and notes by S. Drake. Garden City: Doubleday.

Gamow, George. 1947. *One, Two, Three . . . Infinity.* New York: Viking.

García Canclini, Néstor. 1993. *Transforming Modernity: Popular Culture in Mexico.* Trans. L. Lozano. Austin: University of Texas Press.

———. 1995. *Hybrid Cultures.* Trans. C. L. Chiappari and S. L. López. Minneapolis: University of Minnesota Press.

García Lorca, Federico. 1980. *Deep Song and Other Prose.* Trans. C. Maurer. Boston: Marion Boyars.

Gardner, Howard. 1987. *The Mind's New Science.* New York: Basic Books.

Gardner, Martin. 1983. *Wheels, Life, and Other Mathematical Amusements.* New York: W. H. Freeman.

Gibson, William. 1983. *Neuromancer.* New York: Ace Books.

Gibson, William, and Bruce Sterling. 1991. *The Difference Engine.* New York: Bantam.

Gleick, James. 1987. *Chaos: Making of a New Science.* New York: Viking.

Gombrich, Ernst H. 1985. *Symbolic Images: Studies in the Art of the Renaissance.* 3rd ed. Chicago: University of Chicago Press.

Goodman, Nelson. 1976. *Languages of Art.* Indianapolis: Hackett.

———. 1978. *Ways of Worldmaking.* Indianapolis: Hackett.

———. 1979. *Fact, Fiction and Forecast.* Indianapolis: Bobbs-Merrill.

Gregory, Bruce. 1988. *Inventing Reality: Physics as Language.* New York: John Wiley.

Gregory, Richard L. 1966. *Eye and Brain: The Psychology of Seeing.* New York: McGraw-Hill.

———. 1970. *The Intelligent Eye.* New York: McGraw-Hill.

Gregory, Richard L., and Ernst H. Gombrich, eds. 1973. *Illusion in Nature and Art.* New York: Scribner's.

Greimas, A. J., and J. Courtés. 1982. *Semiotics and Language: An Analytical Dictionary.* Trans. L. Crist et al. Bloomington: Indiana University Press.

Grünbaum, Adolf. 1967. *Modern Science and Zeno's Paradoxes.* Middletown: Wesleyan University Press.

Guattari, Félix. 1984. *Molecular Revolution: Psychiatry and Politics.* Trans. R. Sheed. New York: Penguin.

Hacking, Ian. 1965. *Logic of Statistical Inference.* Cambridge: Cambridge University Press.

Hall, Nina, ed. 1992. *Exploring Chaos: A Guide to the New Science of Disorder.* London: Penguin.

Hanson, Norwood R. 1958. *Patterns of Discovery.* Cambridge: Cambridge University Press.

———. 1969. *Perception and Discovery.* San Francisco: Freeman, Cooper.

Hardin, C. L. 1984. "A New Look at Color." *American Philosophical Quarterly* 21:125–33.

———. 1988. *Color for Philosophers: Unweaving the Rainbow.* Indianapolis: Hackett.

Harris, Paul A. 1993. "Fractal Faulkner: Scaling Time in *Go Down, Moses.*" *Poetics Today* 14 (4):625–51.

Harvey, David. 1989. *The Condition of Postmodernity: An Enquiry into the Origins of Cultural Change.* Cambridge: Basil Blackwell.

Hass, Robert. 1985. *Twentieth Century Pleasures: Prose and Poetry.* New York: Ecco Press.

Hawking, Steven W. 1988. *A Brief History of Time: From the Big Bang to Black Holes.* New York: Bantam.

Hayles, N. Katherine. 1990. *Chaos Bound: Orderly Disorder in Contemporary Literature and Science.* Ithaca: Cornell University Press.

Hayward, Jeremy W. 1984. *Perceiving Ordinary Magic.* Boulder: Shambhala.

———. 1987. *Shifting Worlds, Changing Minds.* Boston: Shambhala.

Hebb, Donald O. 1949. *The Organization of Behavior.* New York: John Wiley.

Heelan, Patrick A. 1983. *Space-Perception and the Philosophy of Science.* Berkeley: University of California Press.

Hempel, Carl G. 1945. "Studies in the Logic of Confirmation." *Mind* 54:1–26, 97–121.

Henderson, Linda Dalrymple. 1983. *The Fourth Dimension: Non-Euclidean Geometry in Modern Art.* Princeton: Princeton University Press.

Herbert, Nick. 1985. *Quantum Reality: Beyond the New Physics.* Garden City: Anchor.

Hicks, Emily D. 1991. *Border Writing: The Multitudinal Text.* Minneapolis: Minnesota University Press.

Hilbert, David R. 1987. *Color and Color Perception: A Study in Anthropocentric Realism.* Stanford: Center for the Study of Language and Information.

Hinton, C. Howard. 1888. *The New Era of Thought.* London: Shonnenschein.

Hofstadter, Douglas R. 1979. *Gödel, Escher, Bach: An Eternal Golden Braid.* New York: Basic Books.

Hookway, Christopher. 1985. *Peirce.* London: Routledge and Kegan Paul.

Hopkins, J. 1973. "Visual Geometry." *Philosophical Review* 82:3–34.

Horwich, Paul. 1987. *Asymmetries in Time: Problems in the Philosophy of Science.* Cambridge: MIT.

Hunt, G. M. K. 1987. "Determinism, Predictability and Chaos." *Analysis* 47:129–33.

Huntington, C. W., Jr. 1989. *The Emptiness of Emptiness.* Honolulu: University of Hawaii Press.

Huntington, Edward V. 1917. *The Continuum, and Other Types of Serial Order.* Cambridge: Harvard University Press.

Hutten, Ernst H. 1962. *The Origins of Science: An Inquiry into the Foundations of Western Thought.* London: George Allen and Unwin.

Ihde, Don. 1973. *Experimental Phenomenology: An Introduction.* New York: Putnam.

James, William. 1948. *Some Problems of Philosophy.* London: Longmans, Green.

———. 1967. *Essays in Radical Empiricism and a Pluralistic Universe.* Gloucester, MA: Peter Smith.

Jameson, D., and L. Horvish. 1989. "Essay Concerning Color Constancy." *Annual Review of Psychology* 40:1–22.

Jantsch, Erich. 1980. *The Self-Organizing Universe: Scientific and Human Implications of the Emergent Paradigm of Evolution.* Oxford: Pergamon.

Jauch, J. M. 1973. *Are Quanta Real?* Bloomington: Indiana University Press.

Jaynes, E. T. 1965. "Gibbs vs. Boltzmann Entropies." *American Journal of Physics* 33 (5):391–98.

Johnson, Mark. 1987. *The Body in the Mind: The Bodily Basis of Meaning, Imagination, and Reason.* Chicago: University of Chicago Press.

Kadanoff, L. P. 1983. "Roads to Chaos." *Physics Today* (December): 46–53.

Kalupahana, David J. 1986. *Nāgārjuna: The Philosophy of the Middle Way.* Albany: State University of New York Press.

Kandinsky, Wassily. 1947. *Concerning the Spiritual in Art.* New York: Wittenborn Art Books.

———. 1979. *Point and Line to Plane.* New York: Dover.

Katz, Jerrold J., and Jerry A. Fodor. 1963. "The Structure of a Semantic Theory." *Language* 39:170–210.

Kellert, Stephen H. 1993. *In the Wake of Chaos: Unpredictable Order in Dynamical Systems.* Chicago: University of Chicago Press.

Kellner, Douglas. 1989. *Jean Baudrillard: From Marxism to Post-Modernism and Beyond.* Stanford: Stanford University Press.

Kern, Stephen. 1983. *The Culture of Time and Space, 1880–1918.* Cambridge: Harvard University Press.

Kevles, Daniel J., and Leroy E. Hood. 1992. *The Code of Codes: Scientific and Human Issues in the Human Genome Project.* Cambridge: Harvard University Press.

Kierkegaard, Søren. 1944. *Concluding Unscientific Postscript.* Trans. D. F. Swanson. Princeton: Princeton University Press.

Klee, Paul. 1959. *The Inward Vision: Watercolors, Drawings and Writings by Paul Klee.* Trans. and ed. N. Gutterman. New York: Abrams. Quoted in Herschel B. Chipp, *Theories of Modern Art: A Source Book by Artists and Critics* (Berkeley: University of California Press, 1959), 184–85.

Kosko, Bart. 1993. *Fuzzy Thinking: The New Science of Fuzzy Logic.* New York: Hyperion.

Lacan, Jacques. 1966. *Ecrits.* Paris: Seuil.

———. 1977. *The Four Fundamental Concepts of Psycho-Analysis.* Trans. A. Sheridan. New York: W. W. Norton.

Laing, Ronald D. 1965. *The Divided Self.* Middlesex: Penguin.

———. 1969. *The Politics of the Family.* New York: Random House.

———. 1970. *Knots.* New York: Random House.

Lakoff, George. 1987. *Women, Fire, and Dangerous Things: What Categories Reveal about the Mind.* Chicago: University of Chicago Press.

Land, Edwin. 1959. "Experiments in Color Vision." *Scientific American* 200 (May): 84–99.

———. 1964. "The Retinex." *American Scientist* 52:247–64.

———. 1977. "The Retinex Theory of Color Vision." *Scientific American* 237 (June): 108–28.

———. 1983. "Recent Advances in Retinex Theory and Some Implications for Cortical Computations: Color Vision and the Natural Image." *Proceedings of the National Academy of Sciences* 80:5163–69.

Landesman, Charles. 1989. *Color and Consciousness.* Philadelphia: Temple University Press.

Landow, George P. 1992. *Hypertext: The Convergence of Contemporary Critical Theory and Technology.* Baltimore: Johns Hopkins University Press.

———, ed. 1994. *Hyper/Text/Theory.* Baltimore: Johns Hopkins University Press.

Langton, C., ed. 1989. *Artificial Life.* Vol. 1. Redwood City, CA: Addison-Wesley.

Langton, C. et al., eds. 1992. *Artificial Life.* Vol. 2. Redwood City, CA: Addison-Wesley.

Lashley, Karl S. 1951. "The Problem of Serial Order in Behavior." In *Cerebral Mechanisms in Behavior: The Hixon Symposium,* ed. L. A. Jeffreys, 112–36. New York: John Wiley.

Layzer, David. 1975. "The Arrow of Time." *Scientific American* 233 (June): 56–69. Quoted in Jeremy Campbell, *Grammatical Man: Information, Entropy, Language and Life* (New York: Simon and Schuster), 89.

———. 1990. *Cosmogenesis: The Growth of Order in the Universe.* Oxford: Oxford University Press.

Lecercle, Jean-Jacques. 1985. *Philosophy through the Looking Glass.* LaSalle: Open Court.

———. 1990. *The Violence of Language.* London: Routledge.

Lee, T. D. 1966. "Space Inversion, Time Reversal and Particle-Anti-Particle Conjugation." *Physics Today* 19 (March): 23–31.

Lefebve, Maurice-Jean. 1964. "La Jolie Tristan ou une estétique de l'infini." *La Nouvelle revue française,* no. 67:102–6.

Levy, Steven. 1992. *Artificial Life.* New York: Random House.

Lewontin, R. C. 1992. "The Dream of the Human Genome." *The New York Review,* May 28, 30–40.

Lindley, David. 1993. *The End of Physics: The Myth of a Unified Theory.* New York: Basic Books.

Lockwood, Michael. 1989. *Mind, Brain, and the Quantum: The Compound "I."* London: Blackwell.

Lorenz, Edward. 1963. "Deterministic Nonperiodic Flow." *Journal of Atmospheric Sciences* 20:131–40.

Lucas, J. R. 1969. "Euclides ab omni naevo vindicatus." *British Journal of Philosophy of Science* 20:1–11.

Lupasco, Stéphane. 1947. *Logique et contradiction.* Paris: Presses Universitaires de France.

McCulloch, Warren S., and Walter Pitts. 1943. "A Logical Calculus of the Ideas Immanent in Nervous Activity." *Bulletin of Mathematical Biophysics* 5:115–33.

McNeill, Daniel, and Paul Freiberger. 1993. *Fuzzy Logic.* New York: Simon and Schuster.

McRobie, Allan, and Michael Thompson. 1992. "Chaos, Catastrophes and Engineering." In *Exploring Chaos: A Guide to the New Science of Disorder,* ed. N. Hall, 149–61. New York: W. W. Norton.

McTaggart, J. M. E. 1927. *The Nature of Existence.* Vol 2. Cambridge: Cambridge University Press.

Mach, Ernst. 1959. *Analysis of Sensations.* Trans. C. M. Williams. New York: Dover.

Malevich, Kazimir. 1959. *The Non-Objective World.* Trans. H. Dearstyne. Chicago: Theobold. Quoted in Herschel B. Chipp, *Theories of Modern Art: A Source book by Artists and Critics* (Berkeley: University of California Press, 1968), 341.

Mandelbrot, Benoit B. 1985. *The Fractal Geometry of Nature.* San Francisco: W. H. Freeman.

Mangel, Anne. 1971. "Maxwell's Demon, Entropy, Information: *The Crying of Lot 49.*" *Triquarterly* 20:194–208.

Matte-Blanco, Ignacio. 1975. *The Unconscious as Infinite Sets: An Essay in Bi-Logic.* London: Duckworth.

———. 1988. *Thinking, Feeling, and Being.* London: Routledge.

Matthaei, Rupprecht, ed. 1971. *Goethe's Color Theory.* New York: Van Nostrand Reinhold.

Maturana, Humberto, and Francisco J. Varela. 1980. *Autopoiesis and Cognition: The Realization of the Living.* Dordrecht: D. Reidel.

———. 1987. *The Tree of Knowledge: The Biological Roots of Human Understanding.* Boston: Shambhala.

Maxwell, James C. 1875. *Theory of Heat.* London: Longmans, Green.

Melhuish, George. 1967. *The Paradoxical Nature of Reality.* Bristol: St. Vincent's Press.

Mellor, D. H. 1981. *Real Time.* Cambridge: Cambridge University Press.

Mercier, Vivian. 1959. "The Mathematical Limit." *Nation* 188 (February): 144–45.

Merleau-Ponty, Maurice. 1962. *Phenomenology of Perception.* New York: Harcourt Brace Jovanovich.

———. 1964. "Eye and Mind." In *The Primacy of Perception and Other Essays,* ed. J. M. Edie. Evanston: Northwestern University Press.

———. 1968. *The Visible and the Invisible.* Evanston: Northwestern University Press.

Merrell, Floyd. 1983. *Pararealities: The Nature of Our Fictions and How We Know Them.* Amsterdam: John Benjamins.

———. 1985. *Deconstruction Reframed.* West Lafayette: Purdue University Press.

———. 1991a. *Unthinking Thinking: Jorge Luis Borges, Mathematics, and the New Physics.* West Lafayette: Purdue University Press.

———. 1991b. *Signs Becoming Signs: Our Perfusive, Pervasive Universe.* Bloomington: Indiana University Press.

———. 1995a. *Semiosis in the Postmodern Age.* West Lafayette: Purdue University Press.

———. 1995b. *Peirce's Semiotics Now: A Primer.* Toronto: Canadian Scholars' Press.

———. 1996. *Signs Grow: Semiosis and Life Processes.* Toronto: University of Toronto Press.

———. 1997. *Peirce, Signs, and Meaning.* Toronto: University of Toronto Press.

Meyer, Leonard. 1967. *Music, the Arts, and Ideas: Patterns and Predictions in Twentieth-Century Culture.* Chicago: University of Chicago Press.

Miller, Arthur I. 1986. *Imagery in Scientific Thought.* Cambridge: MIT.

Mitchell, W. J. T. 1994. *Picture Theory.* Chicago: University of Chicago Press.

Monod, Jacques. 1971. *Chance and Necessity.* New York: Alfred A. Knopf.

Morrissette, Bruce. 1962. "The New Novel of France." *Chicago Review* 15:1–19.

Munsell, Albert H. 1946. *A Color Notation.* Baltimore: Munsell Color Co.

Musil, Robert. 1965. *The Man without Qualities.* Trans. E. Wilkins and E. Kaiser. New York: Capricorn.

Nadeau, Robert L. 1991. *Mind, Machines and Human Consciousness.* Chicago: Contemporary Books.

Nadin, Mihai. 1982. "Consistency, Completeness and the Meaning of Sign Theories." *American Journal of Semiotics* 1 (3):79–98.

———. 1983. "The Logic of Vagueness and the Category of Synechism." In *The Relevance of Charles Peirce,* ed. E. Freeman, 154–66. LaSalle: Monist Library of Philosophy.

Nagel, Thomas. 1974. "What Is It Like to Be a Bat?" *Philosophical Review* 83:435–51.

———. 1986. *The View from Nowhere.* London: Oxford University Press.

Neisser, Ulric. 1967. *Cognitive Psychology.* Englewood Cliffs: Prentice-Hall.

———. 1994. *The Remembering Self: Construction and Accuracy in the Self-Narrative.* Cambridge: Cambridge University Press.

Neville, Robert Cummings. 1992. *The Highroad around Modernism.* Albany: State University of New York Press.

Nicolis, Grégoire, and Ilya Prigogine. 1989. *Exploring Complexity: An Introduction.* New York: W. H. Freeman.

Nietzsche, Friedrich. 1969. *Selected Letters of Friedrich Nietzsche.* Trans. C. Middleton. Chicago: University of Chicago Press.

Nouy, Lecomte du. 1947. *Human Destiny.* New York: Longmans, Green.

O'Brian, Flan. 1974. *The Third Policeman.* London: Picador.

O'Gorman, Edmundo. 1961. *The Invention of America: An Inquiry into the Historical Nature of the New World and the Meaning of Its History.* Bloomington: Indiana University Press.

O'Neill, John. 1992. *Critical Conventions: Interpretation in the Literary Arts and Sciences.* Norman: University of Oklahoma Press.

———. 1995. *The Poverty of Postmodernism.* London: Routledge.

Ortiz, Fernando. 1947. *Cuban Counterpoint: Tobacco and Sugar.* New York: A. A. Knopf.

Ostwald, Wilhelm. 1969. *The Color Primer.* New York: Van Nostrand Reinhold.

Pagels, Heinz R. 1982. *The Cosmic Code: Quantum Physics as the Language of Nature.* New York: Bantam.

———. 1988. *The Dreams of Reason: The Computer and the Rise of the Sciences of Complexity.* New York: Simon and Schuster.

Panofsky, Erwin. 1960. *Renaissance and Renascences in Western Art.* Stockholm: Almqvist and Wiksell.

Park, David. 1980. *The Image of Eternity: Roots of Time in the Physical World.* New York: New American Library.

Pattee, Howard H. 1969. "How Does a Molecule Become a Message?" *Developmental Biology,* suppl. 3:227–33.

———. 1971. "Can Life Explain Quantum Mechanics?" In *Quantum Theory and Beyond,* ed. T. Bastin, 307–19. Cambridge: Cambridge University Press.

———. 1979. "Robert Rosen, Howard Hunt Pattee and Raymond L. Somorja: A Symposium in Theoretical Biology." In *A Question of Physics: Conversations in Physics and Biology,* ed. P. Buckley and F. D. Peat, 84–123. Toronto: University of Toronto Press.

Paz, Octavio. 1961. *The Labyrinth of Solitude: Life and Thought in Mexico.* New York: Grove.

Peirce, Charles Sanders. 1931–35. *Collected Papers of Charles Sanders Peirce,* ed. C. Hartshorne and P. Weiss. Vols. 1–6. Cambridge: Harvard University Press.

Peitgen, H. O., and P. H. Richter. 1986. *The Beauty of Fractals.* New York: Springer-Verlag.

Penrose, Roger. 1989. *The New Emperor's Mind: Concerning Computers, Minds, and the Laws of Physics.* Oxford: Oxford University Press.

Pierce, John Robinson. 1961. *Symbols, Signals, and Noise: The Nature and Process of Communication.* New York: Harper and Row.

Pirenne, M. H. 1970. *Optics, Painting and Photography.* Cambridge: Cambridge University Press.

Poincaré, Henri. 1913. *The Foundations of Science: Science and Hypotheses, the Value of Science, and Science and Method.* Trans. G. B. Halsted. New York: The Science Press.

———. 1914. *Science and Mind.* Trans. F. Maitland. London: Thomas Nelson.

Polanyi, Michael. 1958. *Personal Knowledge.* Chicago: University of Chicago Press.

Pollack, Robert. 1994. *Signs of Life: The Language and Meanings of DNA.* Boston: Houghton Mifflin.

Popper, Karl R. 1963. *Conjectures and Refutations: The Growth of Scientific Knowledge.* Oxford: Oxford University Press.

———. 1972. *Objective Knowledge.* Oxford: Clarendon.

———. 1983. *Realism and the Aim of Science.* Totowa: Rowman and Littlefield.

Porush, David. 1989. *The Soft Machine: Cybernetic Fiction.* New York: Methuen.

Poundstone, William. 1985. *The Recursive Universe: Cosmic Complexity and the Limits of Scientific Knowledge.* New York: William Morrow.

Priest, Graham. 1987. *In Contradiction: A Study of the Transconsistent.* Dordrecht: Martinus Nijhoff.

Priest, Graham, R. Sylvan, and J. Norman, eds. 1989. *Paraconsistent Logic: Essays on the Inconsistent.* Munich: Philosophia Verlag.

Prigogine, Ilya. 1980. *From Being to Becoming: Time and Complexity in the Physical Sciences.* New York: W. H. Freeman.

Prigogine, Ilya, and Isabel Stengers. 1984. *Order out of Chaos: Man's New Dialogue with Nature.* With a foreword by Alvin Toffler. New York: Bantam.

Putnam, Hilary. 1981. *Reason, Truth and History.* Cambridge: Cambridge University Press.

———. 1983. *Realism and Reason.* Cambridge: Cambridge University Press.

———. 1988. *Representation and Reality.* Cambridge: MIT.

———. 1990. *Realism with a Human Face,* ed. J. Conant. Cambridge: Harvard University Press.

Pynchon, Thomas. 1966. *The Crying of Lot 49.* Philadelphia: Lippincott.

———. 1973. *Gravity's Rainbow.* New York: Viking.

Rama, Ángel. 1982. *Transculturación narrativa en América Latina.* Mexico: Siglo XXI.

Ramachandran, V. S. 1987. "Interaction between Colour and Motion in Human Vision." *Nature* 328:645–47.

Ramachandran, V. S., and R. L. Gregory. 1978. "Does Colour Provide an Input to Human Motion Perception?" *Nature* 275:55–56.

Ray, Christopher. 1991. *Time, Space, and Philosophy.* New York: Routledge.

Rescher, Nicholas, and Robert Brandom. 1979. *The Logic of Inconsistency: A Study of Non-Standard Possible World Semantics and Ontology.* Totowa: Rowman and Littlefield.

Richardson, John Adkins. 1971. *Modern Art and Scientific Thought.* Urbana: University of Illinois Press.

Robinson, William S. 1992. *Computers, Minds, and Robots.* Philadelphia: Temple University Press.

Rorty, Richard. 1979. *Philosophy and the Mirror of Nature.* Princeton: Princeton University Press.

———. 1982. *Consequences of Pragmatism.* Minneapolis: Minnesota University Press.

———. 1989. *Contingency, Irony, and Solidarity.* Cambridge: Cambridge University Press.

Rosch, Eleanor. 1973. "On the Internal Structure of Perceptual and Semantic Categories." In *Cognitive Development and the Acquisition of Language,* ed. T. Moore, 111–44. New York: Academic Press.

———. 1974. "Linguistic Relativity." In *Human Communication: Theoretical Explorations,* ed. A. Silverstein, 95–121. New York: Halsted Press.

———. 1975a. "Universals and Cultural Specifics in Human Categorization." In *Cross-Cultural Perspectives on Learning,* ed. R. Brislin, S. Bochner, and W. Lonner, 177–206. New York: Halsted Press.

———. 1975b. "Cognitive Representations of Semantic Categories." *Journal of Experimental Psychology:General* 104:192–233.

———. 1977. "Human Categorization." In *Studies in Cross-Cultural Psychology,* vol. 1, ed. I. N. Warren, 3–49. New York: Academic Press.

Rosenau, Pauline Marie. 1992. *Post-Modernism and the Social Sciences.* Princeton: Princeton University Press.

Roskill, Mark. 1992. *Klee, Kandinsky, and the Thought of Their Time: A Critical Perspective.* Urbana: University of Illinois Press.

Rosso, Stefano. 1983. "A Correspondence with Umberto Eco." Trans. C. Springer. *Boundary 2* 12 (1):1–13.

Rossotti, Hazel. 1983. *Color: Why the World Isn't Grey.* Princeton: Princeton University Press.

Routley, Richard. 1979. *Exploring Meinong's Jungle and Beyond.* Canberra, Australia: Australia National University Departmental Monograph #3.

Rushdie, Salman. 1981. "Calvino." *London Review of Books* (September): 16–17, 17–30.

Russell, Bertrand. 1910. "The Theory of Logical Types." In *Principia Mathematica,* by Alfred North Whitehead and Bertrand Russell, vol. 1, 37–65. Cambridge: Cambridge University Press.

———. 1923. "Vagueness." *Australian Journal of Psychology and Philosophy* 1:84–92.

Sábato, Ernesto. 1945. "Borges: geometrización de la novela." In *Uno y el universe,* 104–10. Buenos Aires: Sudamericana.

Sacks, Oliver. 1995. *An Anthropologist on Mars.* New York: Random House.

Salthe, Stanley. 1993. *Complexity and Change in Biology: Development and Evolution.* Cambridge: MIT.

Sánchez, Luis Rafael. 1981. *Macho Camacho's Beat.* Trans. G. Rabassa. New York: Pantheon.

Savan, David. 1987–88. *An Introduction to C. S. Peirce's Full System of Semeiotic.* Monograph Series of the Toronto Semiotic Circle, no. 1. Toronto: Victoria College.

Sbragia, Albert. 1993. "Italo Calvino's Ordering of Chaos." *Modern Fiction Studies* 39 (2):283–306.

Schlegel, Richard. 1967. *Completeness in Science.* New York: Appleton-Century-Crofts.

Schrödinger, Erwin. 1967. *What Is Life?* and *Matter and Mind.* Cambridge: Cambridge University Press.

Searle, John R. 1980a. "*Las Meninas* and the Paradoxes of Pictorial Representation." *Critical Inquiry* 6 (3):477–88.

———. 1980b. "Minds, Brains and Programs, with Open Peer Commentaries." *Behavioral and Brain Sciences* 3:417–57.

Shannon, Claude E. 1948. "A Mathematical Theory of Information." *Bell Systems Technical Journal* 27:379–423, 623–56.

Shannon, Claude E. and Warren Weaver. 1949. *The Mathematical Theory of Communication.* Urbana: University of Illinois Press.

Sheffer, Henry M. 1913. "A Set of Five Independent Postulates of Boolean Algebra." *Transactions of the American Mathematical Society* 14:481–88.

Shlain, Leonard. 1991. *Art and Physics: Parallel Visions in Space, Time and Light.* New York: William Morrow.

Singh, Jagjit. 1966. *Great Ideas in Information Theory, Language and Cybernetics.* New York: Dover.

Slobodkin, Lawrence B. 1992. *Simplicity and Complexity in Games of the Intellect.* Cambridge: Harvard University Press.

Snyder, J. 1980. "Picturing Vision." *Critical Inquiry* 6:499–526.

Spencer-Brown, G. 1979. *Laws of Form.* New York: E. P. Dutton.

Spiller, Jürg, ed. 1961. *Paul Klee: The Thinking Eye, Notebooks of Paul Klee.* London: Lund Humphries.

Sterling, Bruce. 1988. *Islands in the Net.* New York: Berkeley.

Stewart, Ian. 1989. *Does God Play Dice? The Mathematics of Chaos.* Oxford: Blackwell.

Stewart, Ian, and Martin Golubitsky. 1992. *Fearful Symmetry: Is God a Geometer?* London: Penguin.

Stone, J. D. 1981. "Simplism and the Liar." *Journal of Philosophical Logic* 10:423–51.

Stove, D. C. 1982. *Popper and After: Four Modern Irrationalists.* Oxford: Pergamon.

Svevo, Italo. 1958. *The Confessions of Zeno.* Trans. B. de Zoete. New York: Vintage.

Sypher, Wylie. 1962. *Loss of the Self in Modern Literature and Art.* New York: Random House.

Szamosi, Gésa. 1986. *The Twin Dimensions: Inventing Time and Space.* New York: McGraw-Hill.

Szilard, Leo. 1929. "On the Reduction of Entropy as a Thermodynamic System Caused by Intelligent Beings." *Zeitschrift für Physik* 53:840–56.

Teevan, Richard C., and Robert C. Birney, eds. 1961. *Color Vision.* Princeton: D. van Nostrand.

Thompson, Evan. 1995. *Colour Vision: A Study in Cognitive Science and the Philosophy of Perception.* London: Routledge.

Toulmin, Stephen. 1982. *The Return to Cosmology: Postmodern Science and the Theology of Nature.* Chicago: University of Chicago Press.

———. 1990. *Cosmopolis: The Hidden Agenda of Modernity.* New York: The Free Press.

Tritton, David. 1986. "Chaos in the Swing of a Pendulum." *New Scientist* 14 (July): 37–40.

Turkle, Sherry. 1978. *Psychoanalytic Politics: Freud's French Revolution.* New York: Basic Books.

Uexküll, Jakob von. 1957. "A Stroll through the Worlds of Animals and Men: A Picture Book of Invisible Worlds." In *Instinctive Behavior: The Development of a Modern Concept,* ed. C. H. Scholler, 5–80. New York: International Universities Press.

Ulmer, Gregory L. 1985. *Applied Grammatology: Post(e)-Pedagogy from Jacques Derrida to Joseph Beuys.* Baltimore: Johns Hopkins University Press.

Ushenko, Andrew. 1946. "Zeno's Paradoxes." *Mind* 55:131–43.

Uttal, W. 1978. *Psychobiology of Mind.* Hillsdale: Lawrence Erlbaum.

Valéry, Paul. 1947. *Monsieur Teste.* Trans. J. Matthews. New York: McGraw-Hill.

Varela, Francisco J. 1979. *Principles of Biological Autonomy.* Amsterdam: North-Holland.

———. 1984. "The Creative Circle: Sketches on the Natural History of Circularity." In *The Invented Reality,* ed. P. Watzlawick, 309–23. New York: W. W. Norton.

Varela, Francisco J., Evan Thompson, and Eleanor Rosch. 1993. *The Embodied Mind: Cognitive Science and Human Experience.* Cambridge: MIT.

Vitz, Paul C., and Arnold B. Glimcher. 1984. *Modern Art and Modern Science: The Parallel Analysis of Vision.* New York: Praeger.

Waddington, Conrad H. 1957. *The Strategy of the Genes.* London: George Allen and Unwin.

———. 1970. *Behind Appearances: A Study of the Relations between Painting and Natural Sciences in This Century.* Cambridge: MIT.

Walker, R. C. S. 1978. *Kant: The Arguments of the Philosophers.* London: Routledge and Kegan Paul.

Weaver, Warren. 1949. "The Mathematics of Communication." *Scientific American* 181 (July): 11–15.

Weiss, Rick. 1990. "A Flight of Fancy Mathematics: Chaos Brings Harmony to a Birders' Puzzle." *Science News* 137:172.

Weissert, Thomas P. 1991. "Representation and Bifurcation: Borges's Garden of Chaos Dynamics." In *Chaos and Order: Complex Dynamics in Literature and Science,* ed. N. Katherine Hayles, 223–43. Chicago: University of Chicago Press.

Weizenbaum, Joseph. 1976. *Computer Power and Human Reason: From Judgment to Calculation.* San Francisco: W. H. Freeman.

Weyl, Hermann. 1952. *Symmetry.* Princeton: Princeton University Press.

Wheeler, John Archibald. 1980. "Beyond the Black Hole." In *Some Strangeness in the Proportion: A Centennial Symposium to Celebrate the Achievement of Albert Einstein,* ed. H. Woolf, 341–75. Reading: Addison-Wesley.

Wheelock, Carter. 1969. *The Mythmaker: A Study of Motif and Symbol in the Short Stories of Jorge Luís Borges.* Austin: University of Texas Press.

Wheelwright, Phillip. 1959. *Heraclitus.* Princeton: Princeton University Press.

White, John. 1967. *The Birth and Rebirth of Pictorial Space.* London: Faber and Faber.

Whitehead, Alfred North. 1925. *Science and the Modern World.* New York: Macmillan.

———. 1929. *Process and Reality.* New York: Macmillan.

———. 1938. *Modes of Thought.* New York: Macmillan.

———. 1947. *The Wit and Wisdom of Alfred North Whitehead.* Ed. A. H. Johnson. Boston: Beacon.

Whitrow, G. J. 1980. *The Natural Philosophy of Time.* Oxford: Oxford University Press.

Wicken, Jeffrey S. 1987a. "Entropy and Information: Suggestions for a Common Language." *Philosophy of Science* 54:176–93.

———. 1987b. *Evolution, Thermodynamics, and Information: Extending the Darwinian Program.* New York: Oxford University Press.

Wiener, Norbert. 1948. *Cybernetics: Or, Control and Communication in Animals and the Machine.* Cambridge: MIT.

———. 1950. *The Human Use of Human Beings: Cybernetics and Society.* Boston: Houghton Mifflin.

Wigner, Eugene P. 1965. "Violations of Symmetry in Physics." *Scientific American* 213 (December): 28–36.

Wilson, John Arthur. 1968. "Entropy, Not Negentropy." *Nature* 219:535–36.

Winfree, Arthur T. 1972. "Spiral Waves of Chemical Activity." *Science* 175:634–36.

———. 1987. *When Time Breaks Down: The Three-Dimensional Dynamics of Electrochemical Waves and Cardiac Arrhythmias.* Princeton: Princeton University Press.

Wittgenstein, Ludwig. 1961. *Tractatus Logico-Philosophicus.* Trans. D. F. Pears and B. F. McGuiness. London: Routledge and Kegan Paul.

Wood, David. 1989. *The Deconstruction of Time.* Atlantic Highlands: Humanities Press.

Zadeh, L. 1965. "Fuzzy Sets." *Information and Control* 8:378–53.

———. 1975. "Fuzzy Logic and Approximate Reasoning. (In Memory of Grigore Moisil)." *Synthese* 30:407–28.

Zajonc, Arthur. 1993. *Catching the Light: The Entwined History of Light and Mind.* New York: Bantam.

Zea, Leopoldo. 1963. *The Latin-American Mind.* Norman: University of Oklahoma Press.

———. 1969. *Latin America and the World.* Norman: University of Oklahoma Press.

———. 1992. *The Role of the Americas in History.* Lantham: Rowman and Littlefield.

Zee, Anthony. 1986. *Fearful Symmetry: The Search for Beauty in Modern Physics.* New York: Macmillan.

Zeki, Semir. 1983. "Colour Coding in the Cerebral Cortex: The Reaction of Cells in Monkey Visual Cortex to Wavelengths and Colours." *Neuroscience* 9:741–65.

———. 1985. *Central and Peripheral Mechanisms of Colour.* New York: Macmillan.

———. 1993. *A Vision of the Brain.* Oxford: Blackwell.

Index